Preconstruction Estimating

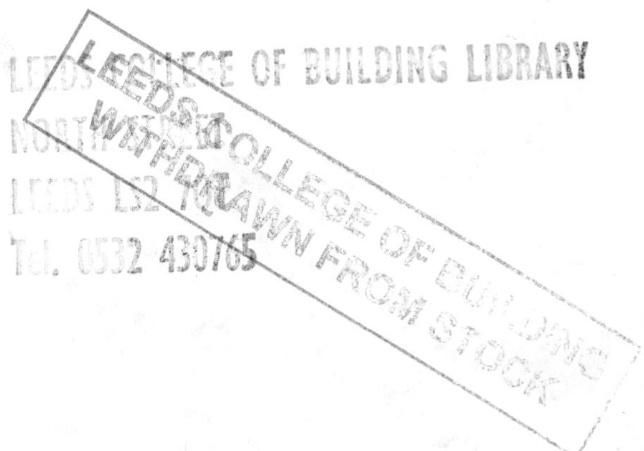

Other Books of Interest from McGraw-Hill

CIVITELLO • *Construction Operations Manual of Policies and Procedures*

COOMBS • *Construction Accounting and Financial Management*

CUSHMAN • *The McGraw-Hill Construction Form Book*

LEVY • *Project Management in Construction*

O'BRIEN • *CPM in Construction Management*

RITZ • *Total Construction Project Management*

Preconstruction Estimating

Budget through Bid

James J. O'Brien
O'Brien-Kreitzberg & Associates, Inc.
Pennsauken, New Jersey

McGraw-Hill, Inc.
New York San Francisco Washington, D.C. Auckland Bogotá
Caracas Lisbon London Madrid Mexico City Milan
Montreal New Delhi San Juan Singapore
Sydney Tokyo Toronto

Library of Congress Cataloging-in-Publication Data

O'Brien, James Jerome.
Preconstruction estimating : budget through bid / James J. O'Brien.
p. cm.
Includes index.
ISBN 0-07-047928-3
1. Building—Estimates. I. Title.
TH435.023 1994
692'.5—dc20 93-30832
CIP

1 2 3 4 5 6 7 8 9 0 DOC/DOC 9 9 8 7 6 5 4 3

ISBN 0-07-047928-3

The sponsoring editor for this book was Larry S. Hager, the editing supervisor was Frank Kotowski, Jr., and the production supervisor was Pamela A. Pelton. It was set in Century Schoolbook by McGraw-Hill's Professional Publishing composition unit.

Printed and bound by R. R. Donnelley & Sons Company.

To Rita G. O'Brien, my inspiration in all things,
and Jessica S. Snyder, who made it happen

ABOUT THE AUTHOR

James J. O'Brien, P.E., CVS, is chairman of the board of O'Brien-Kreitzberg & Associates, Inc., the construction management company that handfled the renovation of San Francisco's cable car system, and is the program manager for the redevelopment of New York's JFK International Airport. Mr. O'Brien is the author of several books on construction-related topics, inlcuding the *Contractor's Managment Handbook*, Second Edition, and *CPM in Construction Management*, also published by McGraw-Hill.

Contents

Preface

This book describes estimating required from the budgetary concept stage of a project through the design process until the project is ready to be bid. The presentation is organized in chronological fashion so that the reader can relate to the estimating process at each milestone in the traditional design process. The estimating process is described at the following major milestone points: budgetary, schematic, design development, and construction document estimates. At the correct chronological points estimating in value engineering is described.

In 9 of the 14 chapters, starting with budgetary, there is a comprehensive case study used to illustrate the estimating process.

Manually prepared estimates are performed at budgetary, design development, and construction document stages. Computerized estimates by two different programs (Composer Gold and I.C.E.) are presented at the schematic, design development, and construction document estimate stages.

At the schematic stage, the limited information requires the estimators to make many more assumptions, and the two programs produce results with a 20% bottom line difference. By the construction document estimate, the computerized estimates are within a percent or two of each other.

Both the manual and computerized estimates follow the Construction Specifications Institute (CSI) format. All three utilized the R. S. Means database for costing. At the design development estimate, R. S. Means Company provided an estimate using a PC spreadsheet approach, which compared very closely with the other three estimates.

Jim O'Brien

Acknowledgments

Organizations

- The American Society of Professional Estimators (ASPE)
- Building Systems Design (BSD)
- Construction Specifications Institute (CSI)
- Management Computer Controls, Inc. (MC^2)
- R. S. Means Company, Inc. (Means)

Individuals

- Kevitt Adler, President, (MC^2)
- Fred M. Seidell, III, Senior Technical Consultant, BSD
- Jessica S. Snyder, Executive Secretary, OK
- Phillip R. Waier, Chief Editor, R. S. Means

Chapter 1

Introduction

Estimating is one of the most important services in the construction industry. While often performed by professionals, it is typically not viewed as a professional service.

The results of estimates can make or break the professional reputations of architects and/or engineers. Good estimating can be the difference between business life and death for contractors and subcontractors. Sound estimating can provide financial success and risk reduction for owners, developers, and their financial backers.

Portions of an estimate can be arrived at with scientific precision. For instance, estimating the construction cost of a list of materials at quantities measured from drawings and extended by unit prices taken from a database is a finite scientific activity. The accuracy of that activity may vary with the accuracy of the takeoff and the suitability and relevance of the database applied. If the database is the proper one to apply and the takeoff is reasonably accurate, the estimate is probably accurate to ±5 to 10%.

Development of that finite scientifically developed number (i.e., construction cost) is important. However, many factors are added to or multiplied against that basic construction estimate. Understanding of these elements is important, because they may increase the budget by as much as 100% and even more.

This book gives an insight to the budgeting-estimating process from concept-budget estimate through schematic estimates and prebid estimates to contractor's estimates, including subcontractor's estimates and change order estimating. Estimating for value engineering is addressed at the design development and cost confidence points during the design process.

Estimating is presented from the owner's, architect-engineer's, value engineer's, project-construction manager's, general contractor's, and subcontractor's viewpoints—and all differ.

Further, the estimates necessary for different types of contracts are discussed, because each has its own special characteristics. Examples are unbalanced bidding in the unit-price-type contract, and the importance of contract pricing strategy in the competitive lump-sum bidding process.

Budgetary estimate. This is the estimate which is often made without consultation by design or estimating professionals, and sets the scene for the entire balance of the project or program. Some owners have described development of this budgetary number as, "picking it out of the air." For instance, in school districts, a state expenditure per student per type of school may be utilized to set a number. Sometimes minimum expenditures per pupil are mandated.

However, setting the budgetary capital number with no input from estimators or design professionals invites an Alice-in-Wonderland situation as the design process proceeds.

A more appropriate approach is to at least utilize similar projects built by other districts, preferably within the same state. If the comparative projects are in other states (or even differing urban vs. suburban areas in the same state), the data must be brought into a comparative condition by utilizing geographical and annual indexes to achieve a parity.

Feasibility estimates. At this stage professional talent is usually utilized. However, there is no design, so the estimator must utilize cost per unit (i.e., student, tenant, patient, etc.) and compare that with a square foot per unit. In addition, the square-foot cost is broken down into gross square feet and net square feet along standard parameters.

Concept and schematic estimate. At this point a designer has been selected, and the estimate is made basically in cost per square foot and/or cubic foot for buildings of similar quality and function.

Design development. At this stage of design, the outline plans and elevations have been developed. Fairly accurate estimates can be made for foundations, structure, exterior, and other architectural features. The engineering portions of the project are in the beginning stages and are still estimated at a square-foot or system basis. At this stage value engineering can profitably be applied.

Fast track and GMP. At what would be design development in other projects, on a fast-track project a guaranteed maximum price (GMP) is estimated at this point in the design. This estimate has to include contingencies for the development of the engineering portion (HVAC, mechanical systems, sprinklers, plumbing, electrical, elevators and escalators, etc.).

Cost confidence. This is an estimate done just prior to the bid process to confirm in the estimator's opinion that the project can be brought in within the budget. If this is not achieved, there may either be a costly (in terms of time) redesign, or the bid packages may be reconfigured into a base contract with alternates to assure that bids can be accepted.

Subcontractor bids. These bids are developed subsequent to the availability of plans and specifications, and prior to the submittal of the general contractor or prime contractor bid. Subcontractors estimate their specialties in great depth, because these are the estimates which will be built up into the general contractor's estimate. Subcontractors may also call upon sub-subcontractors for bids to incorporate into their subcontract bid.

General contractor estimate. General contractors estimate in two parallel paths. One is a detailed estimate of that work which they intend to perform themselves (typically concrete work), and the other a control estimate to have a general idea of the prices which they should receive from each subcontractor. This latter is basically a check estimate to ensure that they are not led astray or that a subcontractor has not made a mistake in understanding the scope of work.

Change order estimates. During the course of construction both the project-construction manager and the general contractor and their subcontractors make estimates of changes in scope in order to negotiate change orders.

Project Cost

This area involves principally the owner and the costs above and beyond the basic construction cost which are still part of the budget, and are very real costs.

Soft costs. These include costs such as design, project management, construction management, and other professional services including estimating.

Land costs. This usually is a "given" in terms of the project budget and is an already established cost as the construction project is designed and constructed.

Financing. This cost is the cost of placing bonds and other professional services involved with developing the finances for the project.

Interest. The interest cost evolves out of the financing costs, and represents the bonding or other loan costs to finance the project.

Equipment. This is the cost of fixed furniture and equipment which the owner requires, and which is part of the capital cost.

Commissioning. This is the cost of moving in, including costs of canceling leases and other costs involved for the project which are incorporated into the capital budget.

Quantities

The estimator (whether for owner-architect, engineer, or contractor) works from the documents provided. These documents will become more detailed as the design evolves.

Manual takeoff. This is the traditional estimating approach in which the estimator "takes off" quantities from the drawings, correlating quantities with the quality as required by the specification.

Quantity survey. This term, more familiar in the United Kingdom, indicates a system in which the quantities are listed in bills of materials (BOM) for use by the contractors.

CSI format. This format of 16 divisions has become very common in the United States. The quantity takeoff is typically organized in the Construction Specifications Institute (CSI) format. It provides a good checklist.

CAD interface. Where the architect-engineer has a CAD system (computer-aided design), bills of material can be generated automatically.

Digitizer. Digitizer can be used for quantity takeoff, for drawings such as contour site drawings and other drawings with irregular shapes.

Allowances. Where the estimator realizes that it will be impossible for the contractor to take off quantities, an allowance category can be established which directs the contractor to bid on a base amount subject to change during the project.

Design Status

Experienced estimators recognize that they will have more complete information as the design evolves. However, to control the design process (i.e., design-to-cost) it is important to have the best available estimates as early as possible.

Concept schematic

At this stage there are usually general floor plans and arrangements as well as architectural elevations. Engineering phases which include structural, HVAC, plumbing, electrical, and conveying systems are usually described in a narrative specification or in criteria.

An experienced estimator can convert this into quantities, particularly in the architectural area. Structural systems can be assumed from bay sizes and loads. Depending on the type of HVAC system, certain of the duct runs and equipment such as chillers can be assumed. The balance of HVAC would be done on a square-foot basis.

Electrical and plumbing systems would be done on a square-foot basis with the exception of specialty equipment which has been identified. This could be priced out on the basis of assumptions by experienced estimators.

Design development

At this stage, floor plans are well developed and often at the same scale as the construction drawings so that they can continue to be utilized in the next stage of design. The schematic estimate can be checked and refined for architectural and structural, while a new estimate would be made for HVAC, plumbing, electrical, and conveying systems. These systems would be only in the early stages of development showing main runs and feeder systems. Secondary supply systems would not be established yet, and assumptions would still have to be made.

Fast track

Fast track design is one in which the foundation and structural design is expedited so that that portion can be bid. That can be done, at the earliest, at a stage between schematic and design development in the traditional approach. At about the same point as design development in the traditional method, packages for exterior skin, roof, and the major portions of the mechanical, electrical, and conveying systems can be estimated, and bids taken.

Construction documents

In the traditional approach, this is the stage at which the drawings and specifications are ready for review and estimating by contractors. The owner and architect-engineer estimate at this stage is often known as a "cost-confidence" estimate. Its main purpose is to pick up areas which have been omitted, and to assure the owner that there is a good probability that the contractor estimates will be within budget.

Schedule

The owner and architect-engineer are in control of the schedule before the construction contract, while the contractor and construction manager for the owner are in control in the construction phase. Time is money, so estimating time is an important cost factor in a project.

Setting schedule. Owners often set a schedule in the most informal manner. Reasons for setting a schedule may include the start of the school year, a sports season, an anniversary date, or other milestone-driven reasons such as the Olympics.

Incentives. If delivery of a project is important, how important is it? Does it deserve incentives as well as liquidated damages?

Claims. An incorrectly estimated schedule can cause claims.

Value Engineering

Estimating plays an important role in the application of value engineering.

Concept. The function analysis to evaluate components of a project and estimate their value vs. cost describes a disciplined method of reviewing program and project to provide the "best bang for the buck." Dramatic savings are possible through the use of value engineering, and estimating is an inherent part of it.

40-hour workshop. The basic method of delivering value engineering in the design process is a 40-hour workshop as defined by the Society of American Value Engineers (SAVE).

Life-cycle costs. The cost of operating a project over its lifetime can often be a significant factor suggesting increased capital costs to arrive at lower life-cycle costs. In fact increasing capital costs is not always a necessity.

VECPs. Value-engineering cost proposals during the construction period even when paid for at 50% of the savings can be a cost-effective way to provide greater value at lower costs.

Types of Contract

The type of contract can directly affect the type of estimate necessary to support the design phase and avoid higher costs during the construction phase.

Cost plus fixed fee. This is one of the most costly methods of implementing a project. Estimating is usually limited because of a lack of information in regard to the final configuration of the project. However, to the extent that estimating can be applied before the cost-plus-fixed-fee contract is awarded, parameters can be set up to contain the cost.

Time and material. This approach is usually utilized during a construction phase to authorize work on a change order when a negotiated price cannot be achieved. Estimating can be utilized to estimate the proper value of the work, and compare it with the cost of time and material. This can be a factor in decisions to utilize or not utilize time and material for change orders.

Fixed price. This is the standard public works contract, and it is very popular in private industry as well. Contractors bid against a completed set of drawings and specifications, and the contract is awarded to the lowest responsible bidder. In the private sector, qualified bidder lists are sometimes utilized and narrow the bidding field with a view toward ensuring quality. Achieving the best fixed price (neither too high nor too low) is the goal of the series of estimates described in this book.

Design-build. A design-build contract offers "whatever you want" for the price you can afford. The role of estimating is to estimate the project independently of the design-build contractor, who has an incentive to provide less than is expected, at a higher cost than is expected. To offset this natural incentive, estimating can be a valuable tool.

Guaranteed maximum price (GMP). This is the result at some point in the fast track process which owners in the private sector utilize to guarantee a cap on their costs. Clearly, continuous estimating of the unfinished drawings which are used to achieve the cap is an important consideration on the owner's side. Failure to have an active estimating support can result in a higher GMP and greater potential for the project to overrun the budget.

Databases

Databases are kept to provide unit costs per unit of quantity which when extended give a cost of doing a component of work. When the components are added together, the result is the base construction cost. Databases can be hard copy or computerized. Computerized databases can be spreadsheet, Lotus 1-2-3, D-Base, or proprietary programs.

R. S. Means. The R. S. Means database is both hard copy and computerized. This database has evolved over the past more than 20 years from a modest database to one very well recognized in the industry. The Means database now has many subsets including:

- Square-foot costs
- Concrete costs
- Structural-steel costs
- Interior costs
- Plumbing costs
- Mechanical costs
- Electrical costs
- Heavy-construction costs
- Site costs

Richardson. Richardson produces a well-recognized database which is hard copy and computerized. It has petrochemical and utility industry components, and also deals with building components, as does the Means database.

In-house. Many organizations have developed their own database, based upon their own experience and other sources. The problem with an in-house database is its lack of completeness or comprehensiveness. Maintaining and annually updating an in-house database is an expensive proposition, probably one which can be financed only by larger firms.

Location indexes. Any database must be adjusted for location. Typically New York is used as 1.00. The ENR location indexes can be utilized, or similar ones developed by R. S. Means, etc.

Annual indexes. Inflation is an important factor in estimating. ENR carries past escalation factors in its ENR Index. There are other indexes also. The problem is in estimating projected future escalation—which is often anyone's best guess.

Computerized Systems

A number of computerized systems can take quantity information, combine it with appropriate unit costs from the database and extend it, and make the appropriate additions by CSI subsection. Some of these are as follows:

Vendor	Software (all trademarks)	Location
Building Systems Design	Composer Gold	Atlanta, GA
CDCI	The Bid Team	Atlanta, GA
MC^2	I.C.E.	Memphis, TN
SoftCost	Success	Atlanta, GA
Timberline	Precision	Beaverton, OR

Contractor Overheads

Contractor overheads make up an important part of the ultimate construction price. Overheads are a reflection of real costs of running the contractor's business and operating the construction contract in the field. The overheads apply to both general contractors and subcontractors. The estimator must be familiar with these markups in order to take the raw construction cost and build it up into a contractor's bid cost.

General conditions. These are the costs in the field of running the office and providing crew facilities, cleanup facilities, toilet facilities, trash removal, site cleanup, housekeeping, security, and other support facilities. They may include shops such as the prefabrication areas and carpenter shop. However, these would best be listed under a separate account number.

Management. This is the cost of noncraft project managers, superintendents, and trade superintendents. It is sometimes lumped in with general conditions, but again is better kept as a separate account.

Equipment. Where equipment is a major cost, such as a tower crane or cranes in a high-rise, these should be kept as a separate line item.

Home office. Home office expense is often charged to the job as a percentage of the revenue generated by the project vs. the overall revenues of the firm. This principle (Eichleay) has been postulated in many claims.

Special factors. The estimator must understand special factors in the project which could drive the costs, such as:

- Required overtime
- Shift work
- Remote location

Construction Equipment Costs

Equipment costs are an important part of the construction estimate. The prebid estimator has to use available information, but only the bidding contractors could know how they will approach pricing their equipment.

Base rates. Equipment prices are available on an industry basis (Green book; Blue book), which gives reasonable rates. However, only the individual contractors know whether they have to buy, rent, or utilize existing equipment.

Depreciation. Those contractors with existing equipment or rental firms with existing equipment which is partially depreciated have the option of using pricing strategies which could improve competition. Only the contractors' estimators will know this information.

Strategy. Different types of equipment can handle a given contract, and contractor strategy is an important part of the estimate, for instance, use of tower crane vs. mobile cranes; also, the use of different types of dirt movers such as scrapers vs. bulldozers, etc.

Productivity

For the contractor's estimate, it is important that the contractor consider productivity if the contract is to produce a favorable result if the bid is achieved. Conversely, if the contractor is too conservative or pessimistic, then the contractor would clearly not be the low bidder. Each contractor or subcontractor should have a database, whether formalized or informal, which describes to the firm their productivity vs. databases such as R. S. Means and/or Richardson. In making a bid, the contractor must consider the following:

- Working conditions
- Overtime required
- Shift work required
- Weather conditions
- Schedule
- Staffing
- Craft availability
- Work rules, union
- Environment
- Logistical support

Pricing Strategies

As the contractor (and subcontractors), particularly in a fixed-price competition, approach the bidding point, the estimators attempt to forecast possible pricing strategies which would impact the overall bid price.

Evaluate competition. Based on who has drawn the plans, the estimators identify the other probable bidders by reputation to determine whether or not they can anticipate "low ball" or break-even bids.

Another important consideration is to identify potential competition that may be already mobilized in the area on another project.

Identify potential errors. Particularly in unit-type price bids, competition is achieved by having the bidders put a unit price against a baseline quantity. If the contractor's estimator identifies an area or areas where the probability is that the quantities will go over or under, the unit price can be unbalanced (i.e., made higher or lower as appropriate).

Qualitative pricing. The estimators have to identify parts of the specification for which a subjective price may be assigned. This would include areas such as exculpatory clauses and liquidated damages.

Escalation. The estimator needs to apply a projected escalation to the midpoint of the project, or on a year-by-year basis.

Change Orders

Estimators should play an important part in the change order process.

Basis and identification. Estimators are in good position to evaluate whether or not, in their opinion, an item of work which is disputed or added is or was part of the scope estimated.

Estimating. When a change order is issued, there has to be an estimate for which the change order negotiation can proceed. This has to have backup worksheets.

Negotiation. The estimator is often part of the negotiation team to support the price requested for change orders.

Time and material. Most contracts permit the owner to direct a change to be done on time and material, if the negotiation cannot be reached. Most contractors do not prefer this arrangement because all of their costs usually are not recovered on the strictly run time and material.

Claims

Estimating is an important part of presenting and/or defending claims.

Disputed work. Where the contractor has a claim for disputed work which they purport is not part of the base contract scope, the estimator can identify included scope from the worksheets of the original estimate. To the extent that disputed claims are for under-

authorization of a change in scope, the estimator's worksheets, again, can provide the basis for asserting the claim.

Delay claims. The estimator's role in delay claims is to identify time-oriented costs such as general conditions, home office overhead, and equipment costs.

The Estimator

As noted at the start of this chapter, estimating is not viewed as a professional service. The American Society of Professional Estimators (ASPE) has recognition of estimating as a profession as a major goal. In 1988, ASPE adopted the following code of ethics:

CANON 1. Professional estimators shall perform services in areas of their discipline and competence.

CANON 2. Professional estimators shall continue to expand their professional capabilities through continuing education programs to better enable them to serve clients, employers, and the industry.

CANON 3. Professional estimators shall conduct themselves in a manner which will promote cooperation and good relations among members of our profession and those directly related to our profession.

CANON 4. Professional estimators shall safeguard and keep in confidence all knowledge of the business affairs and technical procedures of an employer or client.

CANON 5. Professional estimators shall conduct themselves with integrity at all times and not knowingly or willing enter into agreements that violate the laws of the United States of America or of the states in which they practice. They shall establish guidelines for setting forth prices and receiving quotations that are fair and equitable to all parties.

CANON 6. Professional estimators shall utilize their education, years of experience, and acquired skills in the preparation of each estimate or assignment with full commitment to make each estimate or assignment as detailed and accurate as their talents and abilities allow.

CANON 7. Professional estimators shall not engage in the practice of "bid peddling" as defined by this code. This is a breach of moral and ethical standards, and this practice shall not be entered into by a member of the society.

CANON 8. Professional estimators and those in training to be estimators shall not enter into any agreement that may be considered acts of collusion or conspiracy (bid rigging) with the implied or express purpose of defrauding clients. Acts of this type are in direct violation of the Code of Ethics of the American Society of Professional Estimators.

CANON 9. Professional estimators and those in training to be estimators shall not participate in acts, such as the giving or receiving of gifts, that are intended to be or may be construed as being unlawful acts of bribery.

Certification

Five professional societies offer certification. All require certain experience as a minimum, and then successful completion of an examination. They are as follows:

- American Association of Cost Engineers (AACE): Certification is certified cost consultant (CCC) or certified cost engineer (CCE).
- American Society of Professional Estimators (ASPE): Certification is certified professional estimator (CPE).
- Construction Management Association of America (CMAA): Certification is certified construction manager (CCM).
- Project Management Institute (PMI): Certification is project management professional (PMP).
- Society of American Value Engineers (SAVE): Certification is certified value specialist (CVS).

The most relevant certifications for an estimator are the CCC/CCE and CPE.

Chapter

2

Concept Budgetary Estimate

Most projects seem to appear from nowhere; the result is an evolutionary kind of aggregate thinking from many sources which gathers pressure, both political and personal, until the project has been articulated. Projects that evolve in this way include schools, hospitals, public buildings, industrial plants, highways—indeed, almost any identifiable major project. Key characteristics in their evolution are power structure and consensus. Actually, this phase of the project should go through the following stages: establishment of goals, means of accomplishing goals, decision to proceed, identification of funding source, and budget approval.

The decision to go ahead requires the identification of specific projects and the development of preliminary cost estimates, usually on the basis of gross estimating factors such as costs per square foot or cubic foot.

After a project has been given a budget and funding is available to meet the budget, the predesign phase moves to other stages. Site selection, as for a hospital addition or a school replacement or other finite location situation, is often part of the basic decision to go ahead with a project. In many cases, however, a new site should or must be considered. Usually the site consideration precedes the selection of a designer, since the design should be a function of the site. A number of nontechnical factors may funnel the choice of a site into a specific direction. Among the factors to be considered are

1. *Encumbrances.* Are there tenants who will have to be relocated? Are there structures to be removed?
2. *Land costs.* What are the economic values and factors?
3. *Transportation.* Is the location adequately served, and is it served by media suitable to the character of the facility's needs?
4. *Utilities.* What are the availabilities? Where are the potential problems?

5. *Neighborhoods.* Is the environment suitable for the facility? Is the facility suitable for the environment?
6. *Zoning.* Does local zoning conform to the use intended?
7. *Community.* How will the community react to the facility?
8. *Subsurface conditions.* Will the foundations require unusual support? Are there unusual problems to be overcome?

There are other factors, but it is clear that, in choosing one of a number of sites, many factors must be evaluated and considered carefully. Unfortunately, many of them are considered from the viewpoint of hindsight rather than at the proper time in the project.

The last predesign activity should be the development of a specific program to identify the intent of the owner regarding the functional utilization of the project. This philosophical statement is important to the designer, but it is often presented in such a perfunctory, nonspecific fashion that the designer, through a trial-and-error method, ends up establishing the philosophy. It is clearly the owner's responsibility to establish these requirements and to interpret them in terms of cost impact prior to the selection of a designer. Programming is a unique talent, and it will probably require a combination of this knowledge and a consultant's expertise.

Functional planning requires the availability or the assembly of pertinent information regarding the project. Demographic sources such as the U.S. Census, city planning, state planning organizations, and in-house sources should be reviewed. Information should be arranged and stored in a manner which makes it accessible for the review of future projects or for the reconsideration of this one. Often when the information is stored in a computer data bank, such exercises as modeling, gaming, or simulation of various alternatives can be used to test the results of different potential approaches.

The functional programming effort should be tied back to the budgetary estimate, which it should either affirm or revise. Since the functional program incorporates the policy in regard to any project, it should be approved by the appropriate owner or authority.

A concomitant to the functional program is the architectural program, to which the functional program is necessarily related. The architectural program may be incorporated in the schematic design phase by the architect, or it may be furnished to the architect.

Typically, projects do not have formal program documents. The result is uncertainty at the beginning of the design phase. Since designers are not compensated for uncertainty, their only defense is to proceed slowly at the early stages of design and develop a program type of statement which can be confirmed or revised by their clients. Unfortunately, clients often demonstrate a proclivity for changing their minds almost constantly. From the design point of view, this is not only time-consuming but highly expensive. Virtually the only defense designers have is the careful control of the progress of the design—not permitting it to go forward at a normal speed but holding off every activity until a high level of definition has been achieved. This is expensive to designers, and to owners as well, because the true design work is placed in too short a time for economical implementation.

Format

The following format is suggested for the concept-budget estimate:

1. Base construction cost (usually based on square-foot costs)
2. Special equipment and construction (not included in square-foot costs)

Subtotal = construction cost

3. Escalation (to midpoint of construction)
4. Contingency (20% is typical at this early stage)

Subtotal = adjusted construction cost

5. Design cost (6% is typical)
6. Construction management cost (5% is typical)
7. Owner's project management cost (2% is typical)
8. Cost of land

Total budget

Case study

The Widget Housing and Transmission Company (WHATCO) wants the ability in the Philadelphia area to assemble 120 units (minimum) to 180 units (maximum) per year. Company industrial engineers advise that production of one 50-ton unit requires 2 months and 1000 sf of work space. Design will be in 1995 and construction in 1996.

The plant is to be supported by a parts warehouse. Major parts will be stored on pallets on a 4-ft-high forklift-loaded storage area. At maximum pace, 30 units will be in production. Industrial engineering advises that each unit must be supported by 1000 sf of storage. This includes flat storage and bin storage.

Transportation. The siting group must consider rail, highway, air, and deep port access for shipping.

Capital group. This group will select the architect-engineer and construction manager. They specify a 20-year design life (extendable to 30). Class B finishes are to be specified.

Accounting. The project can support a capital budget (total) of $7.5 million (not including interest). Financing will be by either 10-year corporate notes or bonds. A construction bank loan at prime is locked in.

Plant. Assume that the minimum production will set the plant size. Maximum production will be achieved by a second shift:

$$\text{Plant base size} = \frac{120 \text{ units}}{6 \text{ units}} \times 1000 \text{ sf} = 20{,}000 \text{ sf}$$

$$\text{Circulation at } 25\% = 0.25\ (20{,}000 \text{ sf}) = 5{,}000 \text{ sf}$$

$$\text{Staging at each end} = 2 \text{ at } 2500 \text{ sf} = \underline{5{,}000 \text{ sf}}$$

$$\text{Plant} = 30{,}000 \text{ sf}$$

Warehouse

$$\text{Storage base size} = \frac{180 \text{ units} \times 1000 \text{ sf}}{6 \text{ units} \times 4 \text{ levels}} = 7500 \text{ sf}$$

$$\text{Circulation at } 25\% = 0.25\ (7500 \text{ sf}) = \underline{1875 \text{ sf}}$$

$$\text{Warehouse} = 9375 \text{ sf}$$

A regional office is to be incorporated into the complex:

1 regional manager at 400 sf	400 sf
4 district sales at 225 sf	900 sf
1 executive secretary at 225 sf	225 sf
2 sales secretaries at 200 sf	400 sf
Lunchroom at 400 sf	400 sf
Toilet area, 2 at 225 sf	450 sf
Reception area at 400 sf	400 sf
Accounting, 8 at 200 sf	1600 sf
Computer center at 550 sf	550 sf
Net	5325 sf
20% circulation	1065 sf
Gross sf	6390 sf

Industrial engineering provides the following utility and site information:

Utilities

Electric, 5000 kW

Potable water, 500 gpm at 100 psi

Fire water, 5000 gpm at 100 psi

Sewer or septic system

Site

Truck loading and maneuver

Rail siding

Parking, 40 spaces

Concept estimate. Using the Means Facilities Cost Data (1995) index, "square-foot cost" shows two references. Using the first approach (Fig. 2.1), the project cost would be as follows:

At median cost, based on square footage:

Office:

6390 sf at \$76.00 = \$ 485,640

Plant:

30,000 sf at \$36.44 = 1,093,200

Warehouse:

9375 sf at \$36.44 = 341,625

\$1,920,465

At median cost, based on cubic footage:

Office:

6390 sf × 15 ft × \$6.26 = \$ 600,021

Plant:

30,000 sf × 24 ft × \$2.14 = 1,540,800

Warehouse:

9375 sf × 24 ft × \$2.14 = 481,500

\$2,622,321

At ¾ level, based on square footage:

171 / S.F., C.F. and % of Total Costs						
	171 000	/S.F. & C.F. Costs	Unit	1/4	Unit Costs Median	3/4
610	0010	**OFFICES** Low Rise (1 to 4 story)	S.F.	59	76	101
	0020	Total project costs	C.F.	4.37	6.26	8.19
	0100	Sitework	S.F.	4.71	7.55	11.61
	0500	Masonry	S.F.	2.05	4.66	8.84
	1800	Equipment	S.F.	0.75	1.37	3.69
	2720	Plumbing	S.F.	2.26	3.41	4.86
	2770	Heating, ventilating air conditioning	S.F.	4.80	6.71	9.87
	2900	Electrical	S.F.	4.97	6.90	9.48
	3100	Total: Mechanical & Electrical	S.F.	10.38	15.42	22.25
970	0010	**WAREHOUSES** And Storage Buildings	S.F.	26.19	36.44	55.53
	0020	Total project costs	C.F.	1.37	2.14	3.55
	0100	Sitework	S.F.	2.68	5.43	8.13
	0500	Masonry	S.F.	1.66	3.81	8.19
	1800	Equipment	S.F.	0.43	0.88	3.43
	2720	Plumbing	S.F.	0.89	1.52	3.06
	2730	Heating & ventilating	S.F.	1.01	2.63	3.80
	2900	Electrical	S.F.	1.59	2.93	4.94
	3100	Total: Mechanical & Electrical	S.F.	2.90	5.10	11.35

Figure 2.1 Square feet, cubic feet, and percent of total cost table. (*Format adapted from R. S. Means Co., Inc., Kingston, MA.*)

Office:

6390 sf at \$101.00 = \$ 645,390

Plant:

30,000 sf at \$55.53 = 1,665,900

Warehouse:

9375 sf at \$55.53 = 520,594

\$2,831,884

At ¾ level, based on cubic footage:

Office:

6390 sf × 15 ft × \$8.19 = \$ 785,016

Plant:

30,000 sf × 24 ft × \$3.55 = 2,556,000

Warehouse:

9375 sf × 24 ft × \$3.55 = 798,750

\$4,139,762

Average project cost is $2,878,608.

The second approach is as follows:

Means addresses the size of a project versus the sample used to develop the average sf figures. The theory is that, in general, larger buildings will have lower sf costs. This, says Means, is "due mainly to the decreasing (cost) contribution of the exterior walls plus the economy of scale usually achievable in larger buildings."

Figure 2.2 is the R.S. Means explanation of the "Square Foot Project Size Modifier" approach, including a sample calculation by Means. Using this approach, the factors for the WHATCO project are as follows:

Factor

Office:

$$\frac{\text{Proposed sf}}{\text{Typical size sf}} = \frac{6390}{8600} = 0.74$$

Plant:

$$\frac{30{,}000}{25{,}000} = 1.12$$

Warehouse:

$$\frac{9{,}375}{25{,}000} = 0.375$$

From Fig. 2.3 the cost multipliers are

Office	1.05
Plant	0.98
Warehouse	1.125

Using the project size modifiers (Means), the projected costs are shown below:

	Median based on square footage	Median based on cubic footage	¾ based on square footage	¾ based on cubic footage
Office	$ 509,922	$ 630,022	$ 677,660	$ 824,267
Plant	1,060,404	1,494,576	1,615,923	2,479,320
Warehouse	384,328	541,688	585,668	898,594
	1,954,654	2,666,286	2,879,251	4,202,181

Average project cost is $2,925,593.

Using the Means City Cost indexes, the following sampling was selected:

Square-Foot Project Size Modifier (Div. 171)
One factor that affects the S.F. cost of a particular building is the size. In general, for buildings built to the same specifications in the same locality, the larger building will have the lower S.F. Cost. This is due mainly to the decreasing contribution of the exterior walls plus the economy of scale usually achievable in larger buildings. The Area Conversion Scale shown below will give a factor to convert costs for the typical size building to an adjusted cost for the particular project.

The Square-Foot Base Size lists the median costs, most typical project size in our accumulated data and the range in size of the projects.

The Size Factor for your project is determined by dividing your project area in S.F. by the typical project size for the particular Building Type. With this factor, enter the Area Conversion Scale at the appropriate Size Factor and determine the appropriate cost multiplier for your building size.

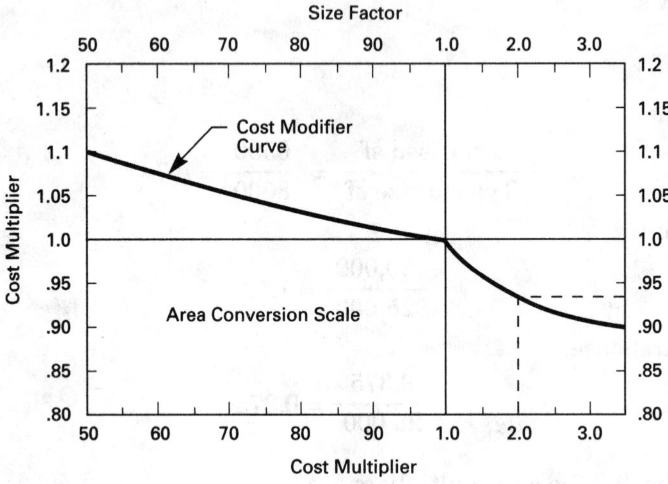

Example: Determine the cost per S.F. for a 100,000 S.F. Midrise apartment building.

$$\frac{\text{Proposed building area} = 100{,}000 \text{ S.F.}}{\text{Typical size from below} = 50{,}000 \text{ S.F.}} = 2.00$$

Enter Area Conversion scale at 2.0, intersect curve, read horizontally the appropriate cost multiplier of 0.94. Size adjusted cost becomes 0.94 x $54.50 = $51.23 based on national average costs.

Note: For Size Factors less than .50, the Cost Multiplier is 1.1
For Size Factors greater than 3.5, the Cost Multiplier is .90

Figure 2.2 Square-Foot Project Size Modifier (R. S. Means Co., Inc., Kingston, MA).

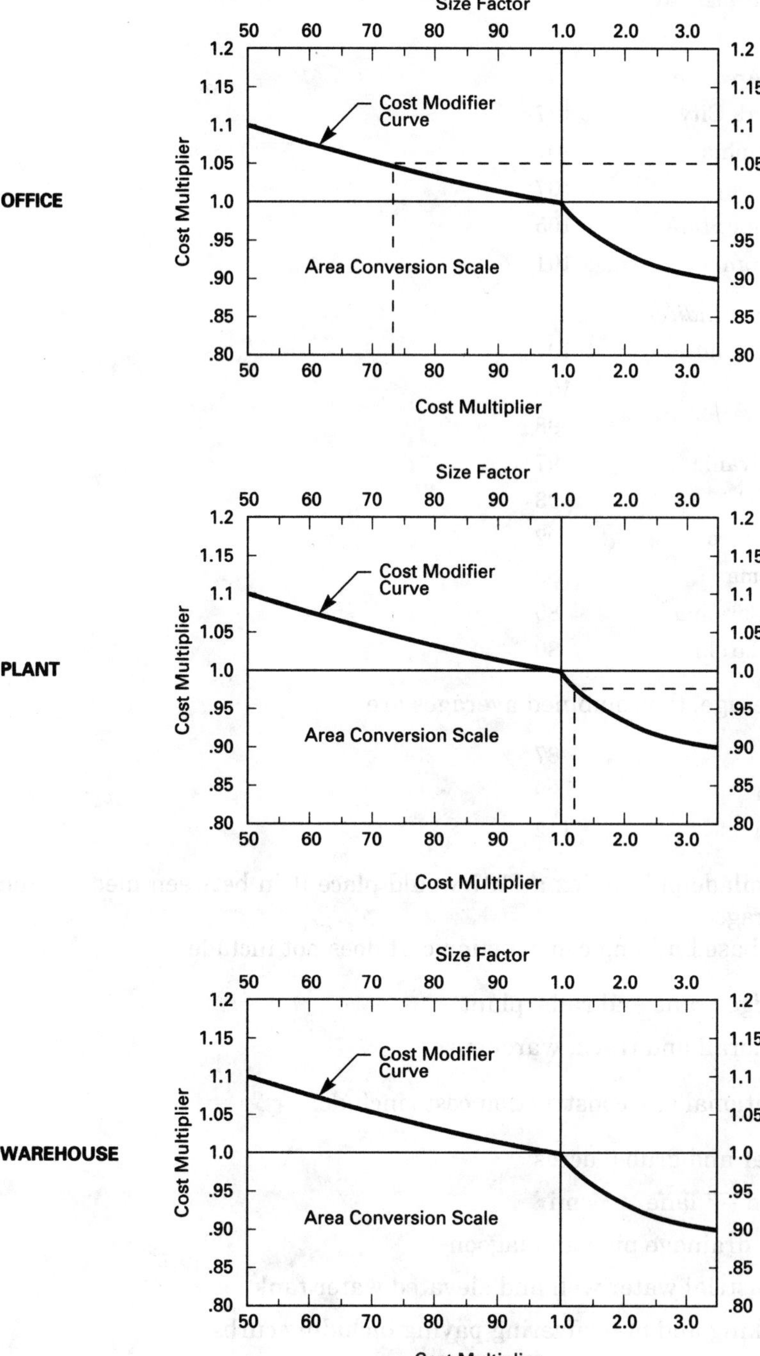

Figure 2.3 Cost modifier factors for WHATCO project.

Urban Areas Indices	
New York City	127
Philadelphia	107
New Jersey	107
New York State	105
Pittsburgh	101

Nonurban Indices	
Rhode Island	101
Oregon	100
Ohio	98
Pennsylvania	97
New York	93
New Mexico	92
Oklahoma	89
North Carolina	80
South Carolina	80

On average, the combined averages are

Low ¼	87
Median	99
High ¾	112

The Philadelphia index at 107 would place it in between median and ¾ average.

The base building construction cost does not include:

- Bridge crane and rails, plant
- Monorail and track, warehouse

Additional site construction costs include:

- Clear and grub 6 acres
- Road—3 lane × ½ mi
- Site drainage pipe and lagoon
- Industrial water well and elevated water tank
- Parking and maneuvering paving including curbs
- Electric service—13.32 kV–3,000 lf
- Railroad siding—1000 lf

Because of the high cube size versus building footprint (particularly in plant and warehouse), the proper cost figure to use is the cost per cubic foot, as adjusted by the size modifier.

Based on the city index, the recommended budget should be prorated between median and ¾ averages, or

$$\text{median index} = 99$$
$$\text{Philadelphia} = 107$$
$$\tfrac{3}{4}\ \text{index} = 112$$

$$\text{Philadelphia} = \text{median} + \frac{(107 - 99)}{(112 - 99)} (\tfrac{3}{4} - \text{median})$$
$$= (\$2{,}666{,}286) + \tfrac{8}{13}\,[4{,}202{,}181 - 2{,}666{,}286]$$
$$= (\$2{,}666{,}286) + \tfrac{8}{13}\,(1{,}535{,}895)$$
$$= (\$2{,}666{,}286) + (\$945{,}166) = \$3{,}611{,}452$$

Equipment costs. Equipment estimates are from the WHATCO Industrial Engineering Group.

Bridge Crane (50 ton × 100-ft span)

Equipment	$250,000
Installation	50,000
Crane rail (assume channel MC18 @ 58 lb/lf)	
600 lf @ $52/lf (erected) =	31,200
(from R.S. Means price for CSI Division 051-3900 W18 ft × 55 lb)	
Total for bridge crane =	$331,200

Monorail (20 ton)

Equipment	$20,000
Crane rail	15,000
Installation	15,0 00
Total for monorail	$50,000

Additional site work

Clear and grub (6 acres).

From Div. 021-100 (Means) medium trees to 12 in diameter:

Cut and chip 6 acres @ $4,212 ea. = $25,272

Grading and planting grass on 6 acres—assume 6 acres @ $8,000 ea. = $48,000

Road—Three lanes (approximately ½ mi). Assume blacktop (from Means):

- Grade Div. (022-100)

  40 ft wide × 5280/2 × 2 ft average = 103,200 = 3822 cy

  Cost @ $3.00/cy = $11,466
- Base Course Div. (022-300)

  13 yd wide × 5280/2 = 1144 sy

  1144 sy @ $12.00/sy = $13,728
- Paving (3 in thick) Div. (025-100)

  1144 sy @ $7.09/sy = $ 8,111

Site drainage and lagoon. Assume this will cost $20,000.

Water

- Elevated water tank 250,000 gal (Div. 132-100) 1 ea. @ $403,000
- Industrial water well—assume a cost of $20,000

Parking and maneuvering

- Parking: 40 spaces—12 ft × 8 in = 3936 sf
- Maneuvering: assume 100 ft × 6 ft = 6000 sf

  Total 9936 sf = 1104 sy

  From the entry road:

  Base course = $12.00/sy

  Paving = $7.09/sy

  This area = $19.09/sy × 1104 sy = $21,075
- Curbs: (100 ft + 60 ft + 60 ft) = 220 lf

  Means (Div. 025-250) concrete curbs @ $8.91/lf × 220 lf = $ 1,960

Electrical service 13.2 kV–3000 lf

- [Poles (Div. 167-100) spaced @ 50 ft = 60] 60 @ $1,000 = $ 60,000
- Cable (Div. 161-100) 3000 lf @ $15 = $ 45,000

R.R. siding (1000 lf)

(Div. 024-520)

#8 turnaround, 1 ea. =	$ 27,000
Track $101/lf @ 1000 lf =	101,000
Fittings at 1 each =	8,000
Subtotal	$136,000
Total additional site work	$788,340

Summary

Base construction cost	$3,611,452
Equipment	381,200
Additional site work	788,340
Subtotal construction cost	$4,780,992
Escalation 5%/2 (construction cost)	119,525
Contingency 20% (construction cost)	956,198
Design 6% (construction cost)	286,860
Construction management 5% (construction cost)	239,050
Owner's project management 2% (construction cost)	95,620
Land cost (per WHATCO real estate)	500,000
Total project cost	$6,978,245
Maximum available	$7,500,000

Chapter

3

Schematic Estimate

Designing a project involves a relatively complex series of activities which become increasingly detailed as the project is moved through the various design phases. These phases are schematic development, preliminary design, and working drawings.

Schematic Design

This is also called the *sketch phase,* during which "concept" plans are developed by the architect and the basic engineering system analysis is made. Design criteria are specified, and schematic drawings are prepared. A set of perspective sketches, or renderings, may be prepared. Exact dimensioning is not expected at the schematic stage.

The schematic design package should include:

Site plan. Plot plan of the site boundary with structures located. The plan may be with, or without, earth contours. To the extent they are identified, utilities would be shown schematically.

Plan view. Buildings are usually laid out to a scale of ¼ in = 1 ft 0 in or ⅛ in = 1 ft 0 in. The program space is blocked out on the plan.

Elevations. The exterior of the project is shown at the same scale as the plan. At schematic, this view has only vertical dimensions or elevations.

Sections. These are taken through the interior at right angles to the exterior walls. At schematic, this view is usually dimensionless.

The schematic level of detail is limited, and the principal takeoff is square footage. Accordingly, the schematic estimate is usually similar to the budgetary or concept schedule.

The schematic design is the first graphical representation of the concept and/or architectural program.

Parts of the schematic estimate format may follow the CSI (Construction Specification Institute) format. However, since major elements are not yet specified, the CSI 16-division format cannot be followed in its entirety.

The MEP (mechanical, electrical, plumbing) portion will be largely undefined at the schematic stage. Typical MEP information would be:

- HVAC heat and cooling design loads
- HVAC air change requirements
- Lighting requirements in footcandles (fc)
- Primary electric service load
- Fire sprinkler requirements
- Water service and storage requirements
- Sanitary sewer requirements

WHATCO project

The schematic graphics for the WHATCO project are as follows:

Fig. 3.1. Site plan

Fig. 3.2. Plant floor plan

Fig. 3.3. Interior section AA, plant

Fig. 3.4. Warehouse floor plan

Fig. 3.5. Interior section BB, warehouse

Fig. 3.6. Office floor plan

Fig. 3.7. Interior section CC, office

The location of the WHATCO project is Philadelphia, Pennsylvania.

Plant

100 ft × 300 ft × 24 ft high

Steel frame

Roof: 100-ft girders with 20-ft LS joist secondary system

Precast plank

20-year 3-ply built-up roof

Siding: Insulated metal panel on girt system

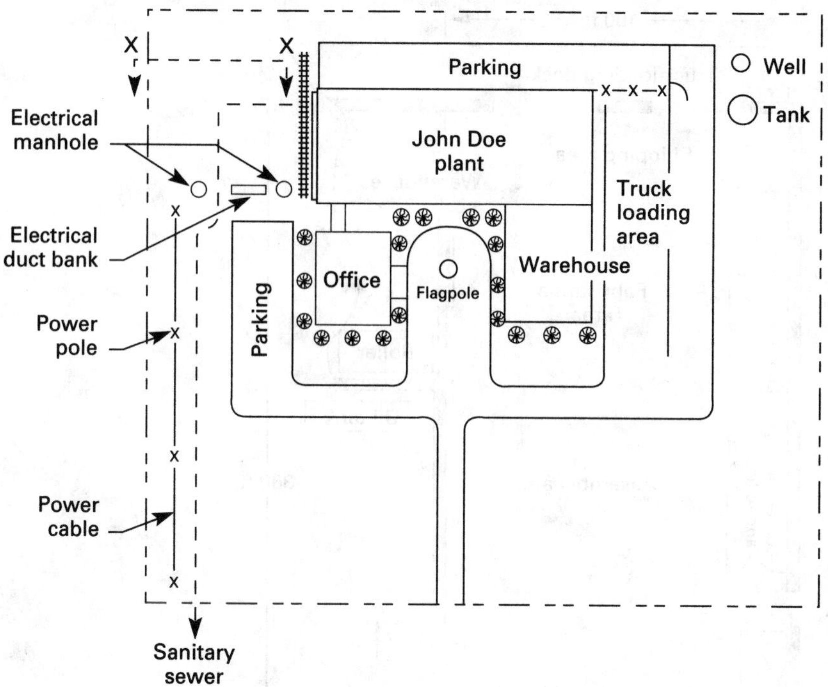

Figure 3.1 Site plan, WHATCO project.

Slab-on-grade 12 in thick; perimeter grade beam 8 in thick × 36 in deep

Railroad siding: 3 roll-up doors 12 ft w × 20 ft ht

Truck loading dock: 3 roll-up doors 12 ft w × 20 ft ht, 3 dock levelers

Crane

50-ton gantry crane, craneway 300 lf

Supported by columns at 30-ft space

Column foundations, 20 cast-in-place piles

Steel frame foundations

- Sides, 20 piles
- End, 4 piles
- Pile caps, 6 ft × 4 ft × 2 ft (common with crane piles)

Switchgear

5000 kW

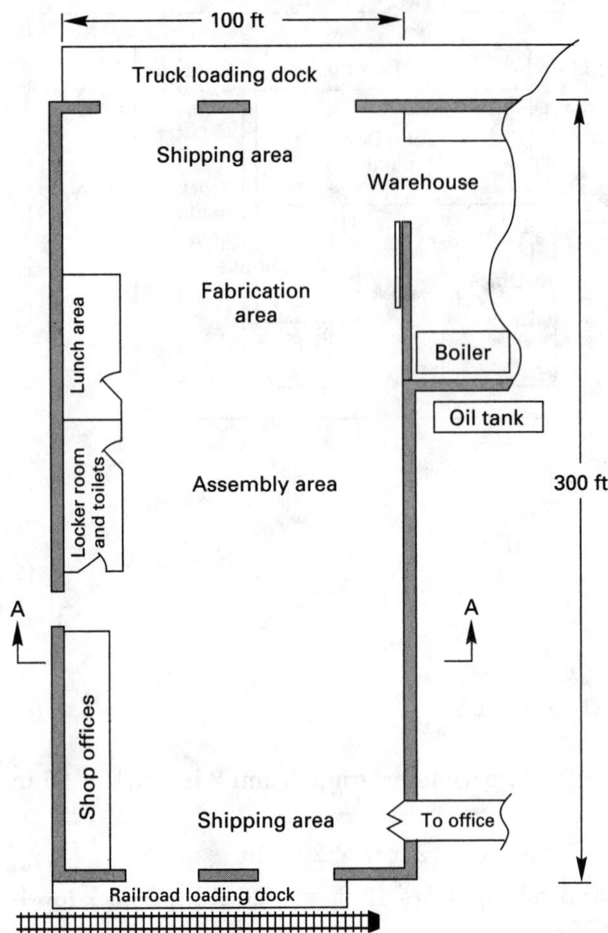

Figure 3.2 Plant floor plan.

Primary 13.2 kV

Secondary 440 V

Fire protection. The hydrant system measures 700 lf × 6 in diameter

Warehouse

120 ft × 80 ft × 24 ft high

Roof and siding: Same as plant

Structural system:

80-ft girders with 20-ft LS joist secondary system

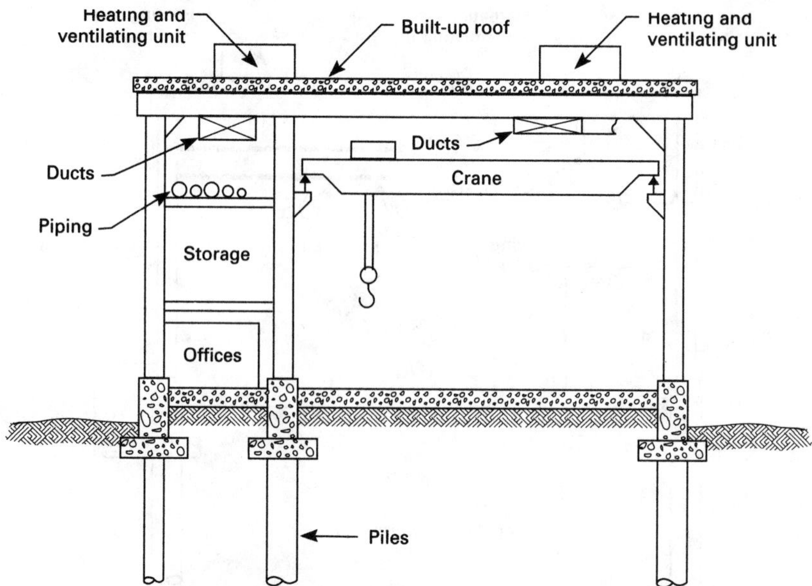

Figure 3.3 Interior section AA, plant (not to scale).

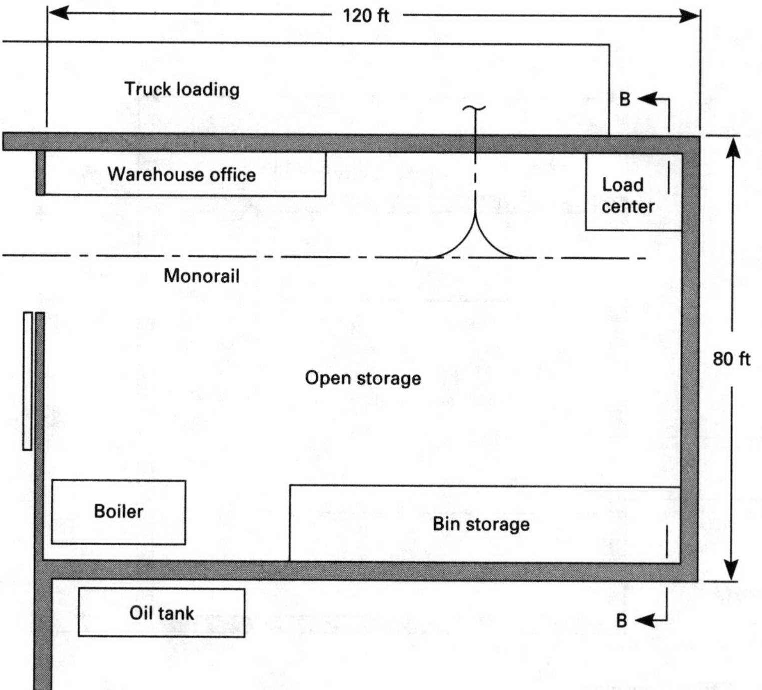

Figure 3.4 Warehouse floor plan.

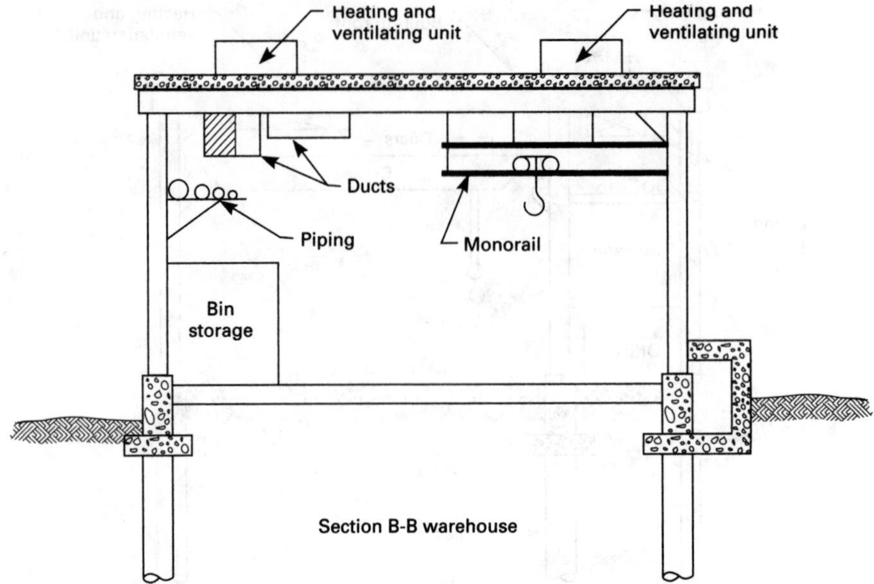

Figure 3.5 Interior section BB, warehouse (not to scale).

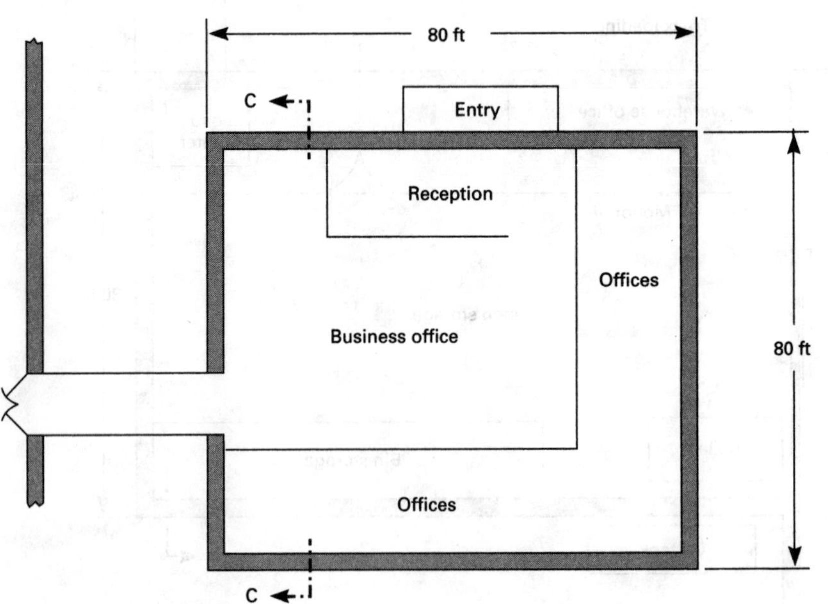

Figure 3.6 Office floor plan.

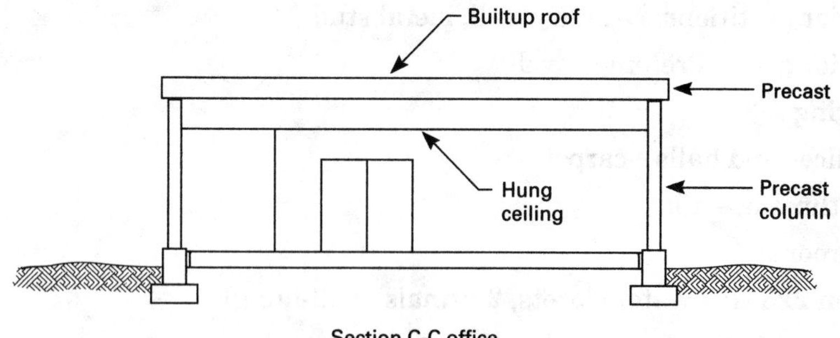

Figure 3.7 Interior section CC, office.

Column foundations

12 cast-in-place piles

Pile caps 4 ft × 4 ft × 3 ft

Grade beams 8 in × 36 in deep

HVAC: Same as plant

Lighting: Same as plant

Power: From plant

Fire Protection: Sprinkler

Rack system: By owner

Bin system: By owner

Bathrooms

Men 500 sf: 4 water closets; 4 urinals (wall-hung)

Women 300 sf, 3 water closets

Office building

80 ft × 80 ft × 15 ft high

Precast frame and roof

Roofing: 20-year 3-ply built-up roof on lightweight fill and 1-in insulation board

Windows: Andersen thermopane (nonoperable)

Doors

Entrance, double plate glass

Exits (2), metal

Interior partitions: ⅜-in dry wall, metal stud

Interior doors: Prehung wood

Flooring

- Offices and halls—carpet
- Bathrooms—tile

Bathrooms:

- Men 225 sf: 2 water closets, 2 urinals (wall-hung)
- Women 225 sf, 3 water closets

Lunchroom: Dishwasher, double sink, stove, microwave, cabinets 400 sf

HVAC

- 750-kBtu heater, air circulation
- 20-ton chiller
- Duct distribution
- Ventilation fans
- Exhaust fans, as necessary

Electric

- 300-kW service
- 100 fc light
- Receptacles at $1/100$ sf $= \dfrac{64}{100} = 64$

Fire protection: Sprinkler

Site

Clear, strip, and grade, 6 acres

Electric 13.2-kV service, 3000-lf pole

Service entrance access, 200-lf duct

Water tower, 250,000 gal at 75-ft head

Other

Codes: BOCA, NFPA, Pennsylvania Department of Labor and Industry

Construction Schedule: 1 year

HVAC

Ventilation: Rooftop fans, 4 air changes/h

Heat:

2 package boilers with hot water (1 million Btu each)

Radiant heaters (6) at doors

Lighting. Industrial 100 fc

Computerized schematic estimate (Composer Gold)

Using the preceding schematic description, Fred M. Seidell III (certified cost consultant) of Building Systems Design (BSD) prepared the following schematic estimate. Mr. Seidell described the process as follows:

> We used Composer Gold to develop the enclosed summary report. The procedure used to get this report out of Gold was to use the Modeling database's resources from this powerful estimating program, and just print out the summary that was developed after entering Philadelphia Davis Bacon wage rates and your design parameters.
>
> Key features and benefits of Gold are highlighted:
>
> - Flexible project and report structures allow adjustable formats for cost estimating and tracking.
> - Latest unit price database from the Corps of Engineers with HTRW data is supplied.
> - Project and site-specific labor and equipment rates may be entered.
> - Assembly database can dramatically increase your estimate production.
> - Modeling database stores historical data for dynamic future access.
> - The system has a networking option for multiuser access.
> - Commercial databases available from R.S. Means, and Richardson Engineering Services provide estimating standards.

O'BRIEN - KREITZBERG & ASSOCIATES, INC.
PROJECT WHATCO: Whatco Manufacturing-Whse-Office - Demonstration Project
Schematic Design Demonstration Report

TITLE PAGE 1

Whatco Manufacturing-Whse-Office
Demonstration Project

Designed By: James J. O'Brien, P.E.
Estimated By: Building Systems Design

Prepared By: Fred M. Seidell III, C.C.C.
Senior Technical Consultant

Date: 10/30/92
Est Construction Time: 360 Days

O'BRIEN - KREITZBERG & ASSOCIATES, INC.
PROJECT WHATCO: Whatco Manufacturing-Whse-Office - Demonstration Project
Schematic Design Demonstration Report

TITLE PAGE 2

BASIS OF ESTIMATE

This estimate was produced from Schematic Design Decription received from -James J O'Brien, P.E. September 12, 1992.

Sketch Plans & Designs	: Dated "NONE RECEIVED"
Original Specifications	: Dated Sept. 11, 1992

This estimate is based upon parametric measurements were used in conjunction with references from similar projects recently estimated by Building Systems Design.

BASIS FOR PRICING

Pricing shown reflects probable construction costs obtainable in the Pennsylvania area on the date of this statement of probable costs.
This estimate is a determination of fair market value for the construction of this project. It is not a prediction of low bid. Pricing assumes competitive bidding for every portion of the construction work for all subcontractors, as well as the general contractor; that is to mean 6 to 7 bids. If less bids are received, bid results can be expected to be higher.

Length of construction is assumed to be 12 months. Any costs for excessive overtime to meet stringent milestone dates are not included in this estimate.

Bid date is assumed to be April 1995. A value of 5% per annual escalation is added to the cost for the construction which is assumed to be finished March 1997.

The General Contractor's Overhead is set at 08% and his Profit margins are set at 07% on all of the work. Bond is set at 1.5%.

A 10% Design and Pricing Contingency has been included.

PROJECT DESCRIPTION

This project consists of : The construction of a Plant Office Warehouse Property in Philadelphia, Pennsylvania.

STATEMENT OF PROBABLE COST

Building Systems Design has no control over the cost of labor and materials, the general contractor's or any subcontractor's method of determining prices, or competitive bidding and market conditions. This opinion of probable cost of construction is made on the basis of experience, qualifications, and best judgement of Building Systems Design familar with the construction industry. We cannot and do not guarantee that proposals, bids or actual construction costs will not vary from this or subsequent cost estimates.

O'BRIEN · KREITZBERG & ASSOCIATES, INC.
PROJECT WHATCO: Whatco Manufacturing-Whse-Office - Demonstration Project
Schematic Design Demonstration Report
10. Plant Building

Substructure	QUANTY	UOM	ManHours	Labor	Equipmnt	Material	TOTAL COST	UNIT COST

Plant Building

SUBSTRUCTURE AND STRUCTURE:

One Story Steel Frame Structure With Heavy Wide Flange Columns On Concrete Foundations Supporting The Roof Structure. The Exterior Walls Are Supported By Concrete Foundation Walls And Footings. The Slab On Grade Has 2 Thicknesses For Different Functions. A Monorail Crane Runway Structure, and A Bridge Crane Runway Structure are also provided for.

ROOFING

The Loading Dock Has A Canopy Roof Structure Detached From The Main Roof. Built-up 20 yr.Roofing Over Rigid Precast Plank Decking. Insulation Covers The Entire Roof Area and Roof sloping is accomplished with the insulation. Gutters And Downspouts Are Used To Channel Rain Water Off The Roof Area.

EXTERIOR WALL

Consists Of Structural Steel and Girt System. Insulated Metal Panels Are Used As A Veneer. The Soffit And Fascia Consists Of Insulated Metal Panels

On Metal Studs. Hollow Metal Doors Are Used To Enter And Exit The Facility. Natural Light Is Transmitted To The Shop And Warehouse Area Through The Use Of Operable Aluminum Windows.

INTERIOR

The Interior Partitions Consist Of some ȧln Cmu And Wire Mesh. The Interior Doors Include Hollow Metal Doors And Wire Mesh. The Interior Partition Finishes Consists Of Paint. The Floors Are Finished With Floor Hardener And Vinyl Composition Tile.

The Ceilings Are 2X4 Suspended Acoustical Ceiling Tiles , Gypsum Wall Board Ceiling And paint At The Exposed Structure.

SPECIALITIES

Specialty Items Include Toilet Accessories, Steel Grating, Trench Drain, Pipe Bollards, Dock Bumpers, Metal Railing And Bridge Cranes.

HEATING

Heating Is Provided By A High Temp To Heating Hot Water Converter.

Heating Hot Water Is Distributed To Four Air Handling Units, Two Heating And Ventilating Units And Twenty Unit Heaters.

Twelve Fans Exhaust Air From The Building.
Domestic Hot Water Is Provided By Two Electric
Water Heaters. Standard Plumbing Fixtures Are Used.

O'BRIEN - KREITZBERG & ASSOCIATES, INC.
PROJECT WHATCO: Whatco Manufacturing-Whse-Office - Demonstration Project
Schematic Design Demonstration Report
20. Warehouse Building

Substructure	QUANTY	UOM	ManHours	Labor	Equipmnt	Material	TOTAL COST	UNIT COST

Warehouse Building

SUBSTRUCTURE and SUPERSTRUCTURE

Building Is Constructed Of A Steel Framed Roof Supported By Cmu Walls, Concrete Tie Beams, Continuous Footings And Deep Slab Turndowns When Required. Foundation Walls Are Used At The Interior Wall Of Electrical Room And At The Building Perimeter. The Slab On Grade Is Designed Using 2 Thicknesses.

ROOFING

The Roof Is Covered With A Precast Plank System and Rigid Insulation. The Roof is a 20 year 3 Ply built-up bitiumnous system. There are Also Gutters And Downspouts.

EXTERIOR WALL

The Exterior Wall Consists Of Two Hour Eight Inch Cmu, Continuous Steel Angles (GIRTS), Vermiculite Insulation, Rebar, Horizontal And Vertical Reinforcement, Bond Beams, And Grouting Of Cavity Walls.

The Exterior is To Be covered with Insulated Metal Siding (PrePainted). The Exterior Doors Are Both Hollow Metal Doors And 5 Overhead Doors With Motors There Are No Exterior Windows. There Are Metal Louvers At The Exterior Walls.

INTERIOR WALLS

Interior Walls The Interior Wall Partitions Consist Of
Two Hour Eight Inch Cmu, Continuous Steel Angles. Bond Beams, Vermiculite Insulation, Vertical And Horizontal Reinforcement And Grouting Of Cavity Walls. However, The Following Occurs Only At The Interior Wall Section Of The Overhead Doors: 3-5/8 In Metal Studs With Two Layers Of 5/8 Inch Fire Rated Gypsum Wallboard Stand On Top Of Two Hour Eight Inch Cmu Wall Partitions, Continuous Steel Angles, Bond Beams, Vermiculite Insulation, Vertical And Horizontal Reinforcement, And Grouting Of Cavity Walls.

INTERIOR DOORS

The Interior Doors Consist Of Both Hollow Metal Doors And Overhead Doors With Motor Operated Chain Hoist. There Are No Interior Windows. The Interior Partitions Are Finished With Epoxy Paint. The Interior Floor Is Finished With Epoxy Coating. There Are No Base Finishes. The Ceiling Is Exposed Structure.

SPECILATIES

The Only Specialty Items Are Pipe Bollards.

H.V.A.C.

Two Package Boilers also Provide Heat To Building water.

ELECTRICAL

O'BRIEN - KREITZBERG & ASSOCIATES, INC.

PROJECT WHATCO: Whatco Manufacturing-Whse-Office - Demonstration Project

Schematic Design Demonstration Report

30. Office Building

Substructure	QUANTY	UOM	ManHours	Labor	Equipmnt	Material	TOTAL COST	UNIT COST

Office Building

This Building is_an office building 80 ft x 80 ft square.

STRUCTURE and SUBSTRUCTURE

The Foundation For This Facility Consists Of Spread Footings And Continous Grade Beams. A Floor Slab Is Poured Monolithically. The Floor Slab Is Of 3000Psi Concrete Poured Over A 6In Rock Capillary Water Barrier, And Vapor Barrier With Rebar Reinforcement - 12In On Center And Construction Joints Approximately 40 ft On Center. The Structural Frame Consists Of Precast Columns Walls and Roof.

ROOFING

Roofing is 20 Year 3 Ply Built Up Roof on Light Weight Fill and 1" Insulation Board.

DOORS and WINDOWS

Entrance Exterior Doors Are Double Plate Glass and Hollow Metal Door Units. Exterior Windows Are Non-Operable, With 1 in Insulated Glass.

INTERIOR PARTITIONS

Interior Walls Are Painted Gypsum Board on Metal Stud Walls.
Interior Windows Are Aluminum Framed With 1/4In Diamond Mesh Wired Glass.

INTERIOR DOORS

Interior Doors Are Also Prehung Wood.

INTERIOR FINISHES

Ceiling Construction Consists Of Suspended Acoustical Tile
Concrete Sealer Is Applied To Storage Area Floors. Ceramic Tile Is Used In Toilet Areas. Office And Halls have Carpet.

SPECIALITIES

Specialties Include Toilet Partitions And Accessories.

EQUIPMENT

Dishwasher, Double Sink, Stove, Microwave, Cabinets in a 400 s.f.

HVAC

Heating Is Provided By A Gas Fired Furnace And Fifty Radiant Heating Panels. Cooling Is Provided By A Room Air Conditioning Unit. A Rooftop Fan Exhausts Air From The Office Area. A Gas Fired Water Heater Is Used To Heat Domestic Hot Water. Standard Plumbing Fixtures Are Used.

FIRE PROTECTION

A Wet Pipe Sprinkler System Runs Throughout The Building. This Building Is Fed Through A 112Kva Pad Mounted Transformer Which Steps The Voltage Down To 480/277V For Use. This High Voltage Is Distributed With A 225A Main Distribution Panel And Used Primarily For Lighting. A 13.2Kva Transformer Steps The Voltage Down To 120/240V, 1 Phase For Use Throughout The Building.

O'BRIEN - KREITZBERG & ASSOCIATES, INC.

PROJECT WHATCO: Whatco Manufacturing-Whse-Office - Demonstration Project

Schematic Design Demonstration Report

SUMMARY PAGE 1

** PROJECT DIRECT SUMMARY - LEVEL 1 (Rounded to 100's) **

	QUANTITY	UOM	ManHours	Labor	Equipmnt	Material	TOTAL COST	UNIT COST
10 Plant Building	30000.00	SF	25,500	920,200	39,100	1,028,300	2,337,600	77.92
20 Warehouse Building	9600.00	SF	7,200	243,800	8,200	226,300	478,300	49.82
30 Office Building	6400.00	SF	4,700	163,600	7,300	214,700	385,500	60.24
40 Site Support	6.00	ACR	800	24,300	14,000	10,800	183,700	30622.86
			-------	----------	--------	----------	----------	
Whatco Manufacturing-Whse-Office	1.00	EA	38,200	1,352,000	68,600	1,480,100	3,385,200	3385192
Overhead @ 8.0 %							270,800	

SUBTOTAL							3,656,000	
Profit @ 7.0 %							255,900	

SUBTOTAL							3,911,900	
Bond @ 1.5 %							58,700	

TOTAL INCL INDIRECTS							3,970,600	
Escalation @ 5.0 % / Annum							595,600	

SUBTOTAL							4,566,200	
Contingency @ 10.0 %							456,600	

TOTAL INCL OWNER COSTS							5,022,800	

O'BRIEN - KREITZBERG & ASSOCIATES, INC.

PROJECT WHATCO: Whatco Manufacturing-Whse-Office - Demonstration Project

Schematic Design Demonstration Report

SUMMARY PAGE 2

** PROJECT DIRECT SUMMARY - LEVEL 2 (Rounded to 100's) **

		QUANTITY	UOM	ManHours	Labor	Equipmnt	Material	TOTAL COST	UNIT COST
10	Plant Building								
10.01	Substructure	30000.00	SF	4,500	145,000	3,300	117,600	265,900	8.86
10.02	Structural Frame	30000.00	SF	3,300	123,300	15,200	282,900	421,400	14.05
10.03	Roofing	30000.00	SF	600	48,300	10,000	43,500	101,700	3.39
10.04	Exterior Closure	30000.00	SF	1,600	54,900	700	85,800	141,400	4.71
10.05	Interior Construction	30000.00	SF	1,100	36,100	300	18,000	54,400	1.81
10.06	Interior Finishes	30000.00	SF	1,100	35,100	500	14,500	50,100	1.67
10.07	Specialties	30000.00	SF	300	9,500	3,700	63,800	77,000	2.57
10.08	Plumbing	30000.00	SF	1,200	41,300	700	45,200	87,200	2.91
10.09	Heating,Ventilation & Air Condit	30000.00	SF	4,300	153,900	2,000	145,900	301,800	10.06
10.10	Special Mechanical Systems	30000.00	SF	200	5,500	100	34,500	40,100	1.34
10.11	Interior Electrical	30000.00	SF	5,300	194,900	800	110,600	306,300	10.21
10.12	Special Interior Electrical Syst	30000.00	SF	400	13,300	0	5,800	19,200	0.64
10.13	Equipment & Conveying	30000.00	SF	1,700	59,100	1,900	60,200	471,200	15.71
	Plant Building	30000.00	SF	25,500	920,200	39,100	1,028,300	2,337,600	77.92
20	Warehouse Building								
20.01	Substructure	9600.00	SF	2,200	70,300	1,400	75,000	146,700	15.28
20.02	Structural Frame	9600.00	SF	400	14,800	2,400	42,000	59,300	6.18

20.03	Roofing	9600.00 SF	200	6,900	100	4,100	11,100	1.16
20.04	Exterior Closure	9600.00 SF	2,700	90,600	700	39,600	130,900	13.63
20.05	Interior Construction	9600.00 SF	100	2,400	0	2,200	4,700	0.48
20.06	Interior Finishes	9600.00 SF	100	3,000	100	1,800	5,000	0.52
20.07	Specialties	9600.00 SF	100	2,000	2,800	5,900	10,700	1.11
20.08	Plumbing	9600.00 SF	100	2,400	0	1,700	4,100	0.43
20.09	Heating,Ventilation & Air Condit	9600.00 SF	500	19,000	200	27,900	47,100	4.90
20.10	Special Mechanical Systems	9600.00 SF	600	20,700	300	7,100	28,200	2.93
20.11	Interior Electrical	9600.00 SF	300	11,700	0	18,800	30,600	3.18
			--------	----------	--------	----------	----------	
	Warehouse Building	9600.00 SF	7,200	243,800	8,200	226,300	478,300	49.82
30	Office Building							
30.01	Substructure	6400.00 SF	900	29,300	1,500	26,100	56,900	8.88
30.02	Structural Frame	6400.00 SF	500	18,200	2,700	54,600	75,500	11.79
30.03	Roofing	6400.00 SF	400	14,500	1,700	30,700	47,000	7.34
30.04	Exterior Closure	6400.00 SF	1,100	39,600	500	44,200	84,300	13.17
30.05	Interior Construction	6400.00 SF	200	5,600	100	5,800	11,500	1.79
30.06	Interior Finishes	6400.00 SF	800	26,200	300	20,400	46,900	7.33
30.07	Specialties	6400.00 SF	0	1,700	400	4,500	6,700	1.04
30.08	Plumbing	6400.00 SF	100	2,600	0	2,700	5,400	0.84
30.09	Heating,Ventilation & Air Condit	6400.00 SF	300	10,000	100	12,600	22,700	3.55
30.10	Special Mechanical Systems	6400.00 SF	0	300	0	2,400	2,700	0.42
30.11	Interior Electrical	6400.00 SF	400	14,400	100	9,800	24,300	3.79
30.12	Special Interior Electrical Syst	6400.00 SF	0	1,100	0	800	1,900	0.30

O'BRIEN - KREITZBERG & ASSOCIATES, INC.

PROJECT WHATCO: Whatco Manufacturing-Whse-Office - Demonstration Project

Schematic Design Demonstration Report

SUMMARY PAGE 3

** PROJECT DIRECT SUMMARY - LEVEL 2 (Rounded to 100's) **

	QUANTITY UOM	ManHours	Labor	Equipmnt	Material	TOTAL COST	UNIT COST
Office Building	6400.00 SF	4,700	163,600	7,300	214,700	385,500	60.24
40 Site Support							
40.01 Site Preparation	6.00 ACR	700	19,000	14,000	9,600	42,600	7093.88
40.10 Electrical Services	3000.00 LF	200	5,400	0	1,200	36,200	12.06
40.20 Water Services		0	0	0	0	105,000	
Site Support	6.00 ACR	800	24,300	14,000	10,800	183,700	30622.86
Whatco Manufacturing-Whse-Office	1.00 EA	38,200	1,352,000	68,600	1,480,100	3,385,200	3385192
Overhead @ 8.0 %						270,800	
SUBTOTAL						3,656,000	

SUBTOTAL		3,911,900
Bond	@ 1.5 %	58,700
		....
TOTAL INCL INDIRECTS		3,970,600
Escalation	@ 5.0 % / Annum	595,600
		....
SUBTOTAL		4,566,200
Contingency	@ 10.0 %	456,600
		....
TOTAL INCL OWNER COSTS		5,022,800

Computerized schematic estimate (I.C.E.)

A computerized cost estimate for the WHATCO project was run using the I.C.E. system by Management Computer Controls, Inc. (MC^2), under the direction of Kevitt Adler, president of MC^2. All of the estimated price was done by the system except Divisions 14, 15, and 16. These were based on square foot costs. The following assumptions were made:

- Escalation assumed to be 5% per year with midpoint of construction in mid-1995. This equals 13%. This was entered as a single line in the estimate but could have just as easily been spread throughout.
- A 10% contingency was figured; this was also entered as a line item.
- All three buildings were figured as free-standing.
- Windows were figured to be 15% of exterior closure in the office area.
- Interior partitions were estimated at 15 lf per 100 sf of building in office.
- Doors were estimated at 1 door per 25 lf of partition in office.
- Structural steel at roof was figured at 6 psf in the plant and 5 psf in the warehouse.
- The crane rail support was estimated to be 125 lb/lf.

Costs excluded were:

- Site costs including railroad siding.
- General conditions, fees, permits, etc.

The I.C.E. schematic estimate is in the following order:

Input data sheets. The *list of takeoff data without diagnostics* shows the raw takeoff, which was entered using MC^2's construction manager module.

Quantity survey report. This is the next step in the audit trail. It lists the raw takeoff from the previous step along with the items and quantities generated.

The estimate by system. This is the format usually favored by MC^2 for cost presentations. You will see detail and recap reports for the plant, the warehouse, and the office. Costs are organized by system (or component), in a Uniformat-style cost reporting structure.

Comparison report. This format lines up the costs for the plant, warehouse, office, and total job.

File - 3256 Estimator - DOUG
Name - Project Name not assigned

Base 1 PC ICE System
LIST OF TAKEOFF DATA
WITHOUT DIAGNOSTICS

Page - 1

ESTIMATING SYSTEMS BY:
Management Computer Controls, Inc.

SEQ NO.	SEC	B	C	D	E	F	G	H	J	K	A	SPEC SELECT 9876543210	ELEM	LOCATIONS S/D ASSY
01.000 FOUNDATIONS														
1	01	6.00	4.00	2.00	2.00						24	23021	0110	1
2	01	20.00									96	00000003	0120	1
3	01	800.00	3.00	.67	3.00						1	20341	0110	1
4	01	800.00	4.00								1	200003	0110	1
02.000 SUBSTRUCTURE														
5	01	300.00	100.00	1.00	.50						1	210012	0210	1
03.000 CONCRETE STRUCTURAL FRAME														
6	01	300.00	100.00								1	201	0320	1
03.005 STEEL & WOOD FRAME SYSTEMS														
7	01	300.00	100.00	6.00							1	11	0320	1
04.000 EXTERIOR WALLS														
8	01	800.00	24.00								1	2005	0410	1
04.005 EXTERIOR DOORS & WINDOWS														
9	01										10	00001	0420	1
10	01	12.00	20.00								6	000002	0420	1
05.000 ROOFING & ROOF ACCESSORIES														
11	01	300.00	100.00								1	21	0500	1
06.210 FINISHES														
12	01	30000.00									1	1009	0210	1
11.000 EQUIPMENT & FURNISHINGS														
13	01										3	1	1110	1
01.000 FOUNDATIONS														
14	02	4.00	4.00	3.00	3.00						14	22021	0110	1
15	02	400.00	3.00	.67	3.00						1	20341	0110	1
16	02	400.00	4.00								1	200003	0110	1

```
02.000  SUBSTRUCTURE
  17   02   120.00    80.00      .50      .50                                   1  210012       0210       1

03.000  CONCRETE STRUCTURAL FRAME
  18   02   120.00    80.00                                                     1  201          0320       1

03.005  STEEL & WOOD FRAME SYSTEMS
  19   02   120.00    80.00     5.00                                            1  11           0320       1
  20   02    55.00    15.00                                                     1  20101        0310       1

04.000  EXTERIOR WALLS
  21   02   400.00    24.00                                                     1  2005         0410       1

04.005  EXTERIOR DOORS & WINDOWS
  22   02                                                                       6  00001        0420       1
  23   02                                                                       2  000001       0420       1

05.000  ROOFING & ROOF ACCESSORIES
  24   02   120.00    80.00                                                     1  21           0500       1

06.200  INTERIOR PARTITIONS
  25   02   166.00    10.00                                                     1  21           0610       1

06.205  INTERIOR DOORS & FRAMES
  26   02     1.00                                                              2  0115         0610       1

06.210  FINISHES
  27   02   166.00              10.00                                           2  27000001     0620       1
  28   02   800.00                                                              1  1000000012   0620       1
  29   02  9600.00                                                              1  1009         0620       1

11.000  EQUIPMENT & FURNISHINGS
  30   02                                                                       2  1            1110       1

01.000  FOUNDATIONS
  31   03   320.00     2.00     1.00     2.00                                   1  21001        0110       1
```

SEQ NO.	SEC	B	C	D	E	F	G	H	J	K	A	SPEC SELECT 9876543210	ELEM	LOCATIONS S/D	ASSY
01.000 FOUNDATIONS															
32	03	4.00	4.00	1.34	2.00						16	22021	0110	1	
33	03	320.00	4.00								1	200002	0110	1	
02.000 SUBSTRUCTURE															
34	03	80.00	80.00	.34	.34						1	210011	0210	1	
35	03	320.00	.67	2.00							1	220303	0210	1	
03.000 CONCRETE STRUCTURAL FRAME															
36	03	80.00	80.00								1	205	0320	1	
04.000 EXTERIOR WALLS															
37	03	320.00	18.00								1	23	0410	1	
38	03	320.00	15.00								1	20002062	0410	1	
04.005 EXTERIOR DOORS & WINDOWS															
39	03										2	00001	0420	1	
40	03										1	0002	0420	1	
41	03	720.00									1	102	0420	1	
05.000 ROOFING & ROOF ACCESSORIES															
42	03	80.00	80.00								1	21	0500	1	
06.200 INTERIOR PARTITIONS															
43	03	960.00	15.00								1	201	0610	1	
06.205 INTERIOR DOORS & FRAMES															
44	03	1.00									39	8001	0610	1	
06.210 FINISHES															
45	03	4950.00									1	10004	0620	1	
46	03	1000.00									1	1001	0620	1	
47	03	450.00									1	101	0620	1	
48	03	6400.00									1	1000000003	0620	1	
49	03	2120.00									1	270002	0620	1	
											1	270003	0620	1	

50	03	120.00						
51	03	1660.00	9.00		1	27000001	0620	1
52	03	550.00	9.00		1	27000004	0620	1
53	03	30.00		8.00	1	2700001	0620	1

06.220 TOILET & BATH ACCESSORIES

54	03	1.00	72.00	42.00	2	000001	0630	1

06.230 MILLWORK, CASEWORK

55	03	6.00			2	5	0630	1
56	03	16.00			1	009	0630	1
57	03	3.00			5	0001	0630	1

MISCELLANEOUS ITEM

SEQ NO.	SEC	ITEM CODE	DESCRIPTION	U/M	QTY	LABOR UNIT	MATERIAL UNIT	SUBCONT/EQUIP UNIT		ELEM	S/D	ASSY
58	03	0610.901	ROOF BLOCKING	BDFT	640	1.250	.450			0500		
59	02	0610.901	ROOF BLOCKING	BDFT	800	1.250	.450			0500		
60	01	0610.901	ROOF BLOCKING	BDFT	1600	1.250	.450			0500		
61	03	0610.902	MISC CARPENTRY	BDFT	1000	1.150	.350			0610		
62	03	0790.903	CAULKING	ALLO				1200.00	SC	0410		
63	02	0790.902	CAULKING	ALLO				1500.00	SC	0410		
64	01	0790.901	CAULKING	ALLO				3000.00	SC	0410		
65	03	1000.903	SPECIALTIES	ALLO				5000.00	SC	0630		
66	02	1000.902	SPECIALTIES	ALLO				6500.00	SC	0630		
67	03	1145.900	KITCHEN EQUIPMENT	ALLO				1800.00	SC	0630		
68	03	1549.903	PLUMBING	SQFT	6400			2.00	SC	0810		
69	02	1549.902	PLUMBING	SQFT	9600			1.30	SC	0810		
70	01	1549.901	PLUMBING	SQFT	30000			.75	SC	0810		
71	03	1599.903	HVAC	SQFT	6400			8.00	SC	0820		
72	02	1599.902	HVAC	SQFT	9600			5.30	SC	0820		
73	01	1599.901	HVAC	SQFT	30000			5.30	SC	0820		
74	03	1559.903	FIRE PROTECTION	SQFT	6400			1.50	SC	0830		
75	02	1559.902	FIRE PROTECTION	SQFT	9600			1.30	SC	0830		
76	01	1559.901	HYDRANT SYSTEM	LNFT	700			14.25	SC	0830		
77	03	1699.903	ELECTRICAL	SQFT	6400			5.75	SC	0900		
78	02	1699.902	ELECTRICAL	SQFT	9600			4.10	SC	0900		

File - 3256 Estimator - DOUG
Name - Project Name not assigned

Base 1 PC ICE System
LIST OF TAKEOFF DATA
WITHOUT DIAGNOSTICS

ESTIMATING SYSTEMS BY:
Management Computer Controls, Inc.

MISCELLANEOUS ITEM

SEQ NO.	SEC	ITEM CODE	DESCRIPTION	U/M	QTY	LABOR UNIT	MATERIAL UNIT	SUBCONT/EQUIP UNIT		ELEM	S/D	ASSY
79	01	1699.901	ELECTRICAL	SQFT	30000			6.50	SC	0900		
80	02	0510.900	STRUCT STL GIRT	CWT	274					0410		
81	01	0510.900	STRUCT STL GIRT	CWT	538					0410		
82	02	0551.902	MISC METALS	ALLO				2500.00	SC	0320		
83	01	0551.901	MISC METALS	ALLO				6500.00	SC	0320		
84	02	1100.901	RACK SYSTEM - BY OWNER							1130		
85	02	1100.902	BIN SYSTEM - BY OWNER							1130		
86	01	0510.901	CRANE RAIL SUPPORT	CWT	750					0700		
87	01	1430.901	BRIDGE CRANE	LS				300000.00	SC	0700		
88	01	1430.902	MONORAIL	LS				50000.00	SC	0700		
89	01	0198.901	ESCALATION @ 13% - MID 95	LS						1300		
90	02	0198.902	ESCALATION @ 13% - MID 95	LS						1300		
91	03	0198.903	ESCALATION @ 13% - MID 95	LS						1300		
92	01	0199.901	CONTINGENCY @ 10%	LS						1400		
93	02	0199.902	CONTINGENCY @ 10%	LS						1400		
94	03	0199.903	CONTINGENCY @ 10%	LS						1400		

File ID - 3256
Client Job No -
Project Name - WHATCO PROJECT
Project Size - 46,000 SQFT
Estimator - DOUG

QUANTITY SURVEY REPORT

Page - 1
Date -
Time -

MANAGEMENT COMPUTER CONTROLS, INC.
2881 DIRECTORS COVE
MEMPHIS, TENNESSEE 38131

1.000 FOUNDATIONS

------------------------DIMENSIONS--SPECIFICATION SELECTION-----------------

DIMENSIONS		SPECIFICATION SELECTION	
B-SQFT,LENGTH (FEET & DECIMAL OF	G-PILASTERS (WHOLE NUMBER)	9-METHOD OF CALCULATING AREA	4-FOUNDATION INSULATION
C-WIDTH,HEIGHT,SHAFT DIA (FEET &	H-PILASTER WIDTH (FEET & DECIMAL	8-TYPE OF FOUNDATION	3-CAISSON
D-DEPTH,THICKNESS (FEET & DECIMA	J-PILASTER PROJECTION (FEET & DE	7-TYPE OF FOUNDATION WALL	2-PILING
E-EXCAVATION DEPTH (FEET & DECIM	K-NOT USED	6-FORMS	
F-REINFORCING STEEL (LBS & DECIM	A-QUANTITY	5-REINFORCING	

WS	LINE SEQ	B	C	D	E	F	G	H	J	K	A	SPEC. SEL. 9876543210	S/D SEC	ELEM	LOC	MARK FIELD
DW	1	6.00	4.00	2.00	2.00						24	23021	01	0110	01	
DW	2	20.00									96	00000003	01	0120	01	
DW	3	800.00	3.00	.67	3.00						1	20341	01	0110	01	
DW	4	800.00	4.00								1	200003	01	0110	01	
DW	14	4.00	4.00	3.00	3.00						14	22021	02	0110	01	
DW	15	400.00	3.00	.67	3.00						1	20341	02	0110	01	
DW	16	400.00	4.00								1	200003	02	0110	01	
DW	31	320.00	2.00	1.00	2.00						1	21001	03	0110	01	
DW	32	4.00	4.00	1.34	2.00						16	22021	03	0110	01	
DW	33	320.00	4.00								1	200002	03	0110	01	

S/D SEC	ELEM	ITEM CODE	DESCRIPTION	U/M	QUANTITY
03	0110	222.300	EXCAVATE CONT. FOOTING	CUYD	48
02	0110	222.301	EXCAVATE COLUMN FOOTING	CUYD	50
03	0110	222.301	EXCAVATE COLUMN FOOTING	CUYD	38
01	0110	222.302	EXCAVATE PILE CAP	CUYD	79
01	0110	222.308	EXCAVATE GRADE BEAM	CUYD	238

02	0110	222.308	EXCAVATE GRADE BEAM	CUYD	119
03	0110	222.320	BACKFILL CONT. FOOTING	CUYD	27
02	0110	222.321	BACKFILL COLUMN FOOTING	CUYD	28
03	0110	222.321	BACKFILL COLUMN FOOTING	CUYD	28
01	0110	222.322	BACKFILL PILE CAP	CUYD	40
01	0110	222.328	BACKFILL GRADE BEAM	CUYD	196
02	0110	222.328	BACKFILL GRADE BEAM	CUYD	98
03	0110	222.330	FINE GRADE CONT. FOOTING	SQFT	640
02	0110	222.331	FINE GRADE COLUMN FOOTING	SQFT	224
03	0110	222.331	FINE GRADE COLUMN FOOTING	SQFT	256
01	0110	222.332	FINE GRADE PILE CAP	SQFT	576
01	0120	232.120	STEEL H PILE	LNFT	1920
02	0110	310.005	COLUMN FTG FORMS	SQFT	672
03	0110	310.005	COLUMN FTG FORMS	SQFT	344
01	0110	310.010	PILE CAP FORMS	SQFT	960
01	0110	310.025	GRADE BEAM FORMS	SQFT	4800
02	0110	310.025	GRADE BEAM FORMS	SQFT	2400
03	0110	321.100	RE-STEEL @ CONTINUOUS FTG	CWT	10
02	0110	321.110	RE-STEEL @ COLUMN FOOTING	CWT	22
03	0110	321.110	RE-STEEL @ COLUMN FOOTING	CWT	12
01	0110	321.120	RE-STEEL @ PILE CAP	CWT	40
01	0110	321.200	RE-STEEL @ GRADE BEAM	CWT	34
02	0110	321.200	RE-STEEL @ GRADE BEAM	CWT	17
03	0110	333.000	CONCRETE @ CONT. FTG	CUYD	25
02	0110	333.005	CONCRETE @ COLUMN FTG	CUYD	27
03	0110	333.005	CONCRETE @ COLUMN FTG	CUYD	14
01	0110	333.010	CONCRETE @ PILE CAP	CUYD	45
01	0110	333.025	CONCRETE @ GRADE BEAM	CUYD	63
02	0110	333.025	CONCRETE @ GRADE BEAM	CUYD	32
03	0110	720.020	1-1/2" FOUNDATION INSULATION	SQFT	1280
01	0110	720.030	2" FOUNDATION INSULATION	SQFT	3200
02	0110	720.030	2" FOUNDATION INSULATION	SQFT	1600

File ID - 3256	QUANTITY SURVEY REPORT	Page - 2
Client Job No -		Date -
Project Name - WHATCO PROJECT		Time -
Project Size - 46,000 SQFT		MANAGEMENT COMPUTER CONTROLS, INC.
Estimator - DOUG		2881 DIRECTORS COVE
		MEMPHIS, TENNESSEE 38131

2.000 SUBSTRUCTURE

---------------------------DIMENSIONS---------------------------SPECIFICATION SELECTION----------

DIMENSIONS		SPECIFICATION SELECTION	
B-SQFT,LENGTH (FEET & DECIMAL OF	G-LNFT SHORING OR LAYBACK (FEET	9-METHOD OF CALCULATING AREA	4-REINFORCING
C-SLAB WIDTH,WALL HEIGHT,EDGE FO	H-PIT,TRENCH-SLAB THICKNESS,STAI	8-TYPE OF CONSTRUCTION	3-WATERPROOFING
D-SLAB/WALL THK, BSMT/PIT/TRENCH	J-PIT,TRENCH-WALL THICKNESS,STAI	7-TYPE OF CONSTRUCTION (CONT.)	2-SHORING BRACING LAYBACK
E-UNDERSLAB FILL THICKNESS (FEET	K-STAIR # OF RISERS	6-FORMS	1-FOUNDATION DRAINAGE
F-REINFORCING STEEL (LBS & DECIM	A-QUANTITY	5-CONCRETE FINISH	0-

WS	LINE SEQ	B	C	D	E	F	G	H	J	K	A	SPEC. SEL. 9876543210	S/D SEC	ELEM	LOC	MARK FIELD
DW	5	300.00	100.00	1.00	.50						1	210012	01	0210	01	
DW	17	120.00	80.00	.50	.50						1	210012	02	0210	01	
DW	34	80.00	80.00	.34	.34						1	210011	03	0210	01	
DW	35	320.00	.67	2.00							1	220303	03	0210	01	

S/D	SEC	ELEM	ITEM CODE	DESCRIPTION	U/M	QUANTITY
	03	0210	222.304	EXCAVATE THICKENED SLAB	CUYD	16
	01	0210	222.334	FINE GRADE SLAB ON GRADE	SQFT	30000
	02	0210	222.334	FINE GRADE SLAB ON GRADE	SQFT	9600
	03	0210	222.334	FINE GRADE SLAB ON GRADE	SQFT	6400
	01	0210	222.350	UNDERSLAB FILL	CUYD	584
	02	0210	222.350	UNDERSLAB FILL	CUYD	187
	03	0210	222.350	UNDERSLAB FILL	CUYD	85
	01	0210	301.019	TROWEL CEMENT FINISH	SQFT	30000
	02	0210	301.019	TROWEL CEMENT FINISH	SQFT	9600

03	0210	301.019	TROWEL CEMENT FINISH	SQFT	6400
03	0210	301.020	POINT & PATCH	SQFT	640
01	0210	301.050	PROTECT & CURE	SQFT	30000
02	0210	301.050	PROTECT & CURE	SQFT	9600
03	0210	301.050	PROTECT & CURE	SQFT	6400
03	0210	310.105	SLAB ON GRADE EDGE FORMS	SQFT	640
03	0210	320.000	6X6-10/10 MESH	SQS	74
01	0210	320.002	6X6-6/6 MESH	SQS	347
02	0210	320.002	6X6-6/6 MESH	SQS	111
03	0210	321.009	RE-STEEL @ SLAB ON GRADE	CWT	7
01	0210	333.035	CONCRETE @ SLAB ON GRADE	CUYD	1145
02	0210	333.035	CONCRETE @ SLAB ON GRADE	CUYD	184
03	0210	333.035	CONCRETE @ SLAB ON GRADE	CUYD	100

QUANTITY SURVEY REPORT

File ID - 3256
Client Job No -
Project Name - WHATCO PROJECT
Project Size - 46,000 SQFT
Estimator - DOUG

Page - 3
Date -
Time -

MANAGEMENT COMPUTER CONTROLS, INC.
2881 DIRECTORS COVE
MEMPHIS, TENNESSEE 38131

3.000 CONCRETE STRUCTURAL SYSTEM

```
------------------------DIMENSIONS-------------------------------------------------------SPECIFICATION SELECTION-----------------

B-SQFT,LENGTH (FEET & DECIMAL OF G-# OF VOIDS (WHOLE NUMBER)       9-METHOD OF CALCULATING AREA      4-MISC.
C-WIDTH (SLAB,BEAM,COLUMN),HEIGH H-LENGTH OF VOID (FEET & DECIMAL 8-CONCRETE SLAB CONSTRUCTION      3-FORMS
D-SLAB & WALL THICKNESS,BEAM DEP J-WIDTH OF VOID (FEET & DECIMAL  7-PRECAST SLAB CONSTRUCTION       2-CONCRETE SLAB FINISH
E-REINFORCING (LBS & DECIMAL OF  K-DIA OF CAPITAL (FEET & DECIMAL 6-BEAMS                           1-UNDER SLAB/BEAM/COLUMN/WALL FI
F-DEPTH OF PAN,DOME (FEET & DECI A-QUANTITY                       5-COLUMNS                         0-REINFORCING
```

WS	LINE SEQ	B	C	D	E	F	G	H	J	K	A	SPEC. SEL. 9876543210	S/D SEC	ELEM	LOC	MARK FIELD
DW	6	300.00	100.00								1	201	01	0320	01	
DW	18	120.00	80.00								1	201	02	0320	01	
DW	36	80.00	80.00								1	205	03	0320	01	

S/D SEC	ELEM	ITEM CODE	DESCRIPTION	U/M	QUANTITY
01	0320	340.100	PRECAST PLANK	SQFT	30000
02	0320	340.100	PRECAST PLANK	SQFT	9600
03	0320	341.000	PC STRUCTURAL SYSTEM	SQFT	6400

File ID - 3256
Client Job No -
Project Name - WHATCO PROJECT
Project Size - 46,000 SQFT
Estimator - DOUG

QUANTITY SURVEY REPORT

Page - 4
Date -
Time -

MANAGEMENT COMPUTER CONTROLS, INC.
2881 DIRECTORS COVE
MEMPHIS, TENNESSEE 38131

3.005 STEEL & WOOD FRAME STRUCTUAL SYSTEM

------------------------DIMENSIONS------------------------SPECIFICATION SELECTION----------------

B-SQFT,LENGTH (FEET & DECIMAL OF
C-WIDTH (FEET & DECIMAL OF FOOT
D-STRUCT STL-LBS/SQFT, WOOD FRM-

A-QUANTITY

9-METHOD OF CALCULATING AREA
8-STEEL FRAME
7-WOOD FRAME
6-METAL DECK
5-WOOD DECK

4-FIREPROOFING
3-PRE-FAB METAL BUILDING

WS	LINE SEQ	B	C	D	E	F	G	H	J	K	A	SPEC. SEL. 9876543210	S/D SEC	ELEM	LOC	MARK FIELD
DW	7	300.00	100.00	6.00								12-11	01	0320	01	
DW	19	120.00	80.00	5.00								1 2-11	02	0320	01	
DW	20	55.00	15.00									1 20101	02	0310	01	

S/D SEC	ELEM	ITEM CODE	DESCRIPTION	U/M	QUANTITY
01	0320	501.011	STRUCTURAL STEEL FRAME	CWT	~~18~~ 1800
02	0320	501.011	STRUCTURAL STEEL FRAME	CWT	~~6~~ 480
02	0310	615.100	WOOD JOIST	SQFT	825
02	0310	615.500	PLYWOOD DECK	SQFT	825

```
   File ID - 3256                QUANTITY SURVEY REPORT                              Page -      5
Client Job No -                                                                      Date -
Project Name   - WHATCO PROJECT                                                      Time -
Project Size   -    46,000 SQFT                                   MANAGEMENT COMPUTER CONTROLS, INC.
Estimator      - DOUG                                             2881 DIRECTORS COVE
                                                                  MEMPHIS, TENNESSEE    38131

      4.000    EXTERIOR WALLS

-----------------------DIMENSIONS--------------------------------------SPECIFICATION SELECTION---------------

B-SQFT,LENGTH (FEET & DECIMAL OF                      9-METHOD OF CALCULATING AREA   4-CORE CONSTRUCTION     (CONT)
C-HEIGHT (FEET & DECIMAL OF FOOT                      8-EXTERIOR FACING              3-INSULATION
D-THICKNESS OF BLOCK (FEET & DEC                      7-EXTERIOR FACING (CONT)       2-EXTERIOR SHEATHING
                                                      6-EXTERIOR FACING (CONT)       1-SOFFIT
                          A-QUANTITY                  5-CORE CONSTRUCTION            0-BALCONY RAILING

WS LINE    B        C       D     E     F     G     H     J     K    A  SPEC. SEL.   S/D     MARK FIELD
   SEQ                                                                  9876543210 SEC  ELEM LOC

DW   8   800.00   24.00                                                1 2005        01    0410 01
DW  21   400.00   24.00                                                1 2005        02    0410 01
DW  37   320.00   18.00                                                1 23          03    0410 01
DW  38   320.00   15.00                                                1 20002062    03    0410 01

            S/D  SEC  ELEM  ITEM CODE       DESCRIPTION                  U/M    QUANTITY

                 03   0410   340.320   PRECAST PANEL ARCH FINISH         SQFT       5760
                 03   0410   720.061   4" BATT INSULATION                SQFT       4800
                 01   0410   740.010   INSULATED METAL SIDING            SQFT      19200
                 02   0410   740.010   INSULATED METAL SIDING            SQFT       9600
                 03   0410   926.580   GYPSUM SHEATHING                  SQFT       4800
                 03   0410   926.601   4"MTL STUD W/GYPSUM-1 SIDE        SQFT       4800
```

File ID - 3256

QUANTITY SURVEY REPORT

Page - 6
Date -
Time -

Client Job No -
Project Name - WHATCO PROJECT
Project Size - 46,000 SQFT
Estimator - DOUG

MANAGEMENT COMPUTER CONTROLS, INC.
2881 DIRECTORS COVE
MEMPHIS, TENNESSEE 38131

4.005 EXTERIOR DOORS & WINDOWS

------------------------DIMENSIONS------------------------ ------------------------SPECIFICATION SELECTION------------------------

B-SQFT,LENGTH (FEET & DECIMAL OF
C-HEIGHT (FEET & DECIMAL OF FOOT
A-QUANTITY

9-METHOD OF CALCULATING AREA
8-EXTERIOR GLASS & GLAZING
7-WINDOWS
6-GLASS DOORS
5-EXTERIOR DOORS
4-OVERHEAD DOORS
3-EXTERIOR SILLS
2-INTERIOR SILLS

WS	LINE SEQ	B	C	D	E	F	G	H	J	K	A	SPEC. SEL. 9876543210	S/D SEC	ELEM	LOC	MARK FIELD
DW	9										10	00001	01	0420	01	
DW	10	12.00	20.00								6	000002	01	0420	01	
DW	22										6	00001	02	0420	01	
DW	23										2	000001	02	0420	01	
DW	39										2	00001	03	0420	01	
DW	40										1	0002	03	0420	01	
DW	41	720.00									1	102	03	0420	01	

S/D SEC	ELEM	ITEM CODE	DESCRIPTION	U/M	QUANTITY
01	0420	805.100	HM FRAME	EACH	10
02	0420	805.100	HM FRAME	EACH	6
03	0420	805.100	HM FRAME	EACH	2
01	0420	805.200	HM DOOR	EACH	10
02	0420	805.200	HM DOOR	EACH	6
03	0420	805.200	HM DOOR	EACH	2

02	0420	831.200	MANUAL OPERATED OH DOOR	EACH	2
01	0420	831.211	12X ELECT OPR OH DR	EACH	6
03	0420	851.110	FIXED WINDOWS	SQFT	720
01	0420	871.027	FINISH HARDWARE ALLOWANCE	OPNG	10
02	0420	871.027	FINISH HARDWARE ALLOWANCE	OPNG	6
03	0420	871.027	FINISH HARDWARE ALLOWANCE	OPNG	2
03	0420	885.205	PAIR ALUM GLASS DOOR	EACH	1
01	0420	990.008	PAINT EXTERIOR DOOR	SIDE	20
02	0420	990.008	PAINT EXTERIOR DOOR	SIDE	12
03	0420	990.008	PAINT EXTERIOR DOOR	SIDE	4
01	0420	990.032	PAINT DOOR FRAME	EACH	10
02	0420	990.032	PAINT DOOR FRAME	EACH	6
03	0420	990.032	PAINT DOOR FRAME	EACH	2

File ID - 3256
Client Job No -
Project Name - WHATCO PROJECT
Project Size - 46,000 SQFT
Estimator - DOUG

QUANTITY SURVEY REPORT

Page - 7
Date -
Time -

MANAGEMENT COMPUTER CONTROLS, INC.
2881 DIRECTORS COVE
MEMPHIS, TENNESSEE 38131

5.000 ROOFING, SHEETMETAL, & ACCESSORIES

------------------------DIMENSIONS------------------------SPECIFICATION SELECTION------------------

B-SQFT, LENGTH (FEET & DECIMAL OF
C-WIDTH (FEET & DECIMAL OF FOOT)

A-QUANTITY

9-METHOD OF CALCULATING AREA
8-TYPE OF ROOF SYSTEM
7-TYPE OF SHINGLE
6-OTHER TYPE OF ROOF
5-SKYLIGHT

4-CEMENTITIOUS DECK
3-FLASHING
2-GUTTER & DOWNSPOUT
1-MISC

WS	LINE SEQ	B	C	D	E	F	G	H	J	K	A	SPEC. SEL. 9876543210	S/D SEC	ELEM	LOC	MARK FIELD
DW	11	300.00	100.00								1	21	01	0500	01	
DW	24	120.00	80.00								1	21	02	0500	01	
DW	42	80.00	80.00								1	21	03	0500	01	

S/D SEC	ELEM	ITEM CODE	DESCRIPTION	U/M	QUANTITY
01	0500	750.100	BUILT-UP ROOF SYSTEM	SQS	300
02	0500	750.100	BUILT-UP ROOF SYSTEM	SQS	96
03	0500	750.100	BUILT-UP ROOF SYSTEM	SQS	64

File ID - 3256
Client Job No -
Project Name - WHATCO PROJECT
Project Size - 46,000 SQFT
Estimator - DOUG

QUANTITY SURVEY REPORT

Page - 8
Date -
Time -

MANAGEMENT COMPUTER CONTROLS, INC.
2881 DIRECTORS COVE
MEMPHIS, TENNESSEE 38131

6.200 INTERIOR PARTITIONS

---DIMENSIONS--- / ---SPECIFICATION SELECTION---

B-SQFT, LENGTH (FEET & DECIMAL O
C-HEIGHT (FEET & DECIMAL OF FOOT
D-THICKNESS OF BLOCK (FEET & DEC

A-QUANTITY

9-METHOD OF CALCULATING
8-MASONARY PARTITION
7-METAL STUD & GYPSUM PARTITION
6-WOOD STUD & GYPSUM PARTITION
5-ADDITIONAL LAYER OF GYPSUM
4-PLASTER PARTITION
3-MISC PARTITION
2-GLASS PARTITION & WINDOWS
1-RAILING
0-MISC. PARTITION

WS	LINE SEQ	B	C	D	E	F	G	H	J	K	A	SPEC. SEL. 9876543210	S/D SEC	ELEM	LOC	MARK FIELD
DW	25	166.00	10.00								1	21	02	0610	01	
DW	43	960.00	15.00								1	201	03	0610	01	

S/D SEC	ELEM	ITEM CODE	DESCRIPTION	U/M	QUANTITY
02	0610	402.105	CONCRETE BLOCK	SQFT	1660
03	0610	926.620	STANDARD DRYWALL PARTITION	SQFT	14400

QUANTITY SURVEY REPORT

File ID - 3256
Client Job No -
Project Name - WHATCO PROJECT
Project Size - 46,000 SQFT
Estimator - DOUG

Page - 9
Date -
Time -

MANAGEMENT COMPUTER CONTROLS, INC.
2881 DIRECTORS COVE
MEMPHIS, TENNESSEE 38131

6.205 INTERIOR DOORS & FRAMES

------------------------DIMENSIONS------------------------SPECIFICATION SELECTION----------------

B-# OF DOORS PER OPENING (WHOLE

A-QUANTITY

9-WOOD DOORS
8-METAL DOORS
7-FRAMES
6-HARDWARE ALLOWANCE
5-THRESHOLD

4-MISC DOORS

WS	LINE SEQ	B	C	D	E	F	G	H	J	K	A	SPEC. SEL. 9876543210	S/D SEC	ELEM	LOC	MARK FIELD
DW	26	1.00									2	0115	02	0610	01	
DW	44	1.00									39	8001	03	0610	01	

S/D SEC	ELEM	ITEM CODE	DESCRIPTION	U/M	QUANTITY
02	0610	805.100	HM FRAME	EACH	2
02	0610	805.200	HM DOOR	EACH	2
03	0610	823.070	PRE-HUNG DOOR	EACH	39
03	0610	871.020	FINISH HARDWARE ALLOWANCE	OPNG	39
02	0610	871.027	FINISH HARDWARE ALLOWANCE	OPNG	2
02	0610	990.031	PAINT INTERIOR DOOR	SIDE	4
03	0610	990.031	PAINT INTERIOR DOOR	SIDE	78
02	0610	990.032	PAINT DOOR FRAME	EACH	2

File ID - 3256 QUANTITY SURVEY REPORT Page - 10
Client Job No - Date -
Project Name - WHATCO PROJECT Time -
Project Size - 46,000 SQFT
Estimator - DOUG

MANAGEMENT COMPUTER CONTROLS, INC.
2881 DIRECTORS COVE
MEMPHIS, TENNESSEE 38131

6.210 FINISHES

------------------------DIMENSIONS------------------------SPECIFICATION SELECTION----------------

DIMENSIONS		SPECIFICATION SELECTION	
B-SQFT OF FINISH AREA,LENGTH OF		9-METHOD OF CALCULATING AREAS	4-BASE
C-LENGTH OF FINISHED PARTITION,W		8-CORRELATE SIDES (USE SS8 WHEN	3-WAINSCOT
D-FINISH WALL HEIGHT (FEET & DEC		7-FLOOR FINISH	2-WALL FINISH
E-WAINSCOT HEIGHT (FEET & DECIMA		6-FLOOR FINISH (CONT)	1-CEILING FINISHES
F-ACT TYPE	A-QUANTITY	5-FLOOR FINISH (CONT)	0-CEILING SYSTEM

WS	LINE SEQ	B	C	D	E	F	G	H	J	K	A	SPEC. SEL. 9876543210	S/D SEC	ELEM	LOC	MARK FIELD
DW	12	30000.00									1	1009	01	0210	01	
DW	27	166.00		10.00							2	27000001	02	0620	01	
DW	28	800.00									1	1000000012	02	0620	01	
DW	29	9600.00									1	1009	02	0620	01	
DW	45	4950.00									1	10004	03	0620	01	
DW	46	1000.00									1	1001	03	0620	01	
DW	47	450.00									1	101	03	0620	01	
DW	48	6400.00									1	1000000003	03	0620	01	
DW	49	2120.00									1	270002	03	0620	01	
DW	50	120.00									1	270003	03	0620	01	
DW	51	1660.00		9.00							1	27000001	03	0620	01	
DW	52	550.00		9.00							1	27000004	03	0620	01	
DW	53	30.00			8.00						1	2700001	03	0620	01	

S/D	SEC	ELEM	ITEM CODE	DESCRIPTION	U/M	QUANTITY
	01	0210	301.053	FLOOR HARDENER	SQFT	30000
	02	0620	301.053	FLOOR HARDENER	SQFT	9600
	02	0620	926.515	GYPSUM BOARD @ CEILING	SQFT	800
	03	0620	931.002	CERAMIC TILE BASE	LNFT	120
	03	0620	931.010	CERAMIC TILE FLOOR	SQFT	450
	03	0620	931.011	CERAMIC TILE WALL	SQFT	240
	03	0620	950.400	ACOUST CEIL SYS-EXPOSED GRID	SQFT	6400
	03	0620	965.010	VINYL COMPOSITION TILE	SQFT	1000
	03	0620	965.026	VINYL BASE	LNFT	2120
	03	0620	968.200	CARPET	SQYD	550
	02	0620	992.100	PAINT WALL	SQS	34
	03	0620	992.100	PAINT WALL	SQS	150
	02	0620	992.110	PAINT CEILING	SQS	8
	03	0620	995.000	VINYL WALL COVERING	SQFT	4950

```
      File ID - 3256                  Q U A N T I T Y   S U R V E Y   R E P O R T                  Page -       11
Client Job No -                                                                                    Date -
Project Name  - WHATCO PROJECT                                                                     Time -
Project Size  -    46,000 SQFT                                                  MANAGEMENT COMPUTER CONTROLS, INC.
Estimator     - DOUG                                                            2881 DIRECTORS COVE
                                                                                MEMPHIS, TENNESSEE    38131

          6.220     TOILET & BATH ACCESSORIES - COMM./INDUSTRIAL/HOSPT

------------------------DIMENSIONS------------------------------------------------------SPECIFICATION SELECTION-----------------

B-PCS REQUIRED PER PLACE (WHOLE                                    9-COMPARTMENTS               4-MIRRORS
C-LENGTH OF SHELF (FEET)                                           8-GRAB BAR/TOWEL BAR/SHELF   3-HOSPITAL ACCESSORIES
D-WIDTH OF MIRROR (INCHES)                                         7-TOILET PAPER HOLDER        2-MISC.
E-HEIGHT OF MIRROR (INCHES)                                        6-DISPENSER/DISPOSALS
                             A-QUANTITY                            5-SOAP DISPENSERS

WS LINE    B       C       D       E       F       G       H       J       K     A  SPEC. SEL.   S/D       MARK FIELD
   SEQ                                                                              9876543210 SEC  ELEM LOC

DW  54    1.00            72.00   42.00                                          2  000001     03    0630 01

                          S/D  SEC  ELEM  ITEM CODE        DESCRIPTION             U/M     QUANTITY

                               03   0630   880.750   FRAMELESS 1/4" PLATE MIRROR   SQFT       42
```

File ID - 3256
Client Job No -
Project Name - WHATCO PROJECT
Project Size - 46,000 SQFT
Estimator - DOUG

QUANTITY SURVEY REPORT

Page - 12
Date -
Time -

MANAGEMENT COMPUTER CONTROLS, INC.
2881 DIRECTORS COVE
MEMPHIS, TENNESSEE 38131

6.230 MILLWORK/CASEWORK/BUILT-IN FITTINGS

------------------------DIMENSIONS------------------------SPECIFICATION SELECTION----------------

B-LENGTH (FEET & DECIMAL OF FOOT
C-$ ALLOWANCE/LNFT (USE W/SS1)
D-SQFT OF RAISED PLATFORM
A-QUANTITY

9-VANITIES
8-COUNTERS
7-CABINETS
6-SHELVING
5-OFFICE MILLWORK

4-HOTEL/MOTEL REST/LOUNGE MILLWO
3-MEDICAL MILLWORK & CASEWORK
2-MISC
1-LNFT ALLOWANCE $ = C DIM.

WS	LINE SEQ	B	C	D	E	F	G	H	J	K	A	SPEC. SEL. 9876543210	S/D SEC	ELEM	LOC	MARK FIELD
DW	55	6.00									2	5	03	0630	01	
DW	56	16.00									1	009	03	0630	01	
DW	57	3.00									5	0001	03	0630	01	

S/D SEC	ELEM	ITEM CODE	DESCRIPTION	U/M	QUANTITY
03	0630	622.112	VANITY TOP	LNFT	12
03	0630	622.208	KITCHEN CABINETS	LNFT	16
03	0630	622.230	CLOSET ROD & SHELF	LNFT	15

```
    File ID - 3256              Q U A N T I T Y   S U R V E Y   R E P O R T                    Page -      13
Client Job No -                                                                                Date -
Project Name  - WHATCO PROJECT                                                                 Time -
Project Size  -    46,000 SQFT                                              MANAGEMENT COMPUTER CONTROLS, INC.
Estimator     - DOUG                                                        2881 DIRECTORS COVE
                                                                            MEMPHIS, TENNESSEE    38131
                -

      11.000     EQUIPMENT/FURNISHINGS

--------------------DIMENSIONS---------------------------------------------------SPECIFICATION SELECTION-----------------

B-LENGTH (FEET & DECIMAL OF_FOOT                        9-DOCK EQUIPMENT           4-RECEPTACLES & PLANTERS
C-WIDTH (FEET & DECIMAL OF FOOT)                        8-DETENTION EQUIPMENT      3-SEATING
D-$ ALLOWANCE, W/SS1                                    7-BANK EQUIPMENT           2-MISC
                                                        6-X-RAY                    1-ALLOWANCES (LS)
                          A-QUANTITY                    5-FURNISHINGS              0-RESIDENTIAL KITCHEN EQUIP.

WS LINE    B     C     D     E     F     G     H     J     K    A    SPEC. SEL.   S/D        MARK FIELD
    SEQ                                                               9876543210 SEC  ELEM LOC

DW   13                                                          3   1          01   1110 01
DW   30                                                          2   1          02   1110 01
```

S/D	SEC	ELEM	ITEM CODE	DESCRIPTION	U/M	QUANTITY
	01	1110	222.600	EXCAV FOR DOCK LEVELER PIT	CUYD	18
	02	1110	222.600	EXCAV FOR DOCK LEVELER PIT	CUYD	12
	01	1110	222.601	BACKFILL @ DOCK LEVELER PIT	CUYD	7
	02	1110	222.601	BACKFILL @ DOCK LEVELER PIT	CUYD	5
	01	1110	301.019	TROWEL CEMENT FINISH	SQFT	48
	02	1110	301.019	TROWEL CEMENT FINISH	SQFT	32
	01	1110	301.020	POINT & PATCH	SQFT	276
	02	1110	301.020	POINT & PATCH	SQFT	184
	01	1110	310.310	DOCK LEVELER PIT FORMS	SQFT	276
	02	1110	310.310	DOCK LEVELER PIT FORMS	SQFT	184
	01	1110	321.006	RE-STEEL @ DOCK LEVELER PIT	CWT	7
	02	1110	321.006	RE-STEEL @ DOCK LEVELER PIT	CWT	5
	01	1110	332.800	CONC IN DOCK LEVELER PIT	CUYD	10
	02	1110	332.800	CONC IN DOCK LEVELER PIT	CUYD	7
	01	1110	550.011	IMBEDDED CURB ANGLE	LNFT	102
	02	1110	550.011	IMBEDDED CURB ANGLE	LNFT	68
	01	1110	1116.000	DOCK LEVELER	EACH	3
	02	1110	1116.000	DOCK LEVELER	EACH	2

File ID - 3256
Client Job No -
Project Name - WHATCO PROJECT
Project Size - 46,000 SQFT
Estimator - DOUG

M I S C E L L A N E O U S   I T E M S   R E P O R T

Page - 1
Date -
Time -

MANAGEMENT COMPUTER CONTROLS, INC.
2881 DIRECTORS COVE
MEMPHIS, TENNESSEE 38131

S D	Sec	Elem	Item Code	Description	Unit Meas	Quantity	Labor	Material	Equipment	Subcontract
	01	1300	198.901	ESCALATION @ 13% - MID 95	LS					
	02	1300	198.902	ESCALATION @ 13% - MID 95	LS					
	03	1300	198.903	ESCALATION @ 13% - MID 95	LS					
	01	1400	199.901	CONTINGENCY @ 10%	LS					
	02	1400	199.902	CONTINGENCY @ 10%	LS					
	03	1400	199.903	CONTINGENCY @ 10%	LS					
	01	0410	510.900	STRUCT STL GIRT	CWT	538				
	02	0410	510.900	STRUCT STL GIRT	CWT	274				
	01	0700	510.901	CRANE RAIL SUPPORT	CWT	750				
	01	0320	551.901	MISC METALS	ALLO					6500.000
	02	0320	551.902	MISC METALS	ALLO					2500.000
	01	0500	610.901	ROOF BLOCKING	BDFT	1,600	1.250	.450		
	02	0500	610.901	ROOF BLOCKING	BDFT	800	1.250	.450		
	03	0500	610.901	ROOF BLOCKING	BDFT	640	1.250	.450		
	03	0610	610.902	MISC CARPENTRY	BDFT	1,000	1.150	.350		
	01	0410	790.901	CAULKING	ALLO					3000.000
	02	0410	790.902	CAULKING	ALLO					1500.000
	03	0410	790.903	CAULKING	ALLO					1200.000
	02	0630	1000.902	SPECIALTIES	ALLO					6500.000
	03	0630	1000.903	SPECIALTIES	ALLO					5000.000
	02	1130	1100.901	RACK SYSTEM - BY OWNER						
	02	1130	1100.902	BIN SYSTEM - BY OWNER						
	03	0630	1145.900	KITCHEN EQUIPMENT	ALLO					1800.000
	01	0700	1430.901	BRIDGE CRANE	LS					300000.000
	01	0700	1430.902	MONORAIL	LS					50000.000

01	0810	1549.901	PLUMBING	SQFT	30,000	.750
02	0810	1549.902	PLUMBING	SQFT	9,600	1.300
03	0810	1549.903	PLUMBING	SQFT	6,400	2.000
01	0830	1559.901	HYDRANT SYSTEM	LNFT	700	14.250
02	0830	1559.902	FIRE PROTECTION	SQFT	9,600	1.300
03	0830	1559.903	FIRE PROTECTION	SQFT	6,400	1.500
01	0820	1599.901	HVAC	SQFT	30,000	5.300
02	0820	1599.902	HVAC	SQFT	9,600	5.300
03	0820	1599.903	HVAC	SQFT	6,400	8.000
01	0900	1699.901	ELECTRICAL	SQFT	30,000	6.500
02	0900	1699.902	ELECTRICAL	SQFT	9,600	4.100
03	0900	1699.903	ELECTRICAL	SQFT	6,400	5.750

PROJECT SCOPE AND STATUS INFORMATION   Page 1

WHATCO PROJECT

```
File ID. : 3256                     Estimate Type: SCHEMATIC
Job Type : MANUFACTURING            Estimate No. :  1
Job Loc. : ANYWHERE, USA            Estimate Date: 11/11/92
Primary  :    46,000 SQFT
Job Start: 07/01/95  (Estimated)                               (Estimated)

For      : O'BRIEN - KREITZBERG & ASSOCIATES, INC.

By       : MANAGEMENT COMPUTER CONTROLS, INC.
           2881 DIRECTORS COVE
           MEMPHIS, TENNESSEE    38131

Scope    : 30,000 SQFT PLANT
            9,600 SQFT WAREHOUSE
            6,400 SQFT OFFICE

Proj Status :
```

File ID. 3256
Project No.

MANAGEMENT COMPUTER CONTROLS, INC.
2881 DIRECTORS COVE
MEMPHIS, TENNESSEE 38131
WHATCO PROJECT

Page 2

Description	Qty	Unit	Unit Price	Total Price
PLANT				1,960,778
WAREHOUSE				495,722
OFFICE				596,118
***ESTIMATE TOTAL				3,052,618

File ID. 3256
Project No.
30,000 SQFT

MANAGEMENT COMPUTER CONTROLS, INC.
2881 DIRECTORS COVE
MEMPHIS, TENNESSEE 38131
WHATCO PROJECT
PLANT

Page 1

Description	Qty	Unit	Unit Price	Total Price	30,000 $/SQFT
FOUNDATIONS				$76,939	2.56
STANDARD FOUNDATIONS				$39,605	1.32
SPECIAL FOUNDATIONS				$37,334	1.24
SUBSTRUCTURE				$108,299	3.61
SLAB ON GRADE				$108,299	3.61
SUPERSTRUCTURE				$273,461	9.11
ROOF CONSTRUCTION				$273,461	9.11
EXTERIOR CLOSURE				$213,240	7.10
EXTERIOR WALLS				$184,827	6.16
EXTERIOR DOORS & WINDOWS				$28,413	.94
ROOFING				$108,339	3.61
CONVEYING SYSTEM				$396,566	13.21
MECHANICAL				$191,475	6.38
PLUMBING				$22,500	.75
H.V.A.C.				$159,000	5.30
FIRE PROTECTION				$9,975	.33

ELECTRICAL	$195,000	6.50
EQUIPMENT	$14,137	.47
FIXED & MOVABLE EQUIPMENT	$14,137	.47
ESCALATION	$225,576	7.51
CONTINGENCY	$157,746	5.25
** SECTION TOTAL	$1,960,778	65.35

File ID. 3256 MANAGEMENT COMPUTER CONTROLS, INC. Page 1
Project No. 2881 DIRECTORS COVE
30,000 SQFT MEMPHIS, TENNESSEE 38131
WHATCO PROJECT
PLANT

Description	Qty	Unit	Unit Price	Total Price	30,000 $/SQFT
FOUNDATIONS					
STANDARD FOUNDATIONS					
EXCAVATE PILE CAP	79	CUYD	6.10	482	.01
EXCAVATE GRADE BEAM	238	CUYD	5.95	1,417	.04
BACKFILL PILE CAP	40	CUYD	9.62	385	.01
BACKFILL GRADE BEAM	196	CUYD	9.62	1,887	.06
FINE GRADE PILE CAP	576	SQFT	.27	158	
PILE CAP FORMS	960	SQFT	4.01	3,853	.12
GRADE BEAM FORMS	4,800	SQFT	4.10	19,690	.65
RE-STEEL @ PILE CAP	40	CWT	43.42	1,737	.05
RE-STEEL @ GRADE BEAM	34	CWT	43.41	1,476	.04
CONCRETE @ PILE CAP	45	CUYD	57.06	2,568	.08
CONCRETE @ GRADE BEAM	63	CUYD	58.41	3,680	.12
2" FOUNDATION INSULATION	3,200	SQFT	.71	2,272	.07
** TOTAL STANDARD FOUNDATIONS				39,605	1.32
SPECIAL FOUNDATIONS					
STEEL H PILE	1,920	LNFT	19.44	37,334	1.24
** TOTAL SPECIAL FOUNDATIONS				37,334	1.24
*** TOTAL FOUNDATIONS				76,939	2.56
SUBSTRUCTURE					
SLAB ON GRADE					
FINE GRADE SLAB ON GRADE	30,000	SQFT	.19	5,760	.19
UNDERSLAB FILL	584	CUYD	12.48	7,291	.24
TROWEL CEMENT FINISH	30,000	SQFT	.29	8,910	.29
PROTECT & CURE	30,000	SQFT	.11	3,390	.11
FLOOR HARDENER	30,000	SQFT	.25	7,710	.25
6X6-6/6 MESH	347	SQS	28.19	9,783	.32
CONCRETE @ SLAB ON GRADE	1,145	CUYD	57.16	65,455	2.18
** TOTAL SLAB ON GRADE				108,299	3.61
*** TOTAL SUBSTRUCTURE				108,299	3.61

Item	Quantity	Unit Cost	Total	Cost/SF
SUPERSTRUCTURE				
ROOF CONSTRUCTION				
PRECAST PLANK	30,000 SQFT	5.00	150,120	5.00
STRUCTURAL STEEL FRAME	1,800 CWT	64.91	116,841	3.89
MISC METALS	ALLO		6,500	.21
** TOTAL ROOF CONSTRUCTION			273,461	9.11
*** TOTAL SUPERSTRUCTURE			273,461	9.11
EXTERIOR CLOSURE				
EXTERIOR WALLS				
STRUCT STL GIRT	538 CWT	68.48	36,847	1.22
INSULATED METAL SIDING	19,200 SQFT	7.55	144,980	4.83
CAULKING	ALLO		3,000	.10
** TOTAL EXTERIOR WALLS			184,827	6.16
EXTERIOR DOORS & WINDOWS				
HM FRAME	10 EACH	96.80	968	.03
HM DOOR	10 EACH	186.90	1,869	.06
12X20 ELECT OPR OH DR	6 EACH	3,640.66	21,844	.72
FINISH HARDWARE ALLOWANCE	10 OPNG	313.50	3,135	.10
PAINT EXTERIOR DOOR	20 SIDE	21.05	421	.01
PAINT DOOR FRAME	10 EACH	17.60	176	
** TOTAL EXTERIOR DOORS & WINDOWS			28,413	.94
*** TOTAL EXTERIOR CLOSURE			213,240	7.10
ROOFING				
ROOFING				
ROOF BLOCKING	1,600 BDFT	2.08	3,339	.11
BUILT-UP ROOF SYSTEM	300 SQS	350.00	105,000	3.50

File ID. 3256
Project No.
30,000 SQFT

MANAGEMENT COMPUTER CONTROLS, INC.
2881 DIRECTORS COVE
MEMPHIS, TENNESSEE 38131
WHATCO PROJECT
PLANT

Description	Qty	Unit	Unit Price	Total Price	30,000 $/SQFT
ROOFING					
ROOFING					
* TOTAL ROOFING				108,339	3.61
*** TOTAL ROOFING				108,339	3.61
CONVEYING SYSTEM					
CONVEYING SYSTEM					
CRANE RAIL SUPPORT	750	CWT	62.08	46,566	1.55
BRIDGE CRANE		LS		300,000	10.00
MONORAIL		LS		50,000	1.66
* TOTAL CONVEYING SYSTEM				396,566	13.21
*** TOTAL CONVEYING SYSTEM				396,566	13.21
MECHANICAL					
PLUMBING					
PLUMBING	30,000	SQFT	.75	22,500	.75
** TOTAL PLUMBING				22,500	.75
H.V.A.C.					
HVAC	30,000	SQFT	5.30	159,000	5.30
** TOTAL H.V.A.C.				159,000	5.30
FIRE PROTECTION					
HYDRANT SYSTEM	700	LNFT	14.25	9,975	.33
** TOTAL FIRE PROTECTION				9,975	.33
*** TOTAL MECHANICAL				191,475	6.38

Description	Quantity	Unit Cost	Total	Cost/SF
ELECTRICAL				
ELECTRICAL				
ELECTRICAL	30,000 SQFT	6.50	195,000	6.50
* TOTAL ELECTRICAL			195,000	6.50
*** TOTAL ELECTRICAL			195,000	6.50
EQUIPMENT				
FIXED & MOVABLE EQUIPMENT				
EXCAV FOR DOCK LEVELER PIT	18 CUYD	6.88	124	
BACKFILL @ DOCK LEVELER PIT	7 CUYD	5.85	41	
TROWEL CEMENT FINISH	48 SQFT	.29	14	
POINT & PATCH	276 SQFT	.10	30	
DOCK LEVELER PIT FORMS	276 SQFT	3.56	984	.03
RE-STEEL @ DOCK LEVELER PIT	7 CWT	43.85	307	.01
CONC IN DOCK LEVELER PIT	10 CUYD	60.30	603	.02
IMBEDDED CURB ANGLE	102 LNFT	7.34	749	.02
DOCK LEVELER	3 EACH	3,761.66	11,285	.37
** TOTAL FIXED & MOVABLE EQUIPMENT			14,137	.47
*** TOTAL EQUIPMENT			14,137	.47
ESCALATION				
ESCALATION				
ESCALATION @ 13% - MID 95	LS		225,576	7.51
* TOTAL ESCALATION			225,576	7.51
*** TOTAL ESCALATION			225,576	7.51
CONTINGENCY				
CONTINGENCY				
CONTINGENCY @ 10%	LS		157,746	5.25
* TOTAL CONTINGENCY			157,746	5.25
*** TOTAL CONTINGENCY			157,746	5.25
**** TOTAL PLANT			1,960,778	65.35

File ID. 3256 | MANAGEMENT COMPUTER CONTROLS, INC. | Page 1

Project No. | 2881 DIRECTORS COVE

9,600 SQFT | MEMPHIS, TENNESSEE 38131

WHATCO PROJECT

WAREHOUSE

Description	Qty	Unit	Unit Price	Total Price	9,600 $/SQFT
FOUNDATIONS				$20,999	2.18
STANDARD FOUNDATIONS				$20,999	2.18
SUBSTRUCTURE				$21,762	2.26
SLAB ON GRADE				$21,762	2.26
SUPERSTRUCTURE				$84,077	8.75
FLOOR CONSTRUCTION				$2,381	.24
ROOF CONSTRUCTION				$81,696	8.51
EXTERIOR CLOSURE				$98,818	10.29
EXTERIOR WALLS				$92,755	9.66
EXTERIOR DOORS & WINDOWS				$6,063	.63
ROOFING				$35,270	3.67
INTERIOR CONSTRUCTION				$19,942	2.07
PARTITIONS				$8,641	.90
INTERIOR FINISHES				$4,801	.50
SPECIALTIES				$6,500	.67
MECHANICAL				$74,880	7.80
PLUMBING				$19,200	2.00
H.V.A.C.				$43,200	4.50
FIRE PROTECTION				$12,480	1.30

ELECTRICAL	$33,600	3.50
EQUIPMENT	$9,463	.98
FIXED & MOVABLE EQUIPMENT	$9,463	.98
ESCALATION	$57,030	5.94
CONTINGENCY	$39,881	4.15
** SECTION TOTAL	$495,722	51.63

File ID. 3256
Project No.
9,600 SQFT

MANAGEMENT COMPUTER CONTROLS, INC.
2881 DIRECTORS COVE
MEMPHIS, TENNESSEE 38131
WHATCO PROJECT
WAREHOUSE

Page 1

Description	Qty	Unit	Unit Price	Total Price	9,600 $/SQFT
FOUNDATIONS					
STANDARD FOUNDATIONS					
EXCAVATE COLUMN FOOTING	50	CUYD	6.12	306	.03
EXCAVATE GRADE BEAM	119	CUYD	5.95	708	.07
BACKFILL COLUMN FOOTING	28	CUYD	9.64	270	.02
BACKFILL GRADE BEAM	98	CUYD	9.62	943	.09
FINE GRADE COLUMN FOOTING	224	SQFT	.27	62	
COLUMN FTG FORMS	672	SQFT	3.90	2,625	.27
GRADE BEAM FORMS	2,400	SQFT	4.10	9,845	1.02
RE-STEEL @ COLUMN FOOTING	22	CWT	43.40	955	.09
RE-STEEL @ GRADE BEAM	17	CWT	43.47	739	.07
CONCRETE @ COLUMN FTG	27	CUYD	57.03	1,540	.16
CONCRETE @ GRADE BEAM	32	CUYD	58.43	1,870	.19
2" FOUNDATION INSULATION	1,600	SQFT	.71	1,136	.11
** TOTAL STANDARD FOUNDATIONS				20,999	2.18
*** TOTAL FOUNDATIONS				20,999	2.18
SUBSTRUCTURE					
SLAB ON GRADE					
FINE GRADE SLAB ON GRADE	9,600	SQFT	.19	1,843	.19
UNDERSLAB FILL	187	CUYD	12.48	2,335	.24
TROWEL CEMENT FINISH	9,600	SQFT	.29	2,851	.29
PROTECT & CURE	9,600	SQFT	.11	1,085	.11
6X6-6/6 MESH	111	SQS	28.18	3,129	.32
CONCRETE @ SLAB ON GRADE	184	CUYD	57.16	10,519	1.09
** TOTAL SLAB ON GRADE				21,762	2.26
*** TOTAL SUBSTRUCTURE				21,762	2.26

Item	Quantity	Unit Cost	Total	Cost/SF
SUPERSTRUCTURE				
FLOOR CONSTRUCTION				
WOOD JOIST	825 SQFT	2.07	1,713	.17
PLYWOOD DECK	825 SQFT	.81	668	.07
** TOTAL FLOOR CONSTRUCTION			2,381	.24
ROOF CONSTRUCTION				
PRECAST PLANK	9,600 SQFT	5.00	48,038	5.00
STRUCTURAL STEEL FRAME	480 CWT	64.91	31,158	3.24
MISC METALS	ALLO		2,500	.26
** TOTAL ROOF CONSTRUCTION			81,696	8.51
*** TOTAL SUPERSTRUCTURE			84,077	8.75
EXTERIOR CLOSURE				
EXTERIOR WALLS				
STRUCT STL GIRT	274 CWT	68.48	18,765	1.95
INSULATED METAL SIDING	9,600 SQFT	7.55	72,490	7.55
CAULKING	ALLO		1,500	.15
** TOTAL EXTERIOR WALLS			92,755	9.66
EXTERIOR DOORS & WINDOWS				
HM FRAME	6 EACH	96.83	581	.06
HM DOOR	6 EACH	186.83	1,121	.11
MANUAL OPERATED OH DOOR	2 EACH	1,060.50	2,121	.22
FINISH HARDWARE ALLOWANCE	6 OPNG	313.50	1,881	.19
PAINT EXTERIOR DOOR	12 SIDE	21.08	253	.02
PAINT DOOR FRAME	6 EACH	17.66	106	.01
** TOTAL EXTERIOR DOORS & WINDOWS			6,063	.63
*** TOTAL EXTERIOR CLOSURE			98,818	10.29
ROOFING				
ROOFING				
ROOF BLOCKING	800 BDFT	2.08	1,670	.17
BUILT-UP ROOF SYSTEM	96 SQS	350.00	33,600	3.50

File ID. 3256
Project No.
9,600 SQFT

MANAGEMENT COMPUTER CONTROLS, INC.
2881 DIRECTORS COVE
MEMPHIS, TENNESSEE 38131
WHATCO PROJECT
WAREHOUSE

Page 2

Description	Qty Unit	Unit Price	Total Price	9,600 $/SQFT
ROOFING				
ROOFING				
* TOTAL ROOFING -			35,270	3.67
*** TOTAL ROOFING			35,270	3.67
INTERIOR CONSTRUCTION				
PARTITIONS				
CONCRETE BLOCK	1,660 SQFT	4.41	7,329	.76
HM FRAME	2 EACH	97.00	194	.02
HM DOOR	2 EACH	187.00	374	.03
FINISH HARDWARE ALLOWANCE	2 OPNG	313.50	627	.06
PAINT INTERIOR DOOR	4 SIDE	20.50	82	
PAINT DOOR FRAME	2 EACH	17.50	35	
** TOTAL PARTITIONS			8,641	.90
INTERIOR FINISHES				
FLOOR HARDENER	9,600 SQFT	.25	2,468	.25
GYPSUM BOARD @ CEILING	800 SQFT	.60	487	.05
PAINT WALL	34 SQS	42.79	1,455	.15
PAINT CEILING	8 SQS	48.87	391	.04
** TOTAL INTERIOR FINISHES			4,801	.50
SPECIALTIES				
SPECIALTIES	ALLO		6,500	.67
** TOTAL SPECIALTIES			6,500	.67
*** TOTAL INTERIOR CONSTRUCTION			19,942	2.07
MECHANICAL				
PLUMBING				
PLUMBING	9,600 SQFT	2.00	19,200	2.00
** TOTAL PLUMBING			19,200	2.00

Item	Quantity	Unit Cost	Total	Per SF
H.V.A.C.				
HVAC	9,600 SQFT	4.50	43,200	4.50
** TOTAL H.V.A.C.			43,200	4.50
FIRE PROTECTION				
FIRE PROTECTION	9,600 SQFT	1.30	12,480	1.30
** TOTAL FIRE PROTECTION			12,480	1.30
*** TOTAL MECHANICAL			74,880	7.80
ELECTRICAL				
ELECTRICAL				
ELECTRICAL	9,600 SQFT	3.50	33,600	3.50
* TOTAL ELECTRICAL			33,600	3.50
*** TOTAL ELECTRICAL			33,600	3.50
EQUIPMENT				
FIXED & MOVABLE EQUIPMENT				
EXCAV FOR DOCK LEVELER PIT	12 CUYD	6.91	83	
BACKFILL @ DOCK LEVELER PIT	5 CUYD	5.80	29	
TROWEL CEMENT FINISH	32 SQFT	.31	10	
POINT & PATCH	184 SQFT	.10	20	
DOCK LEVELER PIT FORMS	184 SQFT	3.57	657	.06
RE-STEEL @ DOCK LEVELER PIT	5 CWT	43.80	219	.02
CONC IN DOCK LEVELER PIT	7 CUYD	60.28	422	.04
IMBEDDED CURB ANGLE	68 LNFT	7.35	500	.05
DOCK LEVELER	2 EACH	3,761.50	7,523	.78
** TOTAL FIXED & MOVABLE EQUIPMENT			9,463	.98
SPECIAL CONSTRUCTION				
RACK SYSTEM - BY OWNER				
BIN SYSTEM - BY OWNER				
** TOTAL SPECIAL CONSTRUCTION				
*** TOTAL EQUIPMENT			9,463	.98

File ID. 3256
Project No.
9,600 SQFT

MANAGEMENT COMPUTER CONTROLS, INC.
2881 DIRECTORS COVE
MEMPHIS, TENNESSEE 38131
WHATCO PROJECT
WAREHOUSE

Description	Qty	Unit	Unit Price	Total Price	9,600 $/SQFT
ESCALATION					
ESCALATION					
ESCALATION @ 13% - MID 95		LS		57,030	5.94
* TOTAL ESCALATION				57,030	5.94
*** TOTAL ESCALATION				57,030	5.94
CONTINGENCY					
CONTINGENCY					
CONTINGENCY @ 10%		LS		39,881	4.15
* TOTAL CONTINGENCY				39,881	4.15
*** TOTAL CONTINGENCY				39,881	4.15
**** TOTAL WAREHOUSE				495,722	51.63

File ID. 3256
Project No.
6,400 SQFT

MANAGEMENT COMPUTER CONTROLS, INC.
2881 DIRECTORS COVE
MEMPHIS, TENNESSEE 38131
WHATCO PROJECT
OFFICE

Description	Qty	Unit	Unit Price	Total Price	6,400 $/SQFT
FOUNDATIONS				$6,707	1.04
STANDARD FOUNDATIONS				$6,707	1.04
SUBSTRUCTURE				$14,114	2.20
SLAB ON GRADE				$14,114	2.20
SUPERSTRUCTURE				$80,000	12.50
ROOF CONSTRUCTION				$80,000	12.50
EXTERIOR CLOSURE				$144,039	22.50
EXTERIOR WALLS				$129,924	20.30
EXTERIOR DOORS & WINDOWS				$14,115	2.20
ROOFING				$23,736	3.70
INTERIOR CONSTRUCTION				$95,784	14.96
PARTITIONS				$44,785	6.99
INTERIOR FINISHES				$40,618	6.34
SPECIALTIES				$10,381	1.62
MECHANICAL				$78,400	12.25
PLUMBING				$17,600	2.75
H.V.A.C.				$51,200	8.00
FIRE PROTECTION				$9,600	1.50
ELECTRICAL				$36,800	5.75
ESCALATION				$47,958	7.49
CONTINGENCY				$68,580	10.71
** SECTION TOTAL				$596,118	93.14

File ID. 3256
Project No.
6,400 SQFT

MANAGEMENT COMPUTER CONTROLS, INC.
2881 DIRECTORS COVE
MEMPHIS, TENNESSEE 38131
WHATCO PROJECT
OFFICE

Page 1

Description	Qty	Unit	Unit Price	Total Price	6,400 $/SQFT
FOUNDATIONS					
STANDARD FOUNDATIONS					
EXCAVATE CONT. FOOTING	48	CUYD	5.70	274	.04
EXCAVATE COLUMN FOOTING	38	CUYD	6.10	232	.03
BACKFILL CONT. FOOTING	27	CUYD	9.63	260	.04
BACKFILL COLUMN FOOTING	28	CUYD	9.64	270	.04
FINE GRADE CONT. FOOTING	640	SQFT	.27	176	.02
FINE GRADE COLUMN FOOTING	256	SQFT	.27	70	.01
COLUMN FTG FORMS	344	SQFT	3.90	1,343	.21
RE-STEEL @ CONTINUOUS FTG	10	CWT	43.50	435	.06
RE-STEEL @ COLUMN FOOTING	12	CWT	43.41	521	.08
CONCRETE @ CONT. FTG	25	CUYD	56.72	1,418	.22
CONCRETE @ COLUMN FTG	14	CUYD	57.07	799	.12
2" FOUNDATION INSULATION	1,280	SQFT	.71	909	.14
** TOTAL STANDARD FOUNDATIONS				6,707	1.04
*** TOTAL FOUNDATIONS				6,707	1.04
SUBSTRUCTURE					
SLAB ON GRADE					
EXCAVATE THICKENED SLAB	16	CUYD	5.68	91	.01
FINE GRADE SLAB ON GRADE	6,400	SQFT	.19	1,229	.19
UNDERSLAB FILL	85	CUYD	12.48	1,061	.16
TROWEL CEMENT FINISH	6,400	SQFT	.29	1,901	.29
POINT & PATCH	640	SQFT	.10	69	.01
PROTECT & CURE	6,400	SQFT	.11	723	.11
SLAB ON GRADE EDGE FORMS	640	SQFT	2.56	1,644	.25
6X6-10/10 MESH	74	SQS	18.54	1,372	.21
RE-STEEL @ SLAB ON GRADE	7	CWT	43.85	307	.04
CONCRETE @ SLAB ON GRADE	100	CUYD	57.17	5,717	.89
** TOTAL SLAB ON GRADE				14,114	2.20
*** TOTAL SUBSTRUCTURE				14,114	2.20

Item	Quantity	Unit Cost	Total	Cost/SF
SUPERSTRUCTURE				
ROOF CONSTRUCTION				
PC STRUCTURAL SYSTEM	6,400 SQFT	12.50	80,000	12.50
** TOTAL ROOF CONSTRUCTION			80,000	12.50
*** TOTAL SUPERSTRUCTURE			80,000	12.50
EXTERIOR CLOSURE				
EXTERIOR WALLS				
PRECAST PANEL ARCH FINISH	5,760 SQFT	20.21	116,416	18.19
4" BATT INSULATION	4,800 SQFT	.40	1,959	.30
CAULKING	ALLO		1,200	.18
GYPSUM SHEATHING	4,800 SQFT	.67	3,245	.50
4"MTL STUD W/GYPSUM-1 SIDE	4,800 SQFT	1.48	7,104	1.11
** TOTAL EXTERIOR WALLS			129,924	20.30
EXTERIOR DOORS & WINDOWS				
HM FRAME	2 EACH	97.00	194	.03
HM DOOR	2 EACH	187.00	374	.05
FIXED WINDOWS	720 SQFT	15.00	10,800	1.68
FINISH HARDWARE ALLOWANCE	2 OPNG	313.50	627	.09
PAIR GLASS DOOR	1 EACH	2,000.00	2,000	.31
PAINT EXTERIOR DOOR	4 SIDE	21.25	85	.01
PAINT DOOR FRAME	2 EACH	17.50	35	
** TOTAL EXTERIOR DOORS & WINDOWS			14,115	2.20
*** TOTAL EXTERIOR CLOSURE			144,039	22.50
ROOFING				
ROOFING				
ROOF BLOCKING	640 BDFT	2.08	1,336	.20
BUILT-UP ROOF SYSTEM	64 SQS	350.00	22,400	3.50

File ID. 3256
Project No.
6,400 SQFT

MANAGEMENT COMPUTER CONTROLS, INC.
2881 DIRECTORS COVE
MEMPHIS, TENNESSEE 38131
WHATCO PROJECT
OFFICE

Description	Qty	Unit	Unit Price	Total Price	6,400 $/SQFT
ROOFING					
ROOFING					
* TOTAL ROOFING				23,736	3.70
*** TOTAL ROOFING				23,736	3.70
INTERIOR CONSTRUCTION					
PARTITIONS					
MISC CARPENTRY	1,000	BDFT	1.85	1,851	.28
PRE-HUNG DOOR	39	EACH	132.71	5,176	.80
FINISH HARDWARE ALLOWANCE	39	OPNG	75.56	2,947	.46
STANDARD DRYWALL PARTITION	14,400	SQFT	2.30	33,220	5.19
PAINT INTERIOR DOOR	78	SIDE	20.39	1,591	.24
** TOTAL PARTITIONS				44,785	6.99
INTERIOR FINISHES					
CERAMIC TILE BASE	120	LNFT	3.94	473	.07
CERAMIC TILE FLOOR	450	SQFT	4.98	2,245	.35
CERAMIC TILE WALL	240	SQFT	3.90	936	.14
ACOUST CEIL SYS-EXPOSED GRID	6,400	SQFT	1.14	7,309	1.14
VINYL COMPOSITION TILE	1,000	SQFT	1.17	1,175	.18
VINYL BASE	2,120	LNFT	1.05	2,226	.34
CARPET	550	SQYD	24.31	13,372	2.08
PAINT WALL	150	SQS	42.78	6,418	1.00
VINYL WALL COVERING	4,950	SQFT	1.30	6,464	1.01
** TOTAL INTERIOR FINISHES				40,618	6.34
SPECIALTIES					
VANITY TOP	12	LNFT	45.08	541	.08
KITCHEN CABINETS	16	LNFT	154.00	2,464	.38
CLOSET ROD & SHELF	15	LNFT	11.93	179	.02

Item	Quantity	Unit Cost	Total	Cost/SF
FRAMELESS 1/4" PLATE MIRROR	42 SQFT	9.45	397	.06
SPECIALTIES	ALLO		5,000	.78
KITCHEN EQUIPMENT	ALLO		1,800	.28
** TOTAL SPECIALTIES			10,381	1.62
*** TOTAL INTERIOR CONSTRUCTION			95,784	14.96
MECHANICAL				
PLUMBING				
PLUMBING	6,400 SQFT	2.75	17,600	2.75
** TOTAL PLUMBING			17,600	2.75
H.V.A.C.				
HVAC	6,400 SQFT	8.00	51,200	8.00
** TOTAL H.V.A.C.			51,200	8.00
FIRE PROTECTION				
FIRE PROTECTION	6,400 SQFT	1.50	9,600	1.50
** TOTAL FIRE PROTECTION			9,600	1.50
*** TOTAL MECHANICAL			78,400	12.25
ELECTRICAL				
ELECTRICAL				
ELECTRICAL	6,400 SQFT	5.75	36,800	5.75
* TOTAL ELECTRICAL			36,800	5.75
*** TOTAL ELECTRICAL			36,800	5.75
ESCALATION				
ESCALATION				
ESCALATION @ 13% - MID 95	LS		47,958	7.49
* TOTAL ESCALATION			47,958	7.49
*** TOTAL ESCALATION			47,958	7.49

File ID. 3256
Project No.
6,400 SQFT

MANAGEMENT COMPUTER CONTROLS, INC.
2881 DIRECTORS COVE
MEMPHIS, TENNESSEE 38131
WHATCO PROJECT
OFFICE

Description	Qty	Unit	Unit Price	Total Price	6,400 $/SQFT
CONTINGENCY					
CONTINGENCY					
CONTINGENCY @ 10%		LS		68,580	10.71
* TOTAL CONTINGENCY				68,580	10.71
*** TOTAL CONTINGENCY				68,580	10.71
**** TOTAL OFFICE				596,118	93.14

Page: 1
Date:
at:14035

RECAP OF COMPARISON OF FOUR ESTIMATES (A B C D) IN MULTIPLE LEVEL ELEMENT (SYSTEM OR COMPONENT) SEQUENCE FOR A SECTION OF PROJECT	ESTIMATE (A) WHATCO PROJECT SECTION-01:PLANT ONLY File ID.:3256 Job Type:MANUFACTURING Job Loc.:ANYWHERE, USA Est Date:11/11/92			ESTIMATE (B) WHATCO PROJECT SECTION-02:WAREHOUSE ONLY File ID.:3256 Job Type:MANUFACTURING Job Loc.:ANYWHERE, USA Est Date:11/11/92			ESTIMATE (C) WHATCO PROJECT SECTION-03:OFFICE ONLY File ID.:3256 Job Type:MANUFACTURING Job Loc.:ANYWHERE, USA Est Date:11/11/92			ESTIMATE (D) WHATCO PROJECT TOTAL ESTIMATE File ID.:3256 Job Type:MANUFACTURING Job Loc.:ANYWHERE, USA Est Date:11/11/92		
Description	30,000 $/SQFT	% of Sec.	TOTAL COST	9,600 $/SQFT	% of Sec.	TOTAL COST	6,400 $/SQFT	% of Sec.	TOTAL COST	46,000 $/SQFT	% of Sec.	TOTAL COST
0100-FOUNDATIONS	$2.56	3.9%	$76,939	$2.19	4.2%	$20,999	$1.05	1.1%	$6,707	$2.27	3.4%	$104,645
0200-SUBSTRUCTURE	$3.61	5.5%	$108,299	$2.27	4.3%	$21,762	$2.21	2.3%	$14,114	$3.13	4.7%	$144,175
0300-SUPERSTRUCTURE	$9.12	13.9%	$273,461	$8.76	16.9%	$84,077	$12.50	13.4%	$80,000	$9.51	14.3%	$437,538
0400-EXTERIOR CLOSURE	$7.11	10.8%	$213,240	$10.29	19.9%	$98,818	$22.51	24.1%	$144,039	$9.92	14.9%	$456,097
0500-ROOFING	$3.61	5.5%	$108,339	$3.67	7.1%	$35,270	$3.71	3.9%	$23,736	$3.64	5.4%	$167,345
0600-INTERIOR CONSTRUCTION				$2.08	4.0%	$19,942	$14.97	16.0%	$95,784	$2.52	3.7%	$115,726
0700-CONVEYING SYSTEM	$13.22	20.2%	$396,566							$8.62	12.9%	$396,566
0800-MECHANICAL	$6.38	9.7%	$191,475	$7.80	15.1%	$74,880	$12.25	13.1%	$78,400	$7.49	11.2%	$344,755
0900-ELECTRICAL	$6.50	9.9%	$195,000	$3.50	6.7%	$33,600	$5.75	6.1%	$36,800	$5.77	8.6%	$265,400
1100-EQUIPMENT	$.47	.7%	$14,137	$.99	1.9%	$9,463				$.51	.7%	$23,600
1300-ESCALATION	$7.52	11.5%	$225,576	$5.94	11.5%	$57,030	$7.49	8.0%	$47,958	$7.19	10.8%	$330,564
1400-CONTINGENCY	$5.26	8.0%	$157,746	$4.15	8.0%	$39,881	$10.72	11.5%	$68,580	$5.79	8.7%	$266,207
** SECTION TOTAL	$65.36	100.0%	$1,960,778	$51.64	100.0%	$495,722	$93.14	100.0%	$596,118	$66.36	100.0%	$3,052,618

COMPARISON OF FOUR ESTIMATES (A B C D) IN MULTIPLE LEVEL ELEMENT (SYSTEM OR COMPONENT) SEQUENCE FOR A SECTION OF PROJECT	ESTIMATE (A) WHATCO PROJECT SECTION-01:PLANT ONLY File ID.:3256 Job Type:MANUFACTURING Job Loc.:ANYWHERE, USA Est Date:11/11/92			ESTIMATE (B) WHATCO PROJECT SECTION-02:WAREHOUSE ONLY File ID.:3256 Job Type:MANUFACTURING Job Loc.:ANYWHERE, USA Est Date:11/11/92			ESTIMATE (C) WHATCO PROJECT SECTION-03:OFFICE ONLY File ID.:3256 Job Type:MANUFACTURING Job Loc.:ANYWHERE, USA Est Date:11/11/92			ESTIMATE (D) WHATCO PROJECT TOTAL ESTIMATE File ID.:3256 Job Type:MANUFACTURING Job Loc.:ANYWHERE, USA Est Date:11/11/92		
Description	30,000 $/SQFT	% of Sec.	TOTAL COST	9,600 $/SQFT	% of Sec.	TOTAL COST	6,400 $/SQFT	% of Sec.	TOTAL COST	46,000 $/SQFT	% of Sec.	TOTAL COST
0800												
MECHANICAL	$6.38	9.7%	$191,475	$7.80	15.1%	$74,880	$12.25	13.1%	$78,400	$7.49	11.2%	$344,755
PLUMBING	$.75	1.1%	$22,500	$2.00	3.8%	$19,200	$2.75	2.9%	$17,600	$1.29	1.9%	$59,300
H.V.A.C.	$5.30	8.1%	$159,000	$4.50	8.7%	$43,200	$8.00	8.5%	$51,200	$5.51	8.3%	$253,400
FIRE PROTECTION	$.33	.5%	$9,975	$1.30	2.5%	$12,480	$1.50	1.6%	$9,600	$.70	1.0%	$32,055
0900												
ELECTRICAL	$6.50	9.9%	$195,000	$3.50	6.7%	$33,600	$5.75	6.1%	$36,800	$5.77	8.6%	$265,400
1100												
EQUIPMENT	$.47	.7%	$14,137	$.99	1.9%	$9,463				$.51	.7%	$23,600
FIXED & MOVABLE EQUIPMENT	$.47	.7%	$14,137	$.99	1.9%	$9,463				$.51	.7%	$23,600
1300												
ESCALATION	$7.52	11.5%	$225,576	$5.94	11.5%	$57,030	$7.49	8.0%	$47,958	$7.19	10.8%	$330,564
1400												
CONTINGENCY	$5.26	8.0%	$157,746	$4.15	8.0%	$39,881	$10.72	11.5%	$68,580	$5.79	8.7%	$266,207
** SECTION TOTAL	$65.36	100.0%	$1,960,778	$51.64	100.0%	$495,722	$93.14	100.0%	$596,118	$66.36	100.0%	$3,052,618

The system results for the I.C.E. estimate include field overhead (superintendents) but not general conditions, bond, and profit. Using industry factors, assume the following:

General conditions	8.0%
Home office expense	5.0%
Bond	0.5%
Profit	10.0%
Total	23.5%

Applying these factors to the I.C.E. schematic estimate:

Base estimate	$3,052,678
Overhead, profit, and bond at 23.5%	717,379
Adjusted base estimate	$3,770,057

Comparison with budget estimate

In comparing the budget and schematic estimates, certain adjustments are necessary:

1. The schematic estimate used a 1992 database and then escalated at 15%/annum to bring it to 1995.
2. The narrative description did not fully describe the site work.
3. The budget estimate included overhead, profit, and bond costs in the base price.
4. The contingency factor at schematic is properly reduced to 10%.

Comparison of adjusted base construction costs

This was estimated without site work at 1995.

Budget, concept estimate

Base construction cost at 1995	$3,611,452
Equipment	381,200
Subtotal	$3,992,652
20% contingency	798,530
Adjusted base construction	$4,791,182

Schematic, Composer Gold

Base construction—at 1992—with site work	$3,970,600
Site work [$183,700 + (overhead, profit + bond) $183,700] = ($183,700 + 71,780)	(255,480)
Subtotal	$3,715,120
Plus 15% escalation	557,268
Subtotal	$4,272,388
Contingency at 10%	427,239
Adjusted base construction	$4,699,627

Schematic estimate by I.C.E. by MC²

Base construction at 1552 without site work and without overhead, profit + bond	$2,445,850
Add general conditions, profit, home office expense at 23.5%	577,125
Subtotal	$3,032,975
Contingency at 10%	303,298
Subtotal	$3,336,273
Escalation to 1995 at 15%	454,946
Subtotal	$3,791,219

The range between high and low computer estimates is 20% of the high figure. At the concept-budget stage the range between median and ¾ sf costs is 32% of the higher cost on a square-foot basis and 37% on the cubic-foot basis.

At the schematic stage, there can be substantial variations due to differences in assumptions. These will narrow at the design development stage.

Chapter 4

Estimate Format

The schematic estimate format is generally similar to the concept and budget estimate format. Starting with the design development estimate, the estimate format should follow the organization outlined by the specifications. At design development, the specification will only be an outline. It can be expected, almost universally in the U.S. construction industry, that the Construction Specification Institute (CSI) format will be used.

Accordingly, the estimate format should follow the CSI organization for the following reasons:

- Most estimating databases are organized on the CSI format.
- The estimator has a preestablished "chart of accounts" with which the estimate input items can be properly entered into the estimate database.
- The CSI format can be used as a checklist to be certain all items are included in the estimate.
- The CSI format is universally recognized, so a reviewer or user of the estimate knows where to look for specific information.

Development of the Uniform System was described on page 0.1 of the 1966 CSI "Uniform System for Building Construction Specifications, Data Filing and Cost Accounting":

> The Uniform System has been developed in response to pressing needs for better and more rapid classification of technical data. Current technology has created these needs by introducing new materials and techniques at a rate that threatens to outstrip our ability to assimilate essential new information and to correlate it with the old.
>
> Previous data filing systems, through obsolescence, complexity, and inflexibility, have been unable to adapt to an expanding body of knowl-

edge. In recent years, the research/storage/retrieval/application relationships existing between technology and specifications have led many to suggest the creation of a data filing system based on specifications.

The American Institute of Architects recognized that its Standard Filing System, in general use throughout the building industry since 1920, was obsolete, and in January 1962 invited the Construction Specification Institute to join with it in sponsoring a construction industry meeting to develop a more broadly based system. The first Conference on Uniform Indexing Systems was held under joint sponsorship in October 1962 to discuss development of a filing system for building product data based on specifications, a concept later enlarged to embrace a specification outline and a contractors' cost accounting guide as well.

The organizations that have helped develop and now endorse the Uniform System are

American Institute of Architects
American Society of Landscape Architects
Associated General Contractors of America, Inc.
Construction Specifications Institute
Council of Mechanical Specialty Contracting Industries, Inc.
National Society of Professional Engineers
Producers' Council, Inc.

The first widely publicized document advocating a nation-wide approach to uniformity of specification writing was published by the Construction Specification Institute in March 1963 as the *CSI Format for Building Specifications.* It proposed the organization of technical specifications into 16 basic groupings designated "Division," each of which was to be based on an interrelationship of place, trade, function, or material. There has been a widespread acceptance of the principle advocated by this document within Federal and State agencies throughout the USA and Canada and among private practitioners in the architectural and engineering professions.

The format is the divisional breakdown of the project into the following 16 divisions:

1. General requirements
2. Site work
3. Concrete
4. Masonry
5. Metals
6. Wood and plastics
7. Thermal and moisture protection

Doors and windows

ishes

10. Specialties
11. Equipment
12. Furnishings
13. Special construction
14. Conveying systems
15. Mechanical
16. Electrical

In 1972, with the publication of the "Uniform Construction Index," two formats—based on the 16 divisions and broadscope section headings concept—merged into one. Widespread acceptance of the 16-division format has helped contractors and increased the accuracy of bids while reducing the effort normally associated with bidding. Contractors have found products easier to control; architects and engineers have greater assurance that specifications are coordinated and complete. Owners naturally have benefited from reduced expenditures due to the increased efficiency provided by the standardization.

In 1978, CSI published its *MasterFormat™*, which provides an organizational structure for groupings of bidding requirements, contract forms, and conditions of the contract, as well as the merged 16 division listings of broad- and narrowscope section titles and numbering systems published by CSI and Construction Specifications Canada (CSC) subsequent to the 1972 edition of the "Uniform Construction Index."

The basis for the 16 divisions or "broadscope" titles follows. The *MasterFormat™* has a further breakdown into "mediumscope" and "narrowscope" titles.

From the *MasterFormat™*, published by the Construction Specification Institute, 1988, and used with permission, the "broadscope" titles are:

DIVISION 1—GENERAL REQUIREMENTS

01010	Summary of Work	01100	Special Project Procedures
01020	Allowances	01200	Project Meetings
01025	Measurement and Payment	01300	Submittals
01030	Alternates/Alternatives	01400	Quality Control
01035	Modification Procedures	01500	Construction Facilities and Temporary Controls
01040	Coordination		
01050	Field Engineering	01600	Material and Equipment
01060	Regulatory Requirements	01650	Starting of Systems/ Commissioning
01070	Abbreviations and Symbols		
01080	Identification Systems	01700	Contract Closeout
01090	Reference Standards	01800	Maintenance

DIVISION 2—SITE WORK

02010	Subsurface Investigation	02500	Paving and Surfacing
02050	Demolition	02600	Piped Utility Materials
02100	Site Preparation	02660	Water Distribution
02140	Dewatering	02680	Fuel and Steam Distribution
02150	Shoring and Underpinning	02700	Sewerage and Drainage
02160	Excavation Support Systems	02760	Restoration of Underground Pipelines
02170	Cofferdams		
02200	Earthwork	02770	Ponds and Reservoirs
02300	Tunneling	02780	Power and Communications
02350	Piles and Caissons	02800	Site Improvements
02450	Railroad Work	02900	Landscaping
02480	Marine Work		

DIVISION 3—CONCRETE

03100	Concrete Formwork	03500	Cementitious Decks and Cleaning
03200	Concrete Reinforcement		
03250	Concrete Accessories	03600	Grout
03300	Cast-In-Place Concrete	03700	Concrete Restoration and Cleaning
03370	Concrete Curing		
03400	Precast Concrete	03800	Mass Concrete

DIVISION 4—MASONRY

04100	Mortar and Masonry Grout	04550	Refractories
04150	Masonry Accessories	04600	Corrosion Resistant Masonry
04200	Unit Masonry	04700	Simulated Masonry
04400	Stone		
04500	Masonry Restoration and Cleaning		

DIVISION 5—METALS

05010	Metal Materials	05400	Cold-Formed Metal Framing
05030	Metal Coatings	05500	Metal Fabrications
05050	Metal Fastening	05580	Sheet Metal Fabrications
05100	Structural Metal Framing	05700	Ornamental Metal
05200	Metal Joists	05800	Expansion Control
05300	Metal Decking	05900	Hydraulic Structures

DIVISION 6—WOOD AND PLASTICS

06050	Fasteners and Adhesives	06200	Finish Carpentry
06100	Rough Carpentry	06300	Wood Treatment
[illegible]30	Heavy Timber Construction	06400	Architectural Woodwork
[illegible]	Wood-Metal Systems	06500	Structural Plastics
[illegible]	[illegible]efabricated Structural [illegible]od	06600	Plastic Fabrications
		06650	Solid Polymer Fabrications

DIVISION 7—THERMAL AND MOISTURE PROTECTION

07100 Waterproofing
07150 Dampproofing
07180 Water Repellents
07190 Vapor Retarders
07195 Air Barriers
07200 Insulation
07240 Exterior Insulation and Finish Systems
07250 Fireproofing
07270 Firestopping
07300 Shingles and Roofing Tiles
07400 Manufactured Roofing and Siding
07480 Exterior Wall Assemblies
07500 Membrane Roofing
07570 Traffic Topping
07600 Flashing and Sheet Metal
07700 Roof Specialties and Accessories
07800 Skylights
07900 Joint Sealers

DIVISION 8—DOORS AND WINDOWS

08100 Metal Doors and Frames
08200 Wood and Plastic Doors
08250 Door Opening Assemblies
08300 Special Doors
08400 Entrances and Storefronts
08500 Metal Windows
08600 Wood and Plastic Windows
08650 Special Windows
08700 Hardware
08800 Glazing
08900 Glazed Curtain Wall

DIVISION 9—FINISHES

09100 Metal Support Systems
09200 Lath and Plaster
09250 Gypsum Board
09300 Tile
09400 Terrazzo
09450 Stone Facing
09500 Acoustical Treatment
09540 Special Wall Surfaces
09545 Special Ceiling Surfaces
09550 Wood Flooring
09600 Stone Flooring
09630 Unit Masonry Flooring
09650 Resilient Flooring
09680 Carpet
09700 Special Flooring
09780 Floor Treatment
09800 Special Coatings
09900 Painting
09950 Wall Coverings

DIVISION 10—SPECIALTIES

10100 Visual Display Boards
10150 Compartments and Cubicles
10200 Louvers and Vents
10240 Grilles and Screens
10250 Service Wall Systems
10260 Wall and Corner Guards
10270 Access Flooring
10290 Pest Control
10300 Fireplaces and Stoves
10340 Manufactured Exterior Specialties
10350 Flagpoles
10400 Identifying Devices
10450 Pedestrian Control Devices
10500 Lockers
10520 Fire Protection Specialties
10530 Protective Covers
10550 Postal Specialties
10600 Partitions
10650 Operable Partitions
10670 Storage Shelving
10700 Exterior Protection Devices for Openings
10750 Telephone Specialties
10800 Toilet and Bath Accessories
10880 Scales
10900 Wardrobe and Closet Specialties

DIVISION 11—EQUIPMENT

11010	Maintenance Equipment
11020	Security and Vault Equipment
11030	Teller and Service Equipment
11040	Ecclesiastical Equipment
11050	Library Equipment
11060	Theater and Stage Equipment
11070	Instrumental Equipment
11080	Registration Equipment
11090	Checkroom Equipment
11100	Mercantile Equipment
11110	Commercial Laundry and Dry Cleaning Equipment
11120	Vending Equipment
11130	Audio-Visual Equipment
11140	Vehicle Service Equipment
11150	Parking Control Equipment
11160	Loading Dock Equipment
11170	Solid Waste Handling Equipment
11190	Detention Equipment
11200	Water Supply and Treatment Equipment
11280	Hydraulic Gates and Valves
11300	Fluid Waste Treatment and Disposal Equipment
11400	Food Service Equipment
11450	Residential Equipment
11460	Unit Kitchens
11470	Darkroom Equipment
11480	Athletic, Recreational and Therapeutic Equipment
11500	Industrial and Process Equipment
11600	Laboratory Equipment
11650	Planetarium Equipment
11660	Observatory Equipment
11680	Office Equipment
11700	Medical Equipment
11780	Mortuary Equipment
11850	Navigation Equipment
11870	Agricultural Equipment

DIVISION 12—FURNISHINGS

12050	Fabrics
12100	Artwork
12300	Manufactured Casework
12500	Window Treatment
12600	Furniture and Accessories
12670	Rugs and Mats
12700	Multiple Seating
12800	Interior Plants and Planters

DIVISION 13—SPECIAL CONSTRUCTION

13010	Air Supported Structures
13020	Integrated Assemblies
13030	Special Purpose Rooms
13080	Sound, Vibration, and Seismic Control
13090	Radiation Protection
13100	Nuclear Reactors
13120	Pre-Engineered Structures
13150	Aquatic Facilities
13160	Ice Rinks
13170	Kennels and Animal Shelters
13180	Site Constructed Incinerators
[illegible]0	Liquid and Gas Storage Tanks
[illegible]	Filter Underdrains and Media
[illegible]	[illegible]estion Tank Covers and [illegible]ppurtenances
13240	Oxygenation Systems
13260	Sludge Conditioning Systems
13300	Utility Control Systems
13400	Industrial and Process Control Systems
13500	Recording Instrumentation
13550	Transportation Control Instrumentation
13600	Solar Energy Systems
13700	Wind Energy Systems
13750	Cogeneration Systems
13800	Building Automation Systems
13900	Fire Suppression and Supervisory Systems
13950	Special Security Construction

DIVISION 14—CONVEYING SYSTEMS

14100	Dumbwaiters	14600	Hoists and Cranes
14200	Elevators	14700	Turntables
14300	Escalators and Moving Walks	14800	Scaffolding
14400	Lifts	14900	Transportation Systems
14500	Material Handling Systems		

DIVISION 15—MECHANICAL

15050	Basic Mechanical Materials and Methods	15650	Refrigeration
		15750	Heat Transfer
15250	Mechanical Insulation	15850	Air Handling
15300	Fire Protection	15880	Air Distribution
15400	Plumbing	15950	Controls
15500	Heating, Ventilating, and Air Conditioning (HVAC)	15990	Testing, Adjusting, and Balancing
15550	Heat Generation		

DIVISION 16—ELECTRICAL

16050	Basic Electrical Materials and Methods	16500	Lighting
		16600	Special Systems
16200	Power Generation—Built-up Systems	16700	Communications
		16850	Electric Resistance Heating
16300	Medium Voltage Distribution	16900	Controls
16400	Service and Distribution	16950	Testing

A *partial* list of mediumscope listings useful for estimating follows:

02160 EXCAVATION SUPPORT SYSTEMS
02162 Cribbing and Walers
02164 Soil and Rock Anchors
02166 Ground Freezing
02167 Reinforced Earth
02168 Slurry Wall Construction

02170 COFFERDAMS
02172 Double Wall Cofferdams
02174 Cellular Cofferdams
02176 Piling with Intermediate Lagging
02178 Sheet Piling Cofferdams

02200 EARTHWORK
02210 Grading
02220 Excavating, Backfilling, and Compacting
02230 Base Course
02240 Soil Stabilization
02250 Vibro-flotation
02270 Slope Protection and Erosion Control

02300 TUNNELING
02305 Tunnel Ventilation and Compression
02310 Tunnel Excavating
02320 Tunnel Lining
02330 Tunnel Grouting
02340 Tunnel Support Systems

02350 PILES AND CAISSONS
02360 Driven Piles
02370 Bored/Augured Piles
02380 Caissons

02500 PAVING AND SURFACING
02510 Asphaltic Concrete Paving
02515 Unit Pavers
02525 Prefabricated Curbs
02540 Synthetic Surfacing
02580 Pavement Marking

02600 PIPED UTILITY MATERIALS
02605 Utility Structures
02610 Pipe and Fittings
02640 Valves and Cocks
02645 Hydrants

02660 WATER DISTRIBUTION
02665 Water Systems
02670 Water Wells
02675 Disinfection of Water Distributing Systems

02680 FUEL DISTRIBUTION
02685 Gas Distribution System
02690 Oil Distribution System
02695 Steam Distribution System

02700 SEWERAGE AND DRAINAGE
02710 Subdrainage Systems
02720 Storm Sewage Systems
02730 Sanitary Sewage Systems
02735 Combined Wastewater System
02740 Septic Systems

02780 POWER AND COMMUNICATIONS
02785 Electrical Power Transmission
02790 Communication Transmission

03100 CONCRETE FORMWORK
03110 Structural Cast-in-Place Concrete Formwork
03120 Architectural Cast-in-Place Concrete Formwork
03130 Permanent Forms

03300 CAST-IN-PLACE CONCRETE
03310 Structural Concrete
03330 Architectural Concrete
03340 Low Density Concrete
03345 Concrete Finishing
03350 Concrete Finishes
03360 Specially Placed Concrete
03365 Post-tensioned Concrete

3400 PRECAST CONCRETE
Structural Precast Concrete (Plant Cast)
uctural Precast Post-
ned Concrete (Plant Cast)

03430 Structural Precast Concrete (Site Cast)
03450 Architectural Precast Concrete (Plant Cast)
03460 Architectural Precast Concrete (Site Cast)
03470 Tilt-up Precast Concrete
03480 Precast Concrete Specialties

04200 UNIT MASONRY
04210 Clay Unit Masonry
04220 Concrete Unit Masonry
04230 Reinforced Unit Masonry
04235 Preassembled Masonry Panels
04240 Non-reinforced Masonry Systems

05100 STRUCTURAL METAL FRAMING
05120 Structural Steel
05140 Structural Aluminum
05150 Steel Wire Rope
05160 Framing Systems

05200 METAL JOISTS
05210 Steel Joists
05250 Aluminum Joists
05260 Composite Joist System

05300 METAL DECKING
05310 Steel Deck
05320 Raceway Deck Systems
05330 Aluminum Deck

05500 METAL FABRICATIONS
05510 Metal Stairs
05515 Ladders
05520 Handrails and Railings
05530 Gratings and Floor Plates
05540 Castings

05700 ORNAMENTAL METAL
05710 Ornamental Stairs
05715 Prefabricated Spiral Stairs
05720 Ornamental Handrails and Railings
05725 Ornamental Metal Castings
05730 Ornamental Sheet Metal

06100 ROUGH CARPENTRY
06105 Treated Wood Foundations
06110 Wood Framing

06115 Sheathing
06120 Structural Panels
06125 Wood Decking
06128 Mineral Fiber Reinforced-Cement Panels

06200 FINISH CARPENTRY
06220 Millwork
06240 Plastic Laminate
06250 Prefinished Wood Paneling
06255 Prefinished Hardboard Paneling
06260 Board Paneling

06400 ARCHITECTURAL WOODWORK
06410 Custom Casework
06420 Panelwork
06430 Stairwork and Handrails
06440 Miscellaneous Ornamental Items
06450 Standing and Running Trim
06460 Exterior Frames
06470 Screens, Blinds and Shutters

07100 WATERPROOFING
07110 Sheet Membrane Waterproofing
07120 Fluid Applied Waterproofing
07130 Bentonite Waterproofing
07140 Metal Oxide Waterproofing
07145 Cementitious Waterproofing

07150 DAMPPROOFING
07160 Bituminous Dampproofing
07175 Water Repellent Coatings
07180 Water Repellent

07200 INSULATION
07210 Building Insulation
07220 Roof and Deck Insulation
07240 Exterior Insulation and Finish Systems

07250 FIREPROOFING
07255 Cementitious Fireproofing
07260 Intumescent Mastic Fireproofing
07265 Mineral Fiber Fireproofing
07270 Firestopping

07300 SHINGLES AND ROOFING TILES
07310 Shingles
07320 Roofing Tiles

07500 MEMBRANE ROOFING
07510 Built-up Bituminous Roofing
07515 Cold-Applied Bituminous Roofing
07520 Prepared Roll Roofing
07530 Single Ply Membrane Roofing
07540 Fluid Applied Roofing
07550 Protected Membrane Roofing
07560 Roof Maintenance and Repairs

07600 FLASHING AND SHEET METAL
07610 Sheet Metal Roofing
07620 Sheet Metal Flashing and Trim
07630 Sheet Metal Roofing Specialties
07650 Flexible Flashings

08100 METAL DOORS AND FRAMES
08110 Steel Doors and Frames
08120 Aluminum Doors and Frames
08130 Stainless Steel Doors and Frames
08140 Bronze Doors and Frames

08300 SPECIAL DOORS
08305 Access Doors
08310 Sliding Doors and Grilles
08315 Pressure-Resistant Doors
08320 Security Doors
08325 Cold Storage Doors and Grilles
08330 Coiling Doors
08350 Folding Doors and Grilles
08355 Chain Closures
08360 Sectional Overhead Doors
08365 Vertical Lift Doors
08370 Industrial Doors
08380 Traffic Doors
08385 Sound Control Doors
08390 Storm Doors
08395 Screen Doors

08500 METAL WINDOWS
08510 Steel Windows
08520 Aluminum Windows
08530 Stainless Steel Windows
08540 Bronze Windows

08600 WOOD AND PLASTIC WINDOWS
08610 Wood Windows
08630 Plastic Windows

08650 SPECIAL WINDOWS
08655 Roof Windows
08660 Security Windows and Screens
08665 Pass Windows
08670 Storm Windows

08700 HARDWARE
08710 Door Hardware
08740 Electro-Mechanical Hardware
08760 Window Hardware
08770 Door and Window Accessories

08800 GLAZING
08810 Glass
08840 Plastic Glazing
08850 Glazing Accessories

08900 GLAZED CURTAIN WALLS
08910 Glazed Steel Curtain Walls
08920 Glazed Aluminum Curtain Walls
08930 Glazed Stainless Steel Curtain Walls
08940 Glazed Bronze Curtain Walls
08950 Translucent Wall and Skylight Systems
08960 Sloped Glazing System
08970 Structural Glass Curtain Walls

09100 METAL SUPPORT SYSTEMS
09110 Non-load-bearing Wall Framing Systems
09120 Ceiling Suspension Systems
09130 Acoustical Suspension Systems

09200 LATH AND PLASTER
09205 Furring and Lathing
09210 Gypsum Plaster
09215 Veneer Plaster
09220 Portland Cement Plaster
09225 Adobe Finish

09300 TILE
09310 Ceramic Tile
09320 Thin Brick
09330 Quarry Tile
09340 Paver Tile
09350 Glass Mosaics
09360 Plastic Tile
09370 Metal Tile
09380 Cut Natural Stone Tile

09400 TERRAZZO
09410 Portland Cement Terrazzo
09420 Precast Terrazzo
09430 Conductive Terrazzo
09440 Plastic Matrix Terrazzo

09500 ACOUSTICAL TREATMENT
09510 Acoustical Ceilings
09520 Acoustical Wall Treatment
09525 Acoustical Space Units
09530 Acoustical Insulation and Barriers

09650 RESILIENT FLOORING
09660 Resilient Tile Flooring
09665 Resilient Sheet Flooring
09670 Fluid-Applied Resilient Flooring
09675 Static Control Resilient Flooring
09678 Resilient Base and Accessories

09680 CARPET
09682 Carpet Cushion
09685 Sheet Carpet
09690 Carpet Tile
09695 Wall Carpet
09698 Indoor-Outdoor Carpet

09700 SPECIAL FLOORING
09705 Resinous Flooring
09710 Magnesium Oxychloride Floors
09720 Epoxy-Marble-Chip Flooring
09725 Seamless Quartz Flooring
09730 Elastomeric Liquid Flooring
09750 Mastic Fills
09755 Plastic Laminate Flooring
09760 Asphalt Plank Flooring

09800 SPECIAL COATINGS
09810 Abrasion-Resistant Coatings
09815 High-Build Glazed Coatings
09820 Cementitious Coatings
09830 Elastomeric Coatings
09835 Textured Plastic Coatings
09840 Fire-Resistant Paints
09845 Intumescent Paints
09850 Chemical-Resistant Coatings
09860 Graffiti Resistant Coatings

09870 Coating Systems for Steel
09880 Protective Coatings for Concrete

09900 PAINTING
09910 Exterior Painting
09920 Interior Painting
09930 Transparent Finishes
09950 Wall Coverings
09955 Vinyl-Coated Fabric Wall Covering
09960 Vinyl Wall Covering
09965 Cork Wall Covering
09970 Wallpaper
09975 Wall Fabrics
09980 Flexible Wood Sheets and Veneers

At this writing, the 1988 edition is the current edition. The CSI *MasterFormat*™ (master list of titles and numbers for the construction industry) is 178 pages long with the following contents:

Introduction
MasterFormat™—An Overview
MasterFormat™—Its Application
Broadscope Section Titles
Master List of Section Titles, Numbers, and Broadscope Section Explanations
Key Word Index

Figures 4.1 and 4.2 are examples of the explanations and cross references included in the Master List. Figure 4.3 is an example of the Key Word Index.

The book is available from The Construction Specification Institute, 601 Madison Street, Alexandria, VA 22314. The 1993 cost (for non-CSI members) was $60, plus handling charges. *MasterFormat*™ is also available from Construction Specifications Canada (CSC).

DIVISION 8 – DOORS AND WINDOWS *Continued*

Section Number	Title	Broadscope Explanation
08500	**METAL WINDOWS**	**08500 – METAL WINDOWS**
-510	Steel Windows	Fixed and operable metal framed windows used singly and in multiples. Operating hardware, screens, and other accessories included. Involves various methods of operation such as sliding, hung, and projecting. Includes window units with louver blinds integrally set between glass panels and replacement windows specifically designed for retrofit work.
-520	Aluminum Windows	***Related Sections:***
-530	Stainless Steel Windows	*Section 08100 - Metal Doors and Frames: Metal frames for sidelights, transoms, and other openings with field installed fixed glazing.*
-540	Bronze Windows	*Section 08400 - Entrances and Storefronts: Metal framing for fixed glazed panels of storefront systems.*
		Section 08650 - Special Windows: Various special function windows.
		Section 08800 - Glazing: Glazing to be field installed in metal windows.
		Section 08900 - Glazed Curtain Walls: Metal framing for glazed curtain wall systems.
		Notes: *Unless windows are factor glazed, glazing is not specified in this section.*
		Metal sills and stools may be specified in this section or in Section 05500 - Metal Fabrications.
08600	**WOOD AND PLASTIC WINDOWS**	**08600 – WOOD AND PLASTIC WINDOWS**
-610	Wood Windows	Fixed and operable wood framed, metal clad wood frames, plastic clad wood frames, and plastic framed windows used singly or in multiples. Operating hardware, screens, and other accessories included. Involves various methods of operation such as sliding, hung, and projecting. Includes window units with louver blinds integrally set between glass panels and replacement windows specifically designed for retrofit work.
	Metal Clad Wood Windows	***Related Sections:***
	Plastic Clad Wood Windows	*Section 06200 - Finish Carpentry: Field constructed wood frames for sidelights, transoms, and other openings for fixed glazing.*
-630	Plastic Windows	*Section 06400 - Architectural Woodwork: Field constructed wood frames for sidelights, transoms, and other openings for fixed glazing.*
		Section 08650 - Special Windows: Various types of special function windows.
		Section 08800 - Glazing: Glazing to be field installed in wood windows.
		Notes: *Unless windows are factory glazed, glazing is not specified in this section.*
		Sills and stools for wood and plastic windows may be specified in this section or in Section 05500 - Metal Fabrications and Section 06200 - Finish Carpentry.
08650	**SPECIAL WINDOWS**	**086500 – SPECIAL WINDOWS**
-655	Roof Windows	Windows including operating hardware used for a variety of special functions and applications and employing operating methods.
-660	Security Windows and Screens	***Related Sections:***
	Security Windows	*Section 07800 - Skylights: Formed plastic skylights and metal framed skylight assemblies with plastic or glass glazing.*
	Security Screens	*Section 08900 - Glazed Curtain Walls: Translucent wall and skylight systems, sloped glazing systems, and structural glass walls.*
-665	Pass Windows	*Section 10530 - Protective Covers: Window awnings.*
-670	Storm Windows	*Section 10700 - Exterior Protection Devices for Openings: Exterior shutters, storm panels, and sun control*

Figure 4.1 Example of broadscope explanation (from Division 8, CSI), used with permission.

DIVISION 16 - ELECTRICAL

Section Number	Title
16050	**BASIC ELECTRICAL MATERIALS AND METHODS**
-110	Raceways
	Cable Trays
	Conduits
	Surface Raceways
	Indoor Service Poles
	Underfloor Ducts
	Underground Ducts and Manholes
-120	Wires and Cables
	Fiber Optic Cable
	Low Voltage Wire
	600 Volt or Less Wire and Cable
	Medium Voltage Cable
	Undercarpet Cable Systems
-130	Boxes
	Floor Boxes
	Outlet Boxes
	Pull and Junction Boxes
-140	Wiring Devices
	Low Voltage Switching
-150	Manufactured Wiring Systems
-160	Cabinets and Enclosures
-190	Supporting Devices
-195	Electrical Identification
16200	**POWER GENERATION - BUILT-UP SYSTEMS**
-210	Generators
	Hydroelectric Generators
	Nuclear Electric Generators
	Solar Electric Generators
	Steam Electric Generators
-250	Generator Controls
	Instrumentation
	Starting Equipment
-290	Generator Grounding
16300	**MEDIUM VOLTAGE DISTRIBUTION**
-310	Medium Voltage Substations
-320	Medium Voltage Transformers
-330	Medium Voltage Power Factor Correction
-340	Medium Voltage Insulators and Lightning Arrestors
-345	Medium Voltage Switchboards
-350	Medium Voltage Circuit Breakers
-355	Medium Voltage Reclosers
-360	Medium Voltage Interrupter Switches
-365	Medium Voltage Fuses
-370	Medium Voltage Overhead Power Distribution
-375	Medium Voltage Underground Power Distribution
-380	Medium Voltage Converters
	Medium Voltage Frequency Changers
	Medium Voltage Rectifiers
-390	Medium Voltage Primary Grounding

Broadscope Explanation

16050 – BASIC ELECTRICAL MATERIALS AND METHODS

Items that are common to more than one section of Division 16.

Related Sections:
Section 09900 - Painting: Electrical identification painting.

Notes: *There are optional ways to specify basic electrical products and installation requirements which are common to specify both products and their installation in this section. Secondly, the products may be specified in this section and additional installation requirements specified in the sections for the systems to which the products apply. A third option is to specify both the products and installation requirements entirely in other sections. With the first two options, references from affected sections should be made to this section.*

16200 – POWER GENERATION - BUILT-UP SYSTEMS

Equipment and installation related to the generation of electric power in large built-up systems.

Related Sections:
Section 02780 - Power and Communications: Overhead and underground distribution
Section 13100 - Nuclear Reactors.
Section 13300 - Utility Control Systems.
Section 13600 - Solar Energy Systems
Section 13700 - Wind Energy Systems.
Section 16300 - Medium Voltage Distribution.
Section 16950 - Testing.

Notes: *This section is not intended for specifying unitized systems. See Section 16600 - Special Systems for packaged engine generators systems.*

16300 – MEDIUM VOLTAGE DISTRIBUTION

Equipment and installation related to the transmission, distribution, and control of electric power ranging from 601 to 35,000 volts; such as substations, switchgear, circuit breakers, and transformers.

Related Sections:
Section 01650 - Facilities Startup/Commissioning: Administrative procedures for starting, testing, and adjusting systems.
Section 02780 - Power and Communications: Overhead and underground site distribution.
Section 13300 - Utility Control Systems.
Section 16200 - Power Generation - Built Up Systems.
Section 16950 - Testing.

Notes: *For systems over 35,000 volts refer to Section 02780 - Power and Communications.*

Transformer vaults may be specified in Section 03400 - Precast Concrete or in other sections appropriate to the type of construction.

84

Figure 4.2 Example of broadscope explanation (from Division 16, CSI), used with permission.

walk-in freezers – water distribution systems

Figure 4.3 Extract from keyword index (CSI).

Chapter

5

Quantity Takeoff Tools

Quantity takeoff is the foundation of the construction estimate. The estimator does not use elegant mathematics. The skills needed (arithmetic, geometry, and trigonometry) are taught in high school. These are defined by Webster's II (New Riverside Dictionary):

Arithmetic. "The mathematics of integers under addition, subtraction, multiplication, [and] division."

Geometry. "The mathematical study of the properties, measurement, and relationships of points, lines, angles, surfaces, and solids."

Trigonometry. "The study of the properties and applications of trigonometric functions." (*Trigonometric function:* "A function of an angle expressed as a ratio of two of the sides of a right triangle that contains the angle.")

Formulas—Geometric

These formulas provide either area [see formulas 1 on page 124] (for plane figures, i.e., two-dimensionals) or volume [see formulas 2 on page 125] (for three-dimensional solids).

Trigonometric functions

Most construction cross sections involve right angles (i.e., 90°). For a right-angle triangle, see formulas 3 on page 126.

Function values for key angles:

Angle	Sine	Cosine	Tangent
30°	0.50	0.87	0.58
45°	0.707	0.707	1.00
60°	0.87	0.50	1.73

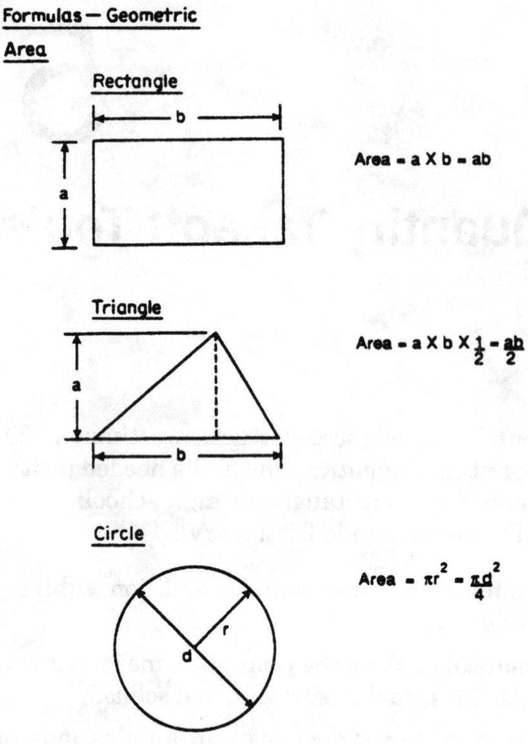

Formulas 1

Mechanics

Generally speaking, a structure as designed will do its job when in place; it is not the usual responsibility of the estimating team to second-guess the structural design. However, when an apparent error is noted, it quite properly may ask the design group for clarification.

Forces

Forces are usually represented for purposes of calculation as *vectors,* that is, arrows that may be drawn to scale in units of force or may just be used to represent the forces. Figure 5.1 shows upward and downward vertical forces and the direction of horizontal forces. The resolution of forces that are neither vertical nor horizontal into their horizontal and vertical components is important for graphical representation, as well as for an understanding of the forces.

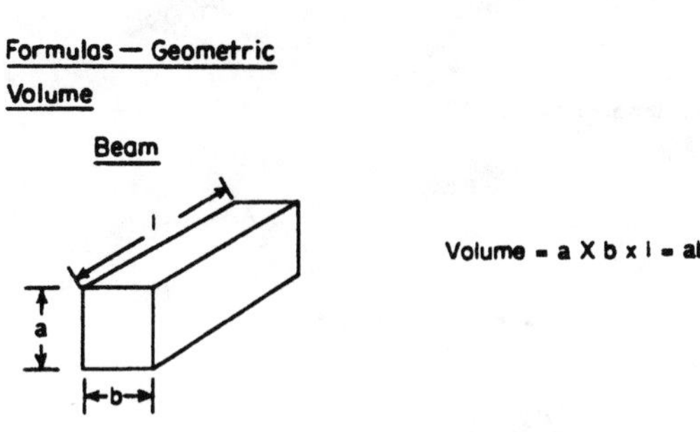

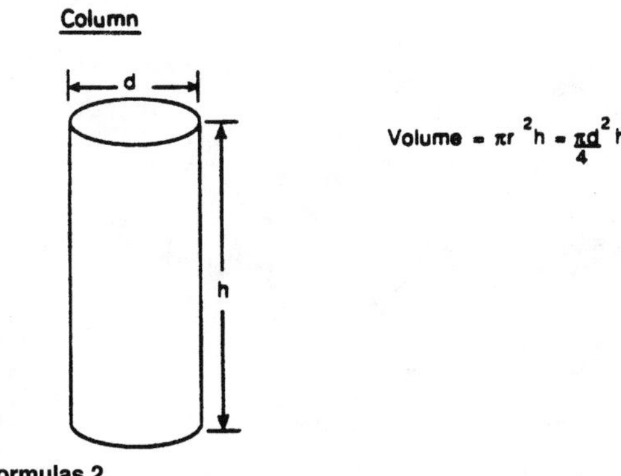

Formulas 2

The resolution works in both directions; that is, a horizontal and vertical force operating on a structure is resolved into one component that is neither vertical nor horizontal. It is this equivalent force or combined force that acts upon the structure and that must be resisted in equal and opposite amounts so that the structure will not be forced to move.

The scientist Sir Isaac Newton, about 300 years ago, posed his three fundamental laws of motion. Two of these deal with equilibrium. Newton's first law states: "A body at rest remains at rest, and a body in motion continues to move at constant speed along a straight line, unless the body is acted upon in either case by an unbalanced force." Either condition is assumed to be equilibrium, but in the case of building structures, stability and rest is the state desired. Newton's third law states: "For every action, there is an equal and opposite reaction." In other words, in a building that is still and at rest, every

Trigonometric Functions

Most construction cross sections involve right angles (i.e. - 90°)

For right angle triangle

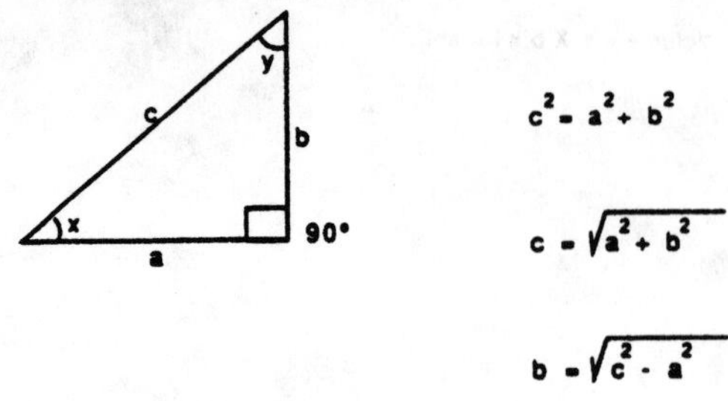

$$\text{Sin X} = \frac{b}{c}$$

$$\text{Cos X} = \frac{a}{c}$$

$$\text{Tan X} = \frac{b}{a}$$

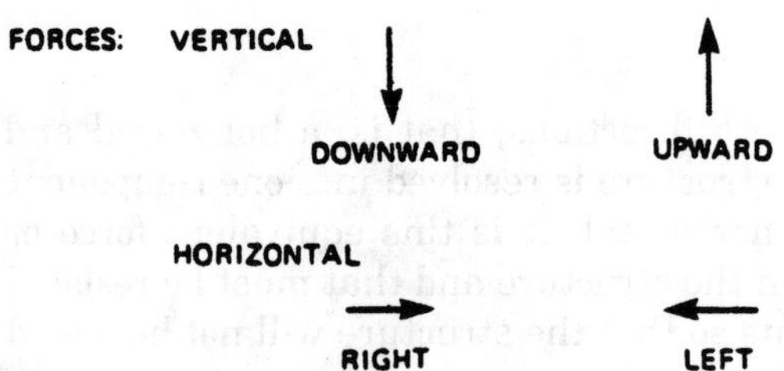

OTHER FORCES CAN BE RESOLVED INTO VERTICAL & HORIZONTAL COMPONENTS:

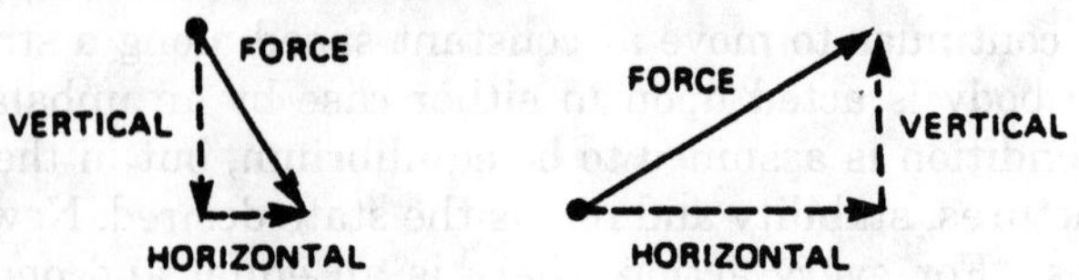

Figure 5.1 Vector presentation of forces by components.

action, either upon the building or resulting from the weight of the building, must be resisted by an equal and opposite reaction if it is to stay in a stable or still condition. When a body is in equilibrium, its resistance to change, either from rest into motion or from motion to come to rest, is known as its *inertia.*

In determining the forces upon a structure at rest, or at stages of rest during erection, all forces added up must equal zero. Thinking in terms of the components, all vertical forces must equal a net of zero, and all horizontal forces must equal a net of zero. It is therefore important to be able to convert components, and the simplest way is to visualize each angular force (i.e., neither horizontal nor vertical) as the hypotenuse of a right triangle. The vertical and horizontal legs of the triangle then become the vertical and horizontal components of that force and, when shown in relation to the structure, can replace the actual force for purposes of calculation. The method of solving, or resolving, forces is shown in Fig. 5.2.

Resolution of the forces depends upon measurement of the angles or sides of triangles, or a combination of both. For instance, in a measurement where both sides are equal, the angles must be 45° angles.

Information on the angles depends upon resolution of the forces on the basis of right triangles. The sum of the three angles within a triangle must equal 180°, and use of a right triangle identifies one of these angles as 90°; so the sum of the other two must be 90° also. The relationship between the sides of a right triangle, where the longest side squared equals the sum of the squares of the opposite two sides, is proved in all trigonometry books.

In order to sum forces, they must be on the same axis, and signs must be given—for instance, downward forces are positive, while upward are negative, and forces to the right positive, while to the left, negative. Any sign convention will work, as long as it is applied uniformly during calculations.

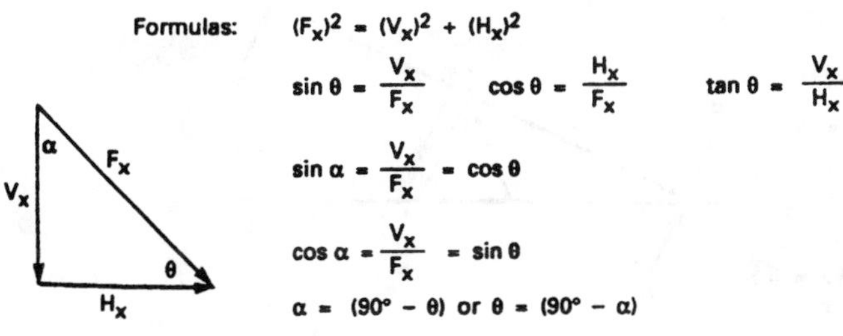

Formulas:
$(F_x)^2 = (V_x)^2 + (H_x)^2$
$\sin\theta = \frac{V_x}{F_x}$
$\cos\theta = \frac{H_x}{F_x}$
$\tan\theta = \frac{V_x}{H_x}$
$\sin\alpha = \frac{V_x}{F_x} = \cos\theta$
$\cos\alpha = \frac{V_x}{F_x} = \sin\theta$
$\alpha = (90° - \theta)$ or $\theta = (90° - \alpha)$
α
F_x
V_x
θ
H_x

Figure 5.2 Resolution of forces.

Moments

The equilibrium of a body is not made complete merely by equalization of its horizontal and vertical forces. For instance, the example in Fig. 5.3 shows a seesaw with no horizontal forces exerted, and a net zero upward and downward force; but obviously the seesaw will move from its horizontal position, tilting to the left-hand side. The unbalanced force causing this motion is the *moment,* which is equal to the arm through which the force acts times the force, in this case 2 times *F*. In Case 2 of Fig. 5.3, the seesaw has come to rest, and again the vertical forces must total zero. One vertical force must resist the downward push at the left-hand side, which is *F*. However, the uplift on the right-hand side must also be resisted, and is resisted by a vertical downward pull of the pivot. A pivot, however, can exert a force only at right angles, so that it sets up both a vertical and a horizontal component. The horizontal component can be resisted only by a pressure in the earth at the grounded end of the seesaw, which is also resisting downward pressure at the left.

The rule for calculating the effect of moments states that moments about any point within the structure, or even a point taken outside of the structure, must be equal and opposite. Solving for the vertical force at the center in Case 2 (see equations in Fig. 5.3), the vertical component must be 2*F*. Accordingly, because the sum of vertical forces must be zero, the force at the left reaction in

Case 1

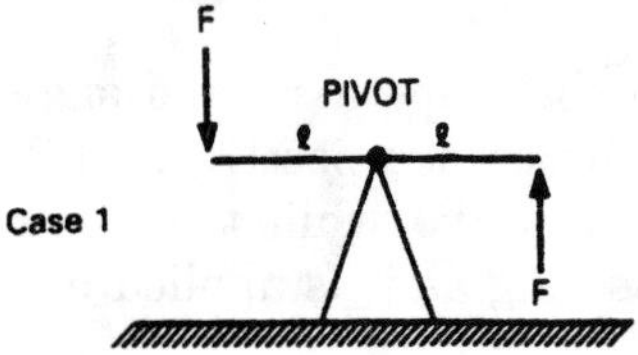

Case 2 Axes shifted to plane of tilted board to simplify results:

Moments at left = 0

$V_c l = 2lF; V_c = 2F$

Moments at pivot = 0

$V_L l - Fl - Fl = 0: V_L = 2F$

Check $\Sigma V_{forces} = 0$

$F - VL + Vc - F = 0$

$F - 2F + 2F - F = 0$

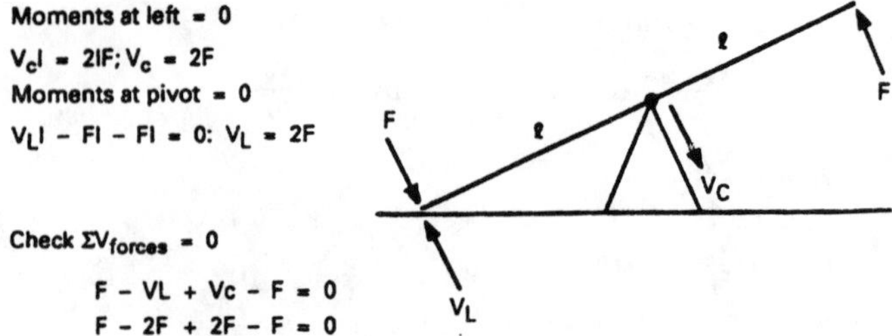

Figure 5.3 Seesaw with no horizontal forces exerted.

the ground must also be $2F$. To simplify calculations, the horizontal and vertical axes were shifted so as to be parallel, and vertical to the seesaw itself. If the force at the right-hand side is not at true right angles to the board, but actually in the vertical, and this condition is not specified, then the angle of the seesaw would have to also be considered, and the height of the pivot above the ground would have to be identified. Further, a horizontal force action to the right at the pivot and to the left at the ground would also be incurred. These horizontal forces would be equal and opposite but would set up a moment equal to the horizontal force times the height of the pivot from the ground and would have to be reacted against by other moments. The other moment would be the horizontal force of F at the right end, times twice the pivot height, if force F is acting truly perpendicular to the board.

In summary, several factors are important in examining a structure:

1. All *forces* must be *equal* and *opposite* and can best be calculated by resolving the forces into their components, which can be added and subtracted arithmetically once the resolution has occurred.
2. In addition to forces being equal and opposite, *moments* about any point in relation to the body must also be *equal* and *opposite.* As demonstrated in Fig. 5.4, this may require reaction to be the result of the type of structure. In this case, a cantilever wall in Case a is providing a reverse or reactive twist. In Case b, a crank arm is providing a reaction torque or twist.
3. The type of *connection* is also important. In Fig. 5.5, various types of connection and their ability to transmit forces are illustrated.

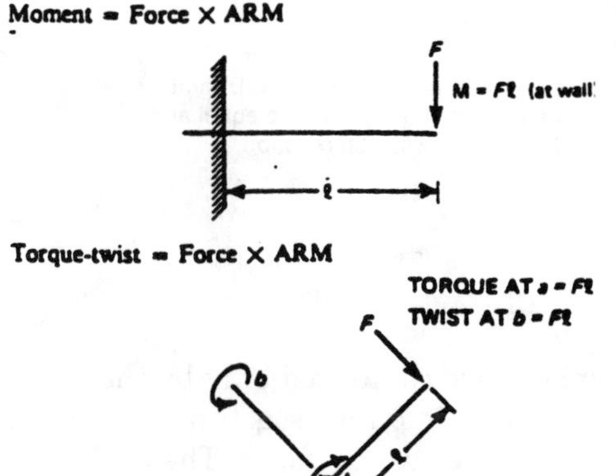

Figure 5.4 Twist or torque.

1. Hanging Cable or Chain—Vertical downward force only, no moment

2. Taut Cable or Chain—Force in direction of pull, no moment

3. Pivot or Rollers: Force at right angle, no moment (up or down)

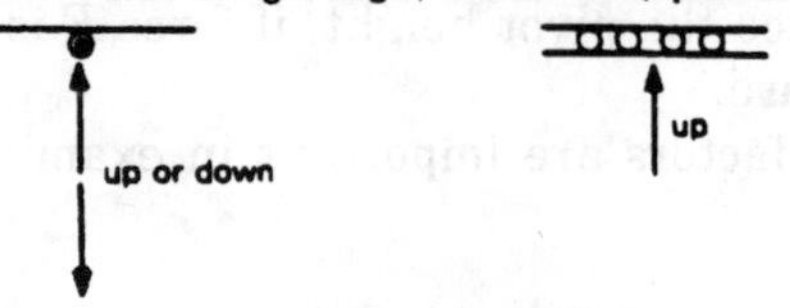

4. Pin Connection: Axial Force + or −; no moment

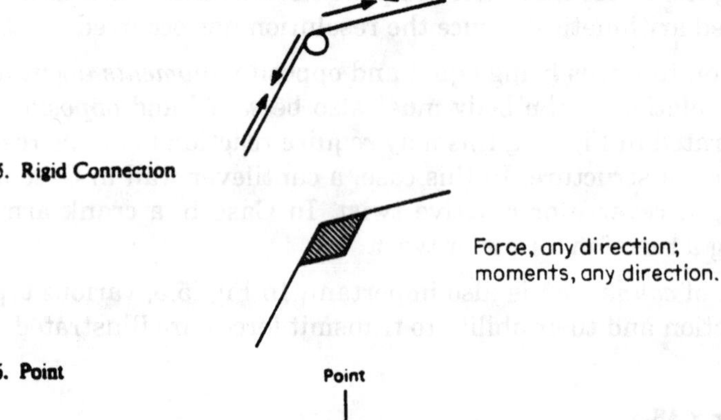

Vertical force in direction of axis; true point to surface transfer has no horizontal component, unless coefficient of friction X area is sufficient to provide equal and opposite reaction, or point "digs in" enough to permit horizontal reaction.

Figure 5.5 Typical connections.

Where two structures or members are connected only by their own relative weight, the ability to transmit force is a function of the roughness or relative smoothness of the two surfaces. The *coefficient of friction* is a measure of the relative ability of the two surfaces to develop a binding or restraining force.

Beams

A beam is a structural member used to transport or carry forces from one point to another, usually horizontally. A beam bridging between two support abutments carries weight between its ends, whereas a cantilever beam is held rigidly at its end or center and carries forces suspended or imposed at its extremities. In carrying these forces, the beam bends or deflects in proportion to the load placed upon it, and this flexure imposes certain internal stresses upon the beam. Its ability to carry load and its amount of deflection are a function of its material and configuration.

Figure 5.6 shows a typical beam in what is normally called positive or downward bending. A cross section of its material is shown, indicating that the bottom fibers are being stretched or are in tension, while the upper fibers are being compressed or squeezed. The neutral axis is somewhere in the central portion of the beam and is a balance point between the ability to resist tension and the ability to resist compression. Usually the neutral axis is stable whether the beam is bent downward or upward.

Using moment analysis, it is obvious that the fibers farthest from the neutral axis are those that must have the greatest ability to resist

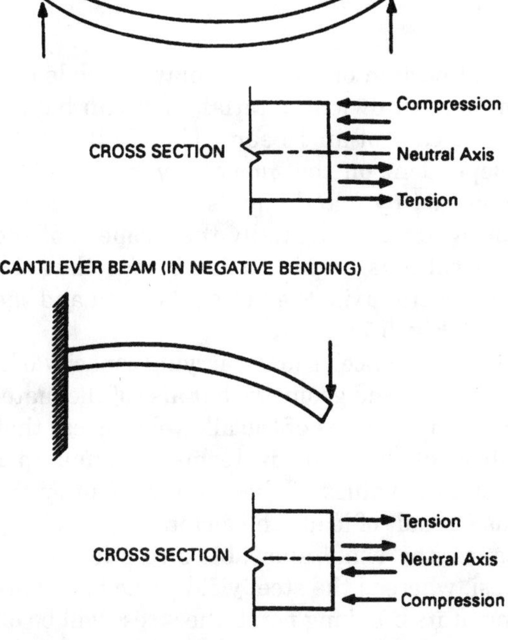

Figure 5.6 Beams in bending.

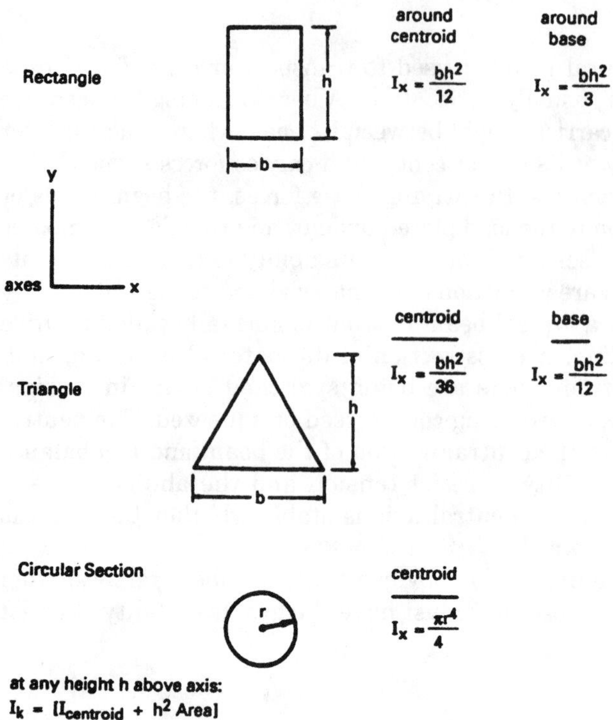

Figure 5.7 Typical moments of inertia.

forces. Accordingly, the central portion of the beam may be made of a lighter material if the beam is a sandwich material, or it can have a smaller cross section, as in the case of an I beam. The ability of the beam to resist bending is dependent on the *moment of inertia* of its cross section, which is symbolized by *I* and equals the cross section times the radius of gyration, which is essentially the moment of the area above or below the neutral axis. Again, it is obvious that the greater the distance from the neutral axis, the larger the area and the greater the ability to resist (see Fig. 5.7).

As illustrated in Fig. 5.8, the resistance or load-carrying capacity of a beam is a function not only of its size and geometry but also of the material. The stresses shown are approximations of the allowable stress that may be developed in the fibers of the beam in terms of pounds per square inch. They do not represent values of the external load or the internal stresses developed as a result of load. The factors of safety vary with the material. For instance, the usual allowable stress in steel is between 20,000 and 25,000 psi, whereas the steel yield value is usually on the order of 36,000 psi, and at its breaking point, the stress will be on the order of 50,000 or more psi. However, this situation does not produce a two-to-one safety factor, but more like a 50% safety factor; for

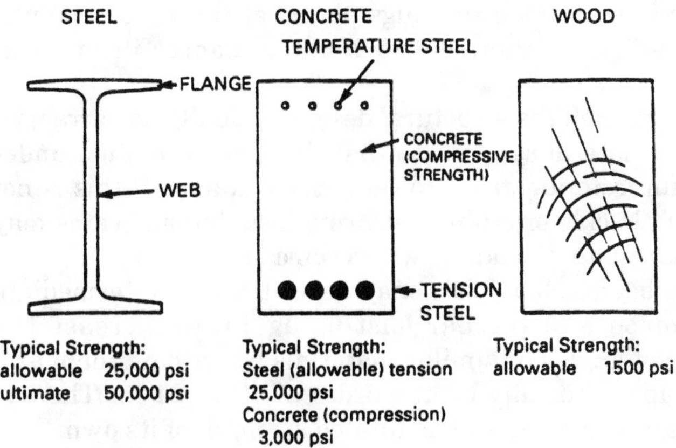

Figure 5.8 Typical beam cross sections.

once the yield point in steel has been passed, it stretches, readily, sharply reducing the area resisting load in a necking phenomenon.

Wood can carry only about 10% of the load of a steel beam. Its working stress is about one-third of the ultimate or breaking stress, because of uncertainties in the makeup of the material. A knot hidden within a timber may cause it to break unexpectedly, thereby requiring a greater safety factor.

Concrete is very strong in compression but has virtually no tensile strength. For this reason, temperature steel is utilized to hold the material together during shrinkage or when subjected to light tensile loads. However, the reinforcing steel is placed in reinforced concrete to carry the entire tension load, while the beam's opposite flange carries the compressive load. Obviously, the placement of the steel reinforcement is quite important. On some occasions, iron workers, not realizing the importance of the placement near the outer fibers, may pull the steel too high into the beam, seriously reducing the moment of inertia of the steel or tensile portion of the beam. In other cases, the steel may be allowed to lie on the ground, as in the case of reinforced-concrete-grade beams, with resultant corrosion and ultimate loss of strength.

With beams, where the emphasis is on developing greater moments of inertia, the narrow or weak axis has limited strength if load is applied from a sideways direction or if a beam is inadvertently carried on its side. Similarly, plates of materials or large sections, such as precast concrete wall sections, if carried flat, may crack badly; whereas, if carried in the upright position, they have a high moment of inertia and can resist handling stresses. Even steel sheets or plates carried in the flat may buckle because of their own dead weight.

Handling of precast members that are to be poststressed is particularly critical. The shop drawings should indicate the methods of handling, which usually involves lifting lugs placed at the ⅓ or ¼ points, so that the dead weight of the member counterbalances itself about the lifting points.

When beams are placed, the structural designer usually has arranged for cross bridging or lateral support, so that there is no rotation under loading of the beam from its strong toward its weak axis. If this is not done, the beam will buckle or crumple. During installation, beams may be subjected to undue lateral loading, which can cause failure.

For lightweight beams, bar joists may be utilized as intermediate members. The purpose of the bar joist is, again, to increase the moment of inertia of angles or similar materials by spacing them at a greater distance apart, usually on the order of 12 to 24 in. The bar material used to space the angles has limited strength of its own.

As a rule of thumb, most steel members are about 1 in deep for each foot to be spanned, so that for a 20-ft span, something on the order of a 20-in I beam could be used. A lightweight beam might be somewhat deeper, whereas a strong heavy section might accomplish this span with 16 in of depth. For larger spans, on the order of 50 ft or more, a different form of beam is utilized—the plate girder. A plate girder is custom-designed for the span and load. It is built up of plates used for both flange and web. For spans over 100 to 150 ft a truss is used. The purpose of a truss is to bridge the space between points, and this is accomplished by placing material along the flanges or outer edge through the use of spacers or chord members.

Figure 5.9 shows some typical truss shapes,. The Fink truss, for instance, might be utilized for the roof beam or span in a building

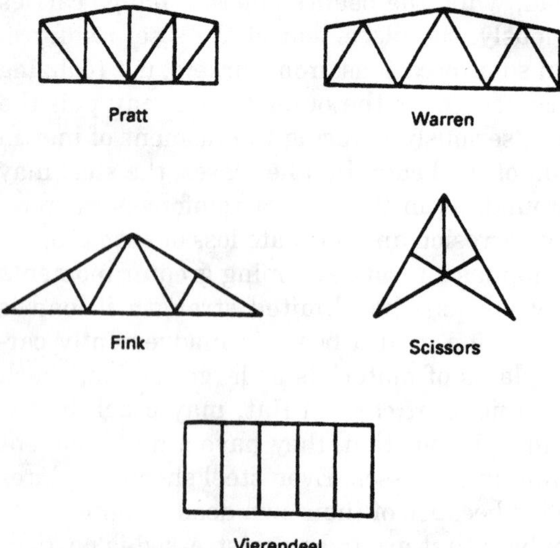

Figure 5.9 Truss types.

with sloped roofs, whereas the Vierendeel is a square truss, very suitable for spanning large openings in buildings. In some cases, the dimensions of the Vierendeel truss may be made to coincide with two or three stories of height, so that lateral finished spaces might go through the truss.

Reinforced concrete beams are larger in dimension when compared with steel beams of equal strength, because of the additional space required to develop the tensile reinforcing steel cross section. However, a wide beam often can be utilized because reinforced concrete beams are incorporated into floor structures.

Columns

Columns are compression members normally installed in a vertical position. Because their primary purpose is the resistance of compressive force, they can support substantial loads with relatively small cross sections. Theoretically, a short column of substantial cross section should be able to support a load equal to the allowable stress times the cross-sectional area. However, columns that fail usually do not do so in direct compression but by buckling. Buckling occurs because the column is relatively slender in terms of its cross section vs. length; and when it is subjected to any small moments, stresses may be set up as a result of the moments and cause a failure. Accordingly, lateral support of columns is an important consideration. This type of support is gained, for instance, by the tie-in of beams at floor level; and in the case of derricks or gin poles, it is achieved by guy wires.

The traditional solution for column loading is found in the Euler equation, which indicates that there is a specific axial load that will hold a column in equilibrium in a bent or curved position when the ends are pin-loaded. This relationship or allowable load P is as follows:

$$P = \pi \frac{EI}{l^2}$$

where E = modulus of elasticity
I = least moment of inertia about any cross-sectional axis
l = length

This equation indicates that the allowable load is directly proportional to both the strength as represented by the modulus of elasticity and the moment of inertia of the cross section. Because the moment of inertia is the least moment, a tube provides the maximum effective placement of material, and the larger the diameter, the more resistant the material. (However, if the material is too thin, local buckling can occur.)

The load is inversely proportional to the square of the length, indicating that the dimension or relationship often considered is the slenderness ratio as a controlling factor. This is the ratio of the length l to the radius of gyration of the cross section.

Column sizes are calculated not only by using the traditional Euler formula, with appropriate variations for end conditions, but also by considering various size and load factors, utilizing families of curves similar to those presented in the *AISC Handbook*.

Conversion Tables

The estimator often must convert quantities to standard measures such as *cubic yards, tons,* or *square feet.* Table 5.1 has English unit conversion factors. Table 5.2 has English to metric and metric to English conversion factors. Table 5.3 has dimensions for concrete reinforcement deformed steel bars. Table 5.4 has weights (psf) for various floors, ceilings, and roofs. Table 5.5 has weights (psf) for walls and partitions.

TABLE 5.1 English Unit Conversion Factors

	Linear Measure	
Inches	Feet	Yards
12.0 =	1.0 =	0.33333
36.0 =	3.0 =	1.0
	Square Measure	
Square inches	Square feet	Square yards
144.0 =	1.0 =	0.111111
1296.0 =	9.0 =	1.0
	Cubic Measure	
Cubic inches	Cubic feet	Cubic yards
1728.0 =	1.0 =	0.037
46,656 =	27.0 =	1.0
	Board feet = 144 in^2 × 1 ft = 0.083 cf	
	Liquid Measure	
U.S. gallons	Cubic feet	Cubic inches
1.0 =	0.1337 =	231
7.48052 =	1.0 =	1728
	Temperature	
	Fahrenheit degrees (less 32°F) × 0.5556°C	
	Centigrade degrees × 1.8 = °F (less 32°F)	

TABLE 5.2 English to Metric, Metric to English Conversion Factors

English to Metric		Metric to English	
1 inch =	25.4 millimeters	1 millimeter =	0.039 inch
	2.54 centimeters		0.0033 foot
	0.0254 meter		
1 foot =	304.8 millimeters	1 centimeter =	0.39 inch
	30.48 centimeters		0.033 foot
	0.305 meter		
1 yard =	0.914 meter	1 meter =	39.4 inches
		3.28 feet	
			1.094 yards
1 cubic inch =	16.39 cubic centimeters	1 cubic centimeter	0.06 cubic inch
			0.000035 cubic foot
1 cubic foot =	28,317 cubic centimeters	1 cubic meter =	35.3 cubic feet
	0.028 cubic meter		1.31 cubic yards
	28.3 liters		
1 cubic yard =	0.765 meter	1 gram =	0.0022 pound
1 pound =	454 grams	1 kilogram =	2.2 pounds
	0.454 kilogram		
	0.00045 metric ton		
pounds/foot =	1.49 kilogram per square meter	kilogram per square centimeters =	14.2 pounds per square inch
pounds/square inch =	0.07 kilogram per square centimeter	kilogram per square meters =	0.20 pound per square inch
	0.0007 kilogram/millimeter		
square feet =	0.093 square meters	liter =	0.26 gallon
			0.035 cubic foot
square inches =	6.45 square centimeters	1 square centimeter	0.155 square inch
	645 square millimeters		
square yards =	0.836 square meter	1 square meter =	10.76 square feet
			1.2 square yards
tons =	907 kilograms		
horsepower (U.S.) =	1.01 metric horsepower		

TABLE 5.3 Dimensions for Concrete Reinforcement Deformed Steel Bars

Deformed bar designation numbers[a]	Unit weight, lb/ft	Nominal dimensions, round sections: Diameter, in	Cross-sectional area, in^2	Perimeter, in
3	0.376	0.375	0.11	1.178
4	0.668	0.500	0.20	1.571
5	1.043	0.625	0.31	1.963
6	1.502	0.750	0.44	2.356
7	2.044	0.857	0.60	2.749
8	2.670	1.000	0.79	3.142
9[b]	3.400	1.128	1.00	3.544
10[b]	4.303	1.270	1.27	3.990
11[b]	5.313	1.410	1.56	4.430

[a]Bar numbers are based on the number of eighths of an inch included in the nominal diameter of the bars.

[b]Bars of designation nos. 9, 10, and 11 correspond to the former 1-in^2, 1⅛-in^2, 1¼-in^2 sizes and are equivalent to those former standard bar sizes in weight and nominal cross-sectional areas.

TABLE 5.4 Weights: Floors, Ceilings, Roofs

Floors	Weight (psf)
Cement finish, per inch of thickness	12
⅞-in hardwood floor on sleepers clipped to concrete without fill	5
1½-in terrazzo floor finish directly on slab	19
1½-in terrazzo finish on 1-in mortar bed	30
¾-in ceramic or quarry tile on ½-in mortar bed	16
¾-in ceramic or quarry tile on 1-in mortar bed	22
¼-in vinyl tile directly on concrete	1
Ceilings	
¾-in plaster on metal lath furring	8
¾-in plaster on metal lath and channel suspended ceiling construction	10
Acoustical fiber tile directly on concrete blocks or tile	1
Acoustical fiber tile on rock lath and channel ceiling construction	5
Roofs	
Three-ply felt and gravel (or slag)	5½
Three-ply felt composition roof, no gravel	3
Slate, ½-in thick	19
Sheathing, ¾-in thick, yellow pine	3½
Skylight with galvanized iron frame, ¼-in wire glass	7
Gypsum, per inch of thickness	4
Light-weight fill or insulation, porous glass, vermiculite, etc., per inch of thickness	1 to 2
Metal deck (20 gauge)	2¼
Metal deck (18 gauge)	3
Corrugated metal (20 gauge)	1½
Flat cement tile, per inch of thickness	13

TABLE 5.5 Weights of Walls and Partitions

Walls	Weight (psf) Unplas-tered	One side plastered	Both sides plastered
4-in brick wall	40	45	50
9-in brick wall	80	85	90
6-in concrete wall	36	41	46
8-in concrete wall	51	56	61
12-in concrete wall	59	64	69
6-in hollow light-weight block (tile or cinder)	22	27	32
8-in hollow light-weight block (tile or cinder)	33	38	43
12-in hollow light-weight block (tile or cinder)	44	49	54
4-in brick, 4-in hollow concrete block backing	68	73	—
4-in brick, 8-in hollow concrete block backing	91	96	—
4-in brick, 12-in hollow concrete block backing	119	124	—
4-in glass block	20	—	—
Windows, glass, frame and sash	8	—	—
Structural glass, per inch of thickness	15	—	—
Partitions			
2 × 4 studs, or metal studs, lath and ¾-in plaster	—	—	18

Reference Library

Cost and pricing databases and references are described in Chap. 7. In addition, the following reference volumes are useful to identify vendors. This is important for solicitation of vendor information, in particular cost:

- *Sweet's Catalog File Series* (annual) by McGraw-Hill:
  - *Products for General Building & Renovation* (18 volumes typical)
  - *Engineering and Retrofit* (6 volumes typical)

  The catalog's information is arranged in CSI format.

- *Thomas Register of American Manufacturers* (annual) by Thomas Publishing Company, New York. Products are listed alphabetically. Volumes are arranged (typically):
  - *Products and Services* (16 volumes, 30,000 pages)
  - *Company Profiles* (2 volumes)
  - *Catalog File* (ThomCat) (7 volumes)

  The catalog portions often provide specification data, sample layouts, and/or general dimensions. Other good references include:

- *Standard Handbook for Civil Engineers* (McGraw-Hill)
- *Masterformat* (Construction Specification Institute)

- *Marks' Standard Handbook for Mechanical Engineers* (McGraw-Hill)
- *Standard Handbook for Electrical Engineers* (McGraw-Hill)
- *Steel Construction Manual* (American Institute of Steel Construction)
- *Standard Details for Fire-Resistive Building Construction* (McGraw-Hill)
- *Graphic Standards* (Wiley)
- *Mechanical and Electrical Equipment for Buildings* (Wiley)

Chapter

6

Design Development Estimate Quantities

This stage is also called preliminary design. After approval of the schematic phase, the drawings are refined to a degree sufficient to permit the development of dimensioned space layouts, heating and ventilating systems main feeders, and electrical main feeders, as well as definitive development of the structural framework. Requirements for utilities are also definitively developed, and specific equipment requirements are determined. The budgetary type cost estimate is revised, and a more firm estimate made. Value engineering (VE) and constructibility reviews are best applied at this stage.

The design development design package should include:

Site plan. Schematic site plan updated to reflect any additional information. Level of completion probably 25 to 50%.

Plan views. Building layouts are usually at a scale of ¼ in = 1 ft 0 in. Bay sizes should be finalized. Room, corridor, equipment rooms, electrical equipment, and other locations should be well developed. Level of completion is approximately 60 to 80%. These drawings are often numbered as the A series (architectural).

Elevations. The exterior view of the building shows the perimeter, again usually at the same vertical and horizontal scale. ¼ in = 1 ft 0 in is typical. At this stage, drawings are probably at 40 to 60% level of completion. These are usually in the A, or architectural, series.

Sections. Interior views of the building showing cross sections, similar to elevations. At this stage, sections are in a preliminary level of completion, probably 20 to 40%. These are usually in the A, or architectural, series.

Foundations. These are shown as an overlay of the architectural plans, at the same scale. Level of completion at this stage is probably 60 to 80%. The foundation plans are often numbered either an F (for foundation) or S (for structure). Sections through the foundation plans are numbered in the same series (F or S). Sections are dimensioned both horizontally and vertically.

Structural. These are shown as an overlay of the architectural plans to the same scale. Beam sizes (structural steel or concrete) are shown. Column sizes may also be shown. Slabs are shown, and slab reinforcing may be shown. Level of completion is probably 70 to 90% for main members. Secondary members, bracing, and connections probably will not be shown. The structural plans are usually in the S series. Sections through the structural plans will show both beam and column sizes. Sections are dimensioned both horizontally and vertically. Sections through the structure are usually in the S series.

HVAC. At this stage, plans would be an overlay of the architectural footprint at the same scale. Level of completion would probably be 0 to 20% for equipment location and ductwork. A flow diagram (not to scale) showing equipment by type (chillers, boilers, fans, etc.), and method of delivery (pipe, ductwork, etc.) should be well developed, probably 40 to 60%. This diagram usually includes tables listing equipment and capacities, such as Btu, horsepower, and kW. HVAC drawings are usually an HVAC or M (mechanical) series.

Mechanical. At this stage, the following would not usually have reached the drawing stage:

- Fire sprinkler
- Equipment
- Plumbing

Mechanical drawings are usually the M series. This may include plumbing and equipment, or these (respectively) could be the P and EQ series.

Electrical. At this stage, plans would be an overlay of the architectural drawings at the same scale. Level of completion would probably be at a level of completion of 0 to 20%. A single-line diagram showing electrical equipment by type (switchgear, motor starters, transformers, power distribution panels, lighting panels, etc.) should be at a level of about 40 to 50%. This diagram would usually have tables listing the equipment and ratings such as kW, voltage, and amperage. This diagram would also have tables listing equipment in the HVAC, mechanical, plumbing, and equipment sections which are electrically powered. These, of course, must follow the development of the HVAC, mechanical, plumbing, and equipment specifications.

Specifications. An outline of the specifications will have been started. This will almost always be in CSI format. The information included will be the architect's and engineers' initial thoughts as to features they expect to require by CSI category. These usually relate to quality rather than quantities. At this point, the design professionals may list examples (i.e., "or equal") of the type of material, fixtures, finishes, etc., they expect to specify in the final design.

Suggestions for the Estimator

The following can be adapted by an estimator regardless of the type of estimate or the discipline of the estimator.

1. Use preprinted or columnar forms for orderly sequence of dimensions and locations. If you create your own form, organize it in columns to avoid errors.
2. Use only the front side of each paper or form except for certain preprinted summary forms. If you must use the reverse side, put an arrow at the bottom of the sheet.
3. Be consistent in listing dimensions; for example, length × width × height. This helps in rechecking to ensure that, say, the total length of partitions is appropriate for the building area.
4. Dimensions:
   a. Use printed (rather than measured) dimensions where given.
   b. Add up multiple printed dimensions for a single entry where possible.
   c. Measure all other dimensions carefully.
   d. Use each set of dimensions to calculate multiple related quantities.
5. Convert foot and inch measurements to decimal feet when listing.
6. Do not "round off" quantities until the final summary.

Procedure

For portions of the design development stage of design, a fairly complete material takeoff can be made. These include:

- Excavation

- Foundations
- Structure
- Concrete
- Roof
- Exterior shell
- Some interior work

For other areas such as mechanical, electrical, and site work, the design has advanced little from the schematic phase, so that the same information can be used.

Design Development, WHATCO Project

The plant and warehouse administrative and support areas of 7000 sf are combined with the regional office space of 6400 sf, in a total figure of 13,000 sf. Some reduction is made possible by reduction of overlapping functions (restrooms, lunchrooms, and net reduction of office space). However, the offices are now two stories, so space is added for two stairwells and one elevator. The design development team identified problems in the schematic layout. In the plant, the shop offices, locker room, toilets, and lunch area intrude 20 ft into the 100-ft width. This reduces the working area by 20%.

It was decided to relocate the following from the plant:

Shop offices	20 ft × 100 ft =	2000 sf
Locker room and toilets	50 ft × 20 ft =	1000 sf
Lunchroom	50 ft × 20 ft =	1000 sf
	Subtotal	4000 sf

The following spaces were removed from the warehouse:

Boiler area	20 ft × 40 ft =	800 sf
Load center	20 ft × 20 ft =	400 sf
Warehouse office	50 ft × 20 ft =	1000 sf
	Toilets =	800 sf
	Subtotal	3000 sf

These relocations permit the warehouse to be reduced to 6000 sf.

The design team reconfigured the space into a reconfigured footprint as shown in Fig. 6.1. The new configuration is more efficient,

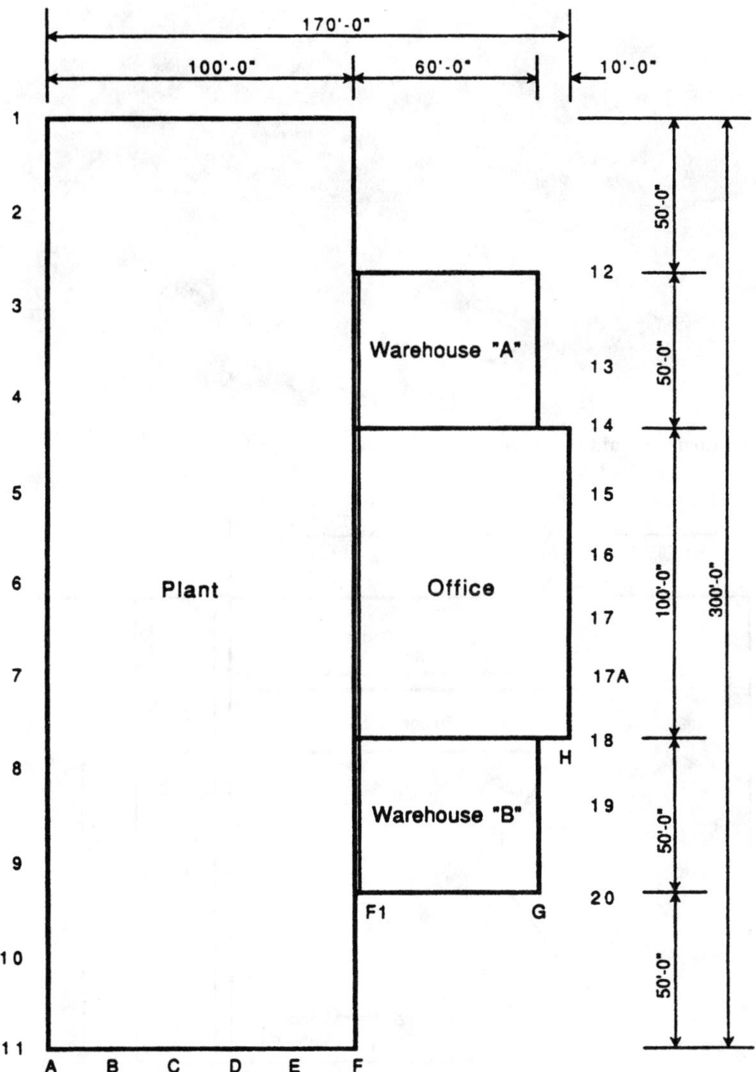

Figure 6.1a Revised plant footprint (i.e., plan), WHATCO project.

improves aesthetics, and makes a better corporate statement (i.e., consolidation vs. diversity).

The rearrangement also restored a clear 100-ft span in the plant as shown in Fig. 6.2 (section). It also deletes a row of piles (see Fig. 6.3), A review of vertical clearances in the plant (Fig. 6.2) demonstrates that the height must be increased from 24 ft to 30 ft.

Figure 6.1*b* Revised plant perspective view, WHATCO project.

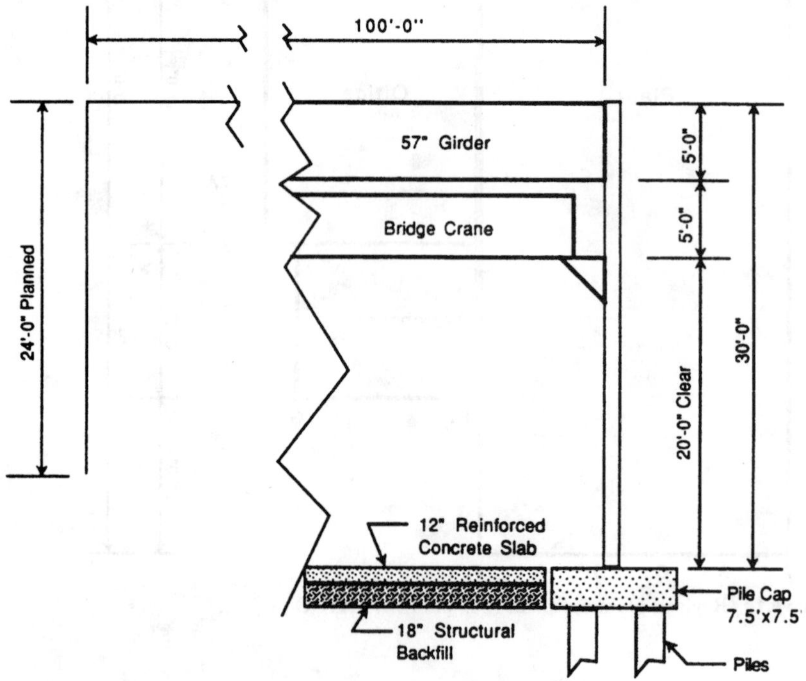

Figure 6.2 Plant section comparing schematic with design development.

Division 1, general requirements

This area can be based on industry experience. In a paper[1] to the ASCE, the author suggested the following factors:

---

[1]"Use of Cost Engineering Tools to Establish Damages," paper at ASCE National Convention, October 1983, by James J. O'Brien, P.E.

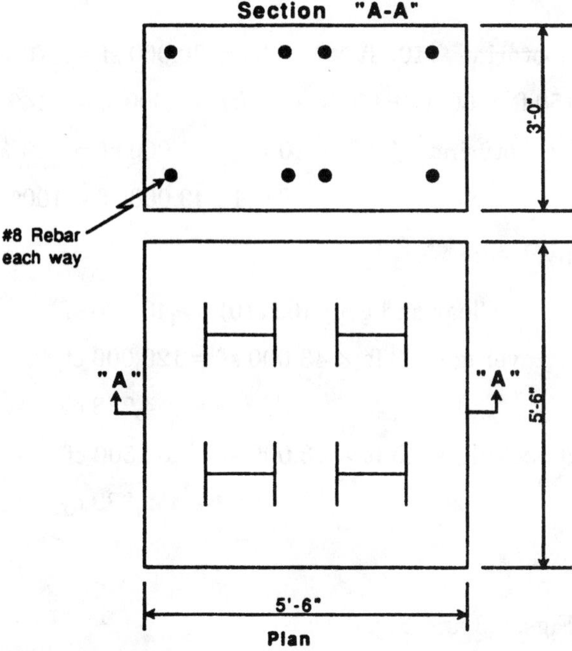

Figure 6.3 Pile cap.

Field overhead. This amounts to 20 % of labor cost plus 10% material cost.

For 40% labor/60% materials = 14%

For 50% labor/50% materials = 15%

For 60% labor/40% materials = 16%

Field expenses and general conditions. These are estimated to be 4 to 8% of the basic construction cost.

Home office expense. This amounts to 3 to 5% of the basic construction cost.

Bond. This is estimated at 0.5 to to 2.0% of the construction cost. Based on this experience, a reasonable range for general requirements would be 21.1 to 29.5% of the base construction cost.

Division 2, site work

Site work design outside of the building footprint has not advanced, so continue to use the schematic site work price.

CSI 02200 earthwork

Plant footprint	(100 ft × 300 ft) = 30,000 sf =	70%
Warehouse footprint	(50 ft × 60 ft) + (50 ft × 60 ft) = 6,000 sf =	14%
Office footprint	(70 ft × 100 ft) = 7,000 sf =	16%
	Total 43,000 sf =	100%

The work that was involved:

Clear and grub (02110) = 43,000 sf

Excavate and level: average of 3 ft × 43,000 sf = 129,000 cf

= 4,778 cy

Structural backfill = 1.5 ft × 43,000 sf = 64,500 cf

= 2,389 cy

Excavation

Plant pile caps (see Figs. 6.3 and Fig. 6.4):

22 × (5.5 ft × 5.5 ft × 3 ft) = 22 (90.75 cf) = 74 cy

Plant spread footers (see Figs. 6.5 and Fig. 6.6):

8 (4 ft × 4 ft × 2.5 ft) = 320 cf = 12 cy

Warehouse spread footings:

6 (4 ft × 4 ft × 2.5 ft) = 240 cf = 9 cy

Office spread footings:

4 (4 ft × 4 ft × 2.5 ft) = 160 cf = 6 cy

Foundations

Piles (02350). Figure 6.3 is a typical pile cap with 4 W14 piles. The piles are W14 - 90 lb. Figure 6.4 shows the 22 pile cap locations. Assume each pile is 20 ft long.

Total pile length = 22 × 4 × 20 ft = 1760 lf

Total pile weight = 1760 lf × 90 lb/lf = 158,400 lb

= 79.2 tons

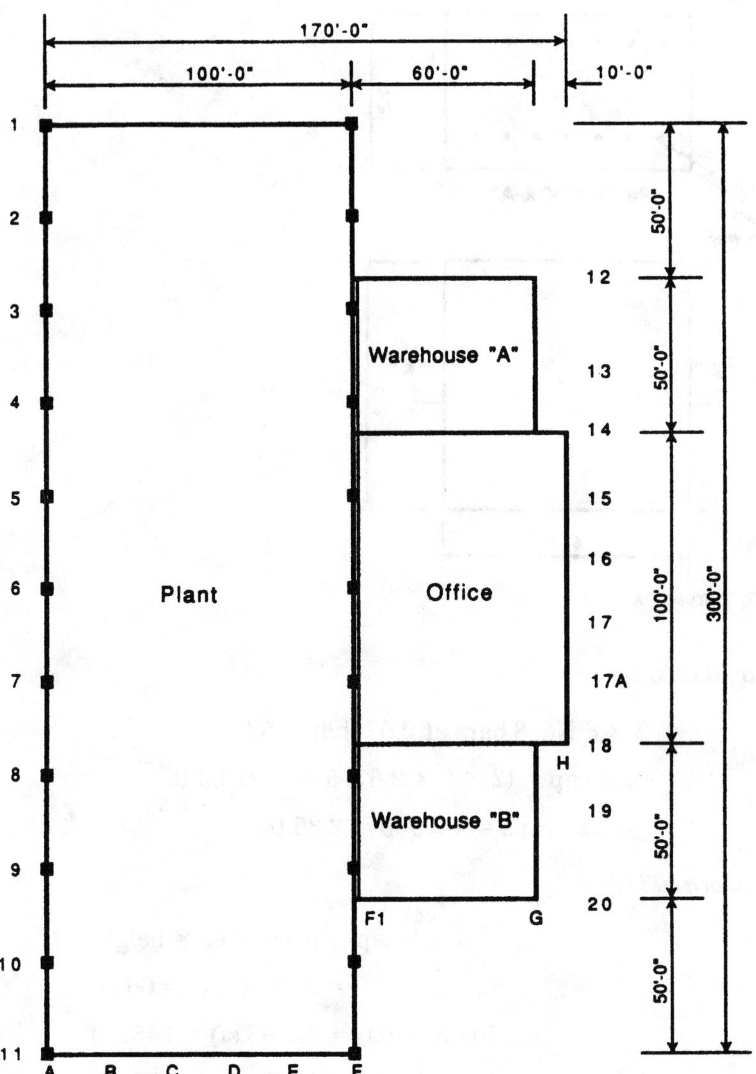

Figure 6.4 Pile cap locations.

Division 3, concrete (03300)

Pile caps (see Figs. 6.3 and 6.4):

Concrete (03300):

$$(5.5 \text{ ft} \times 5.5 \text{ ft} \times 3 \text{ ft})\,(22) = (90.75 \text{ cf})\,(22)$$

$$= 1996.5 \text{ cf} = 74 \text{ cy}$$

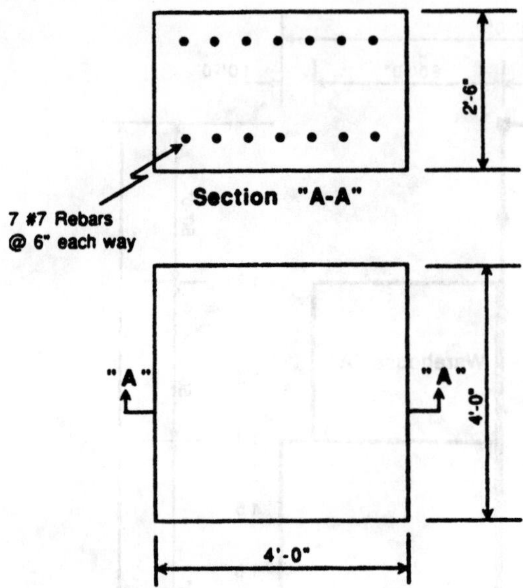

Figure 6.5 Spread footing.

Rebar (03200):

$$2 \times 8 \text{ No. 8 bars at } 2.67 \text{ lb/lf} \times 5 \text{ ft long}$$

$$\text{Each cap} = (2 \times 8 \times 2.67)\, 5 \text{ ft} = 213.6 \text{ lb}$$

$$22 \text{ caps} = 4699 \text{ lb} = 2.35 \text{ tons}$$

Formwork (03100):

$$\text{Each cap} = \text{perimeter} \times \text{height}$$

$$= 22 \text{ lf} \times 3 \text{ ft} = 66 \text{ sf}$$

$$\text{Installation} = 22\,(66 \text{ sf}) = 1452 \text{ sf}$$

$$\text{Material (assume 4 uses/form)} = 1452 \text{ sf}/4 = 363 \text{ sf}$$

Spread footings. See Figs. 6.5 and 6.6.

Concrete (03300):

$$(4.0 \text{ ft} \times 4.0 \text{ ft} \times 2.5 \text{ ft})\,(18) = 40 \text{ cf} \times 18 = 720 \text{ cf} = 27 \text{ cy}$$

$$\text{Plant } 8/18\,(27 \text{ cy}) = 12 \text{ cy}$$

$$\text{Warehouse } 6/18\,(27 \text{ cy}) = 9 \text{ cy}$$

$$\text{Office } 4/18\,(27 \text{ cy}) = 6 \text{ cy}$$

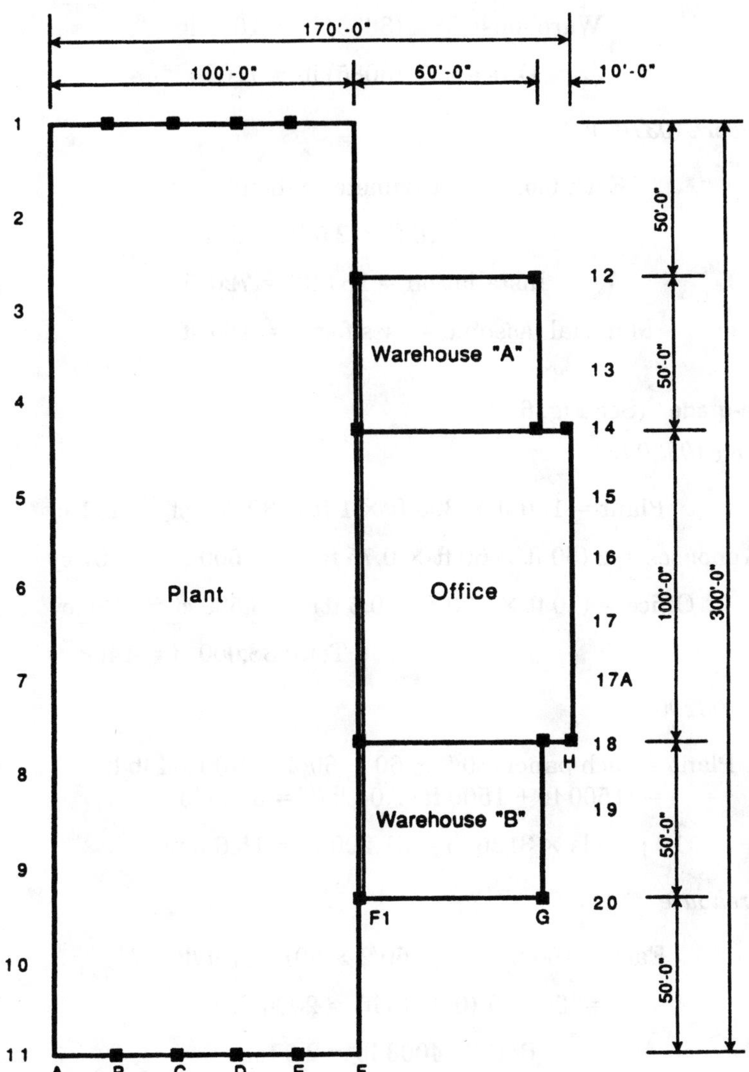

Figure 6.6 Spread footing locations.

Rebar (03200):

$$2 \times 14 \text{ No. 7 bars at } 2.04 \text{ lb/lf} \times 3 \text{ ft long}$$

$$\text{Each footing} = (2 \times 14 \times 3 \text{ ft})\ 2.04 \text{ lb/ft} = 171.4 \text{ lb}$$

$$(18)\,(171.4 \text{ lb}) = 3085 \text{ lb} = 1.54 \text{ tons}$$

$$\text{Plant } 44\%\ (3085) \text{ lb} = 1357 \text{ lb}$$

$$\text{Warehouse } 33\% \ (3085) \text{ lb} = 1018 \text{ lb}$$

$$\text{Office } 23\% \ (3085) \text{ lb} = 710 \text{ lb}$$

Formwork (03100):

$$\text{Each footing} = \text{perimeter} \times \text{height}$$
$$= 16 \text{ ft} \times 2.5 \text{ ft} = 42 \text{ sf}$$
$$\text{Installation} = 18\ (40) = 720 \text{ sf}$$
$$\text{Material (assume 4 uses/form)} = 180 \text{ sf}$$

Slabs-on-grade (See Fig. 6.7)

Concrete (03300):

$$\text{Plant} = 100 \text{ ft} \times 300 \text{ ft} \times 1 \text{ ft} = 30{,}000 \text{ cf} = 1111 \text{ cy}$$
$$\text{Warehouses} = 2\ (50 \text{ ft} \times 60 \text{ ft} \times 0.75 \text{ ft}) = 4{,}500 \text{ cf} = 167 \text{ cy}$$
$$\text{Office} = (70 \text{ ft} \times 100 \text{ ft} \times 0.5 \text{ ft}) = \underline{3{,}500 \text{ cf} = 130 \text{ cy}}$$
$$\text{Total } 38{,}000 \text{ cf} = 1408 \text{ cy}$$

Rebar (03200):

$$\text{Plant} = \text{each panel } (50/2 \times 60 + 60/2 \times 50)\ 1.04 \text{ lb/lf}$$
$$= (1500 \text{ ft} + 1500 \text{ ft})\ 1.04 \text{ lb/lf} = 3120 \text{ lb}$$
$$10 \text{ panels} \times 3120 \text{ lb} = 31{,}200 \text{ lb} = 15.6 \text{ tons}$$

Warehouse

$$\text{Each} = (60/2 \times 50 + 50/2 \times 60)\ 0.67 \text{ lb/lf}$$
$$= 3000 \text{ ft } (0.67 \text{ lb/lf}) = 2000 \text{ lb}$$
$$\text{Both} = 4000 \text{ lb} = 2.0 \text{ tons}$$

Office

$$(70/2 \times 100 + 100/2 \times 70)\ 0.67 \text{ lb/lf} = (3500 + 3500)\ 0.67 = 4690 \text{ lb}$$
$$= 2.35 \text{ tons}$$

Rebar summary:

$$\text{Plant} = 15.6 \text{ tons}$$
$$\text{Warehouses} = 2.0 \text{ tons}$$
$$\text{Office} = \underline{2.35 \text{ tons}}$$
$$\text{Subtotal } 19.95 \text{ tons}$$

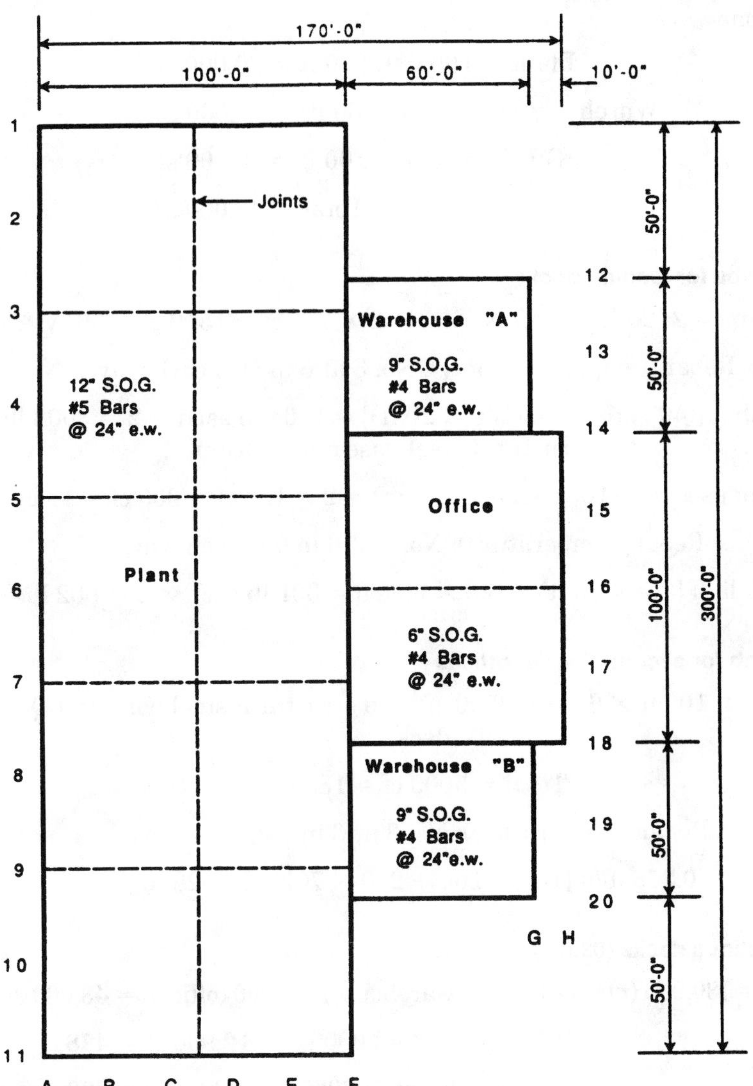

Figure 6.7 Slabs-on-grade.

Subtotal	19.95 tons
Allow 10% for splice overlaps	2.0 tons
	21.95 tons
	(say 22 tons)

Formwork. Assume metal forms 1 ft high, largest pour = 50 ft × 70 ft (office), and perimeter = 240 lf.

Membrane

Plant = 100 ft × 300 ft = 30,000 sf

Warehouses = 2(50 ft × 60 ft) = 6,000 sf

Office = 70 ft × 100 ft = 7,000 sf

Total 43,000 sf

Deck slabs for loading docks

Plant = 2 (20 ft × 50 ft × 9 in) = 2 × 750 cf = 1500 cf = 56 cy

Rebar (temperature only): No. 3 at 6 in O-C each way

0.376 lb/lf [(40 × 50 ft) + (100 × 20 ft)] = 1504 lb each × 2 = 3008 lb (on structural steel base and Q deck)

Warehouses = 2 (10 ft × 20 ft × 9 in) = 2 × 150 cf = 300 cf = 11 cy

Rebar (temperature): No. 3 at 6 in 0-C each way

0.376 lb/lf [(20 × 20 ft) + (40 × 10 ft)] = 301 lb each × 2 = 602 lb

Deck slab for second floor of offices

(70 ft × 100 ft × 6 in) = 3500 cf (on structural-steel base and Q deck)

Total = 5000 cf = 185 cy

Rebar (temperature): No. 3 at 6 in O-C each way

0.376 lb/lf [140 × 100 + 200 × 70] = 10,528 lb

Cementitious decks (03500)

Roof sf = [30,000 (plant) + 6000 (warehouse) + 7000 (office)] = 43,000 sf

Roof fill = (4 in × 43,000 sf) = 12,900 cf = 478 cy

Plant = 3000/43,000 (478) = 333 cy

Warehouse = 6000/43,000 (478) = 67 cy

Office = 7000/43,000 (478) = 79 cy

Grade beams. Figure 6.8 is a cross section of the grade beam for the WHATCO project. The cross-section area is:

Stem 3 ft × 1 ft = 3 sf

Base 2 ft × 1 ft = 2 sf

Subtotal 5 sf

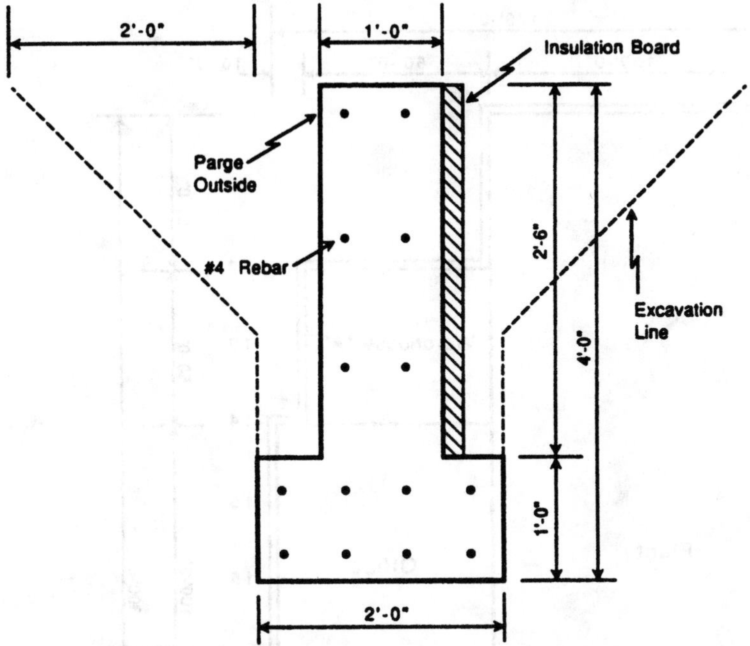

Figure 6.8 Cross section of grade beam.

The grade beam perimeter from Fig. 6.9 is:

Plant (300 ft + 100 ft + 100 ft + 50 ft + 50 ft) = 600 lf

Warehouses (60 ft + 50 ft + 60 ft + 50 ft) = 220 lf

Office (10 ft + 100 ft + 10 ft) = 120 lf

940 lf

Concrete

(Cross section × perimeter) = (5 sf × 940 ft) = 4700 cf

Plant 64% (174 cy) = 111 cy

Warehouse 23% (174 cy) = 40 cy

Office 13% (174 cy) = 23 cy

174 cy

Parging (07160)

3 ft × 940 ft = 2820 sf

Plant 64% = 1805 sf

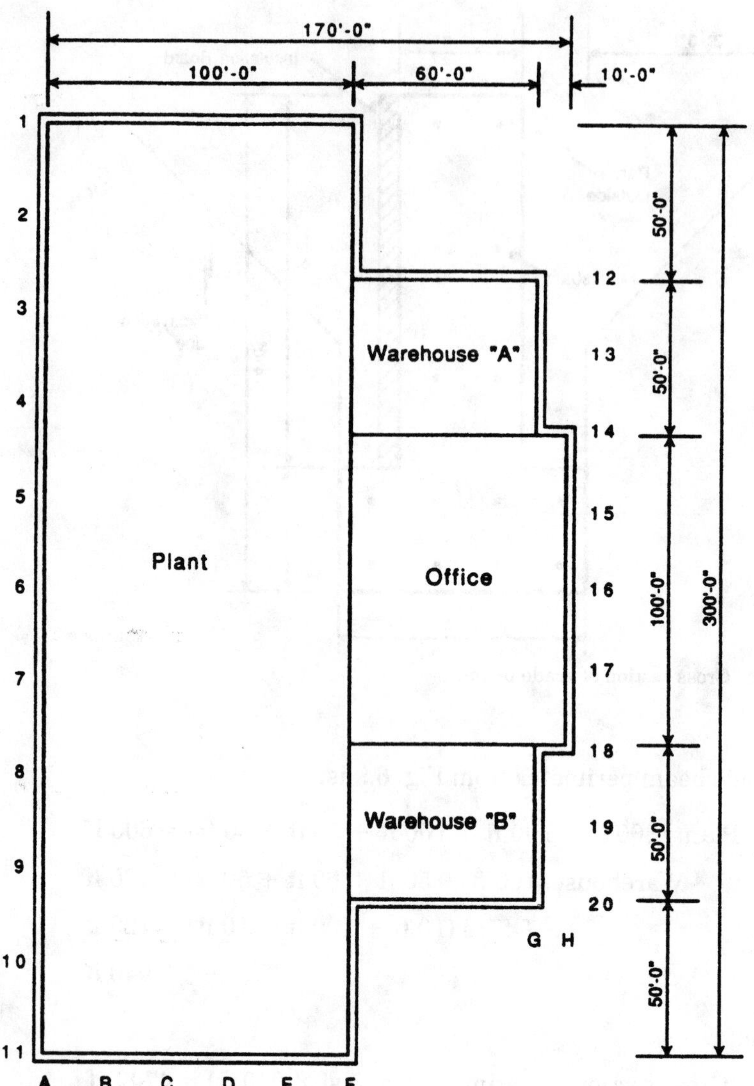

Figure 6.9 Grade beam plan.

Warehouse 23% = 649 sf

Office 13% = 367 sf

Insulation board (07210)

3 ft × 940 ft = 2820 sf

Plant 64% = 1805 sf

Warehouse 23% = 649 sf

Office 13% = 367 sf

Rebar (03200)

14 No. 4 Bars at 0.67 lb/lf × 940 ft perimeter = 8817 lb

= 4.4 tons

Add 10% for startups and splices = 0.4 tons

Subtotal = 4.8 tons

Plant 64% (9600 lb) = 6144 lb

Warehouse 23% (9600 lb) = 2208 lb

Office 13% (9600 lb) = 1248 lb

Excavation (02200)

Bottom 2 ft: 2 ft × 2 ft = 2 sf

Upper 2 ft: (2 ft × 2 ft) + 2[(½) 2 × 2] = 8 sf

Subtotal = 10 sf

Volume = (10 sf × 940 ft) = 9400 cf

= 348 cy

From Fig. 6.9:

Plant 600 lf = 64% = 223 cy

Warehouse 220 lf = 13% = 77 cy

Office 120 lf = 13% = 45 cy

940 lf

Backfill = (10 sf − 5 sf) = 50%

Plant = 112 cy

Warehouse = 39 cy

Office = 22 cy

Office building, stairs

Forms (3 uses) (03110):

(Slant length × width) = 25 ft × 5 ft = 125 sf

Concrete: 3 (125 sf × 1.0 ft) = 375 cf = 14 cy

Division 4, masonry

Unit masonry (04200)

Office facade (10 ft + 100 ft + 10 ft) × H (height)

Height (see Fig. 6.10) = 30 ft 0 in

Facade sf = 120 ft × 30 ft = 3600 sf

Assume 25% glazing:

Brick = 2700 sf

Glazing = 900 sf

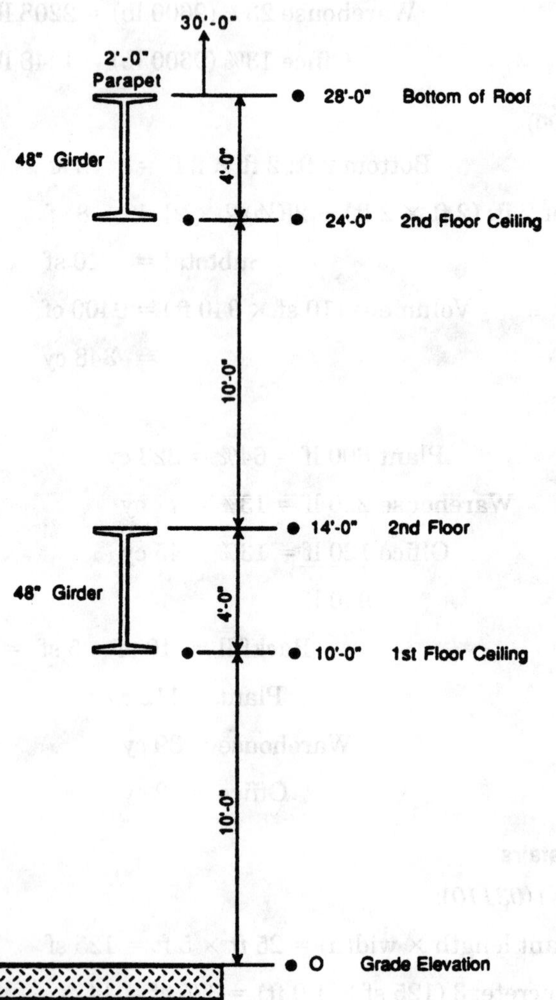

Figure 6.10 Office section.

Brick facade

Scaffolding, 2-story, 36 csf (01525)

Backup masonry block (04220) 8-in thick, reinforced = 2700 sf

Face brick (04210)

Common 8 in face brick 4 in thick = 2700 sf

Coping, precast (04210) 14 in wide = 120 lf

Division 5, metals

Structural metal framing (05100). Figure 6.11 is a plan showing beams and girders. Figure 6.12 is a summary of beams and girders (05120). Figures 6.13 and 6.14 are plans of columns. Figure 6.15 is a summary of columns (05120). Figure 6.16 is a plan showing joists (05210). Figure 6.17 is a joist summary. Figures 6.18 and 6.19 show cross bracing for the plant area. The warehouse and office area rely on moment connections. Figure 6.20 shows structural steel for the three loading docks.

Loading docks

Plant

Structural steel

MC 18 in/58 lb (10 × 20 ft × 58 lb) = 11,600 lb

MC 12 in/31 lb (50 ft + 50 ft) = 3,100 lb

Subtotal = 14,700 lb each

Total = 29,400 lb = 14.7 tons

Panels

(50 ft × 20 ft) = 1000 sf each

Total 2000 sf

Warehouse

Structural steel

MC 15 in/40 lb (5 × 10 ft × 40 lb) = 2000 lb

MC 12 in/31 lb (20 ft + 20 ft) = 1240 lb

Subtotal = 3240 lb

Total = 6480 lb

Panels

(20 ft × 10 ft) = 200 sf each = total = 400 sf

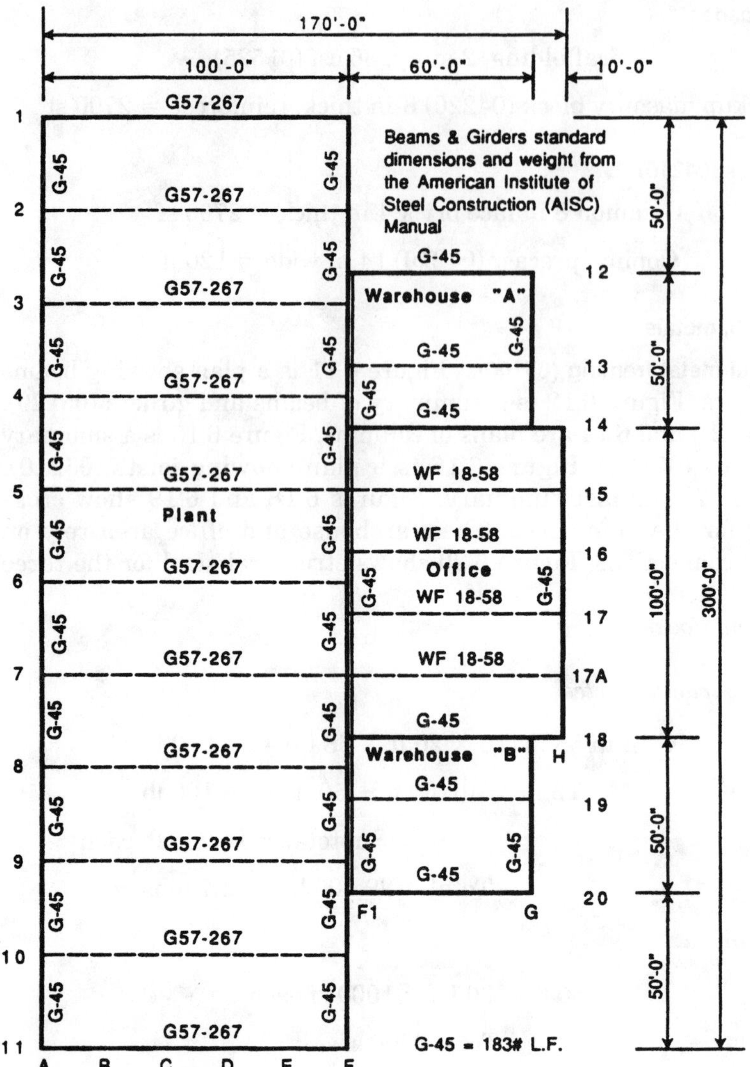

Figure 6.11 Beams and girders.

Piers (steel):

Warehouse 2 × 5 × 3 ft × 33 lb = 990 lb = 0.5 ton

Plant 2 × 10 × 3 ft × 33 lb = 1880 lb = 1.0 ton

Concrete piers:

Warehouse 2 (5 × 4 cf) = 40 cf = 1.5 cy

2 (5 × 9 cf) = 90 cf = 3.3 cy

Girder/Beam Summary (051200)

Location	Column	Line	FT	#/FT	#	No.	Total #	
Plant	1 to 11	G-57	100	267	26,700	11	293,700	
Warehouse	12 - 13	G-45	60	183	10,980	2	21,960	
Office	14	G-45	70	183	12,810	2	25,620	2 Floors
Office	15 - 17A	WF18	70	58	4,060	8	32,480	2 Floors
Office	18	G-45	70	183	12,810	2	25,620	2 Floor
Warehouse	19 - 20	G-45	60	183	10,980	2	21,960	
Plant	A	G-45	30	183	5,490	10	54,900	
Plant	F	G-45	30	183	5,490	10	54,900	
Warehouse	F1	G-45	50	183	9,150	2	18,300	
Office	F1	G-45	100	183	18,300	1	18,300	
Warehouse	G	G-45	50	183	9,150	2	18,300	
Office	H	G-45	100	183	18,300	2	36,600	
	Plant =	403,500# =		202 Tons			622,340#	
	Office =	138,620# =		69 Tons			•	
	Warehouse = 80,520# =			40 Tons			311 Tons	
				311 Tons				

Figure 6.12 Girder and beam summary (05120).

$$\text{Plant } 2\,(10 \times 9 \text{ cf}) = 180 \text{ cf} = 6.6 \text{ cy}$$

$$2\,(10 \times 4 \text{ cf}) = 40 \text{ cf} = 3.0 \text{ cy}$$

Perimeter curb angle:

$$\text{Plant } 2\,(20 \text{ ft} + 50 \text{ ft} + 20 \text{ ft}) = 180 \text{ lf}$$

$$\text{Warehouse } 2\,(10 \text{ ft} + 20 \text{ ft} + 10 \text{ ft}) = 80 \text{ lf}$$

Metal decking (05300). Deck type is Load Master heavy-duty 22-gauge (or equal). Weight (galvanized) is 1.68 psf (see Fig. 6.21).

Deck summary:

Plant 30,000 sf × 1.68 psf = 50,400 lb
[+ (loading dock 1000 sf × 1.68 psf = 1680 lb) = 52,080 lb]

Warehouse 60 (50 + 50) = 6000 sf × 1.68 psf = 10,080 lb
[+ (loading docks 400 sf × 1.68 psf = 672 lb) = 10,752 lb]

Offices (70 × 100) sf × 1.68 psf = 11,760 lb

Total = 72,240 lb

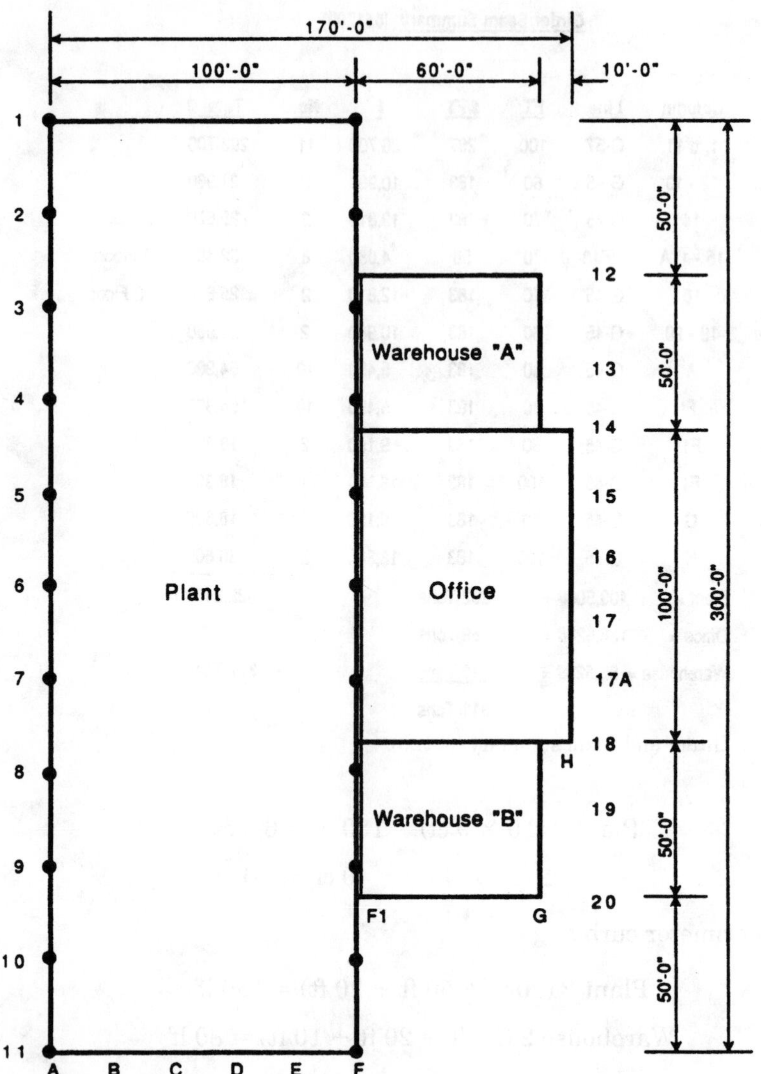

Figure 6.13 Columns, 14W 145#.

Division 6, wood and plastics

None.

Division 7, thermal and moisture protection

Roofing:

Plant	30,000 sf
Warehouse	6,000 sf
Office	7,000 sf
Total	43,000 sf

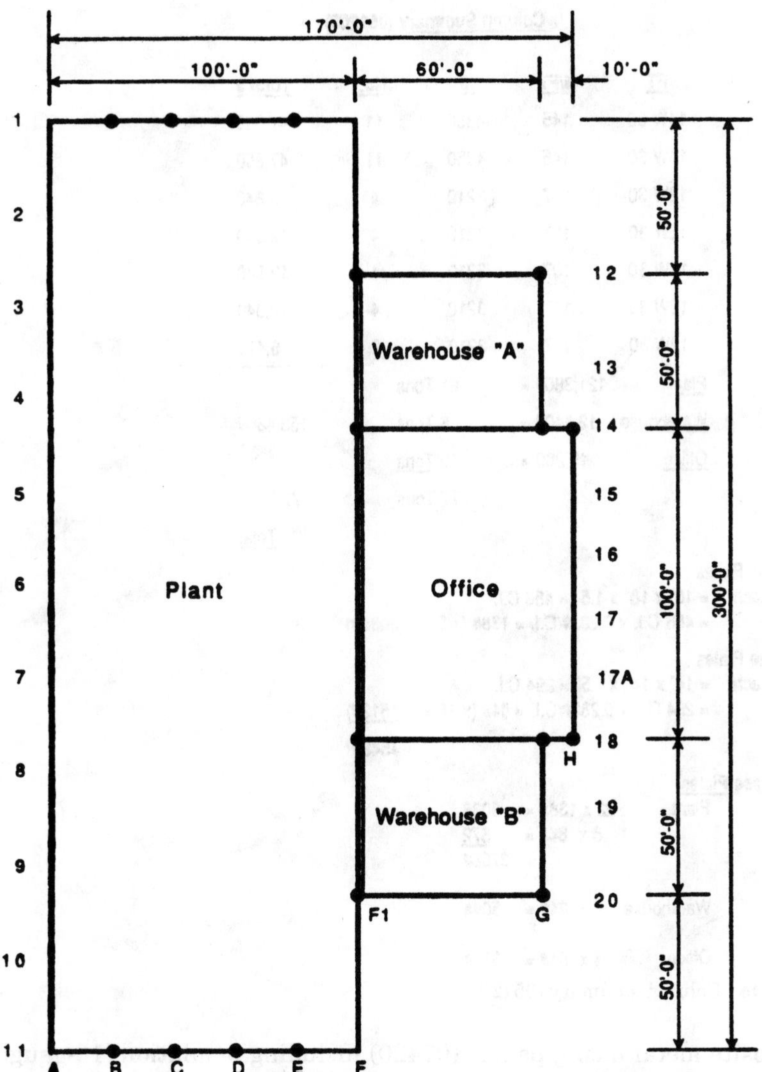

Figure 6.14 Columns, 12W 170#.

Roofing (*07500*): 43,000 sf
Roofing insulation (*07200*): 43,000 sf
Flashing (*07600*):

$$\text{Plant} = (100\text{ ft} + 300\text{ ft} + 100\text{ ft} + 300\text{ ft}) = 800\text{ lf}$$

$$\text{Warehouse} = (50\text{ ft} + 60\text{ ft} + 50\text{ ft} + 60\text{ ft})2 = 440\text{ lf}$$

$$\text{Office} = (100\text{ ft} + 70\text{ ft} + 100\text{ ft} + 70\text{ ft}) = 140\text{ lf}$$

Column Summary (051200)

Column Line	FT	#/FT	#	No.	Total #
A	14W 30	145	4350	11	47,850
F	14W 30	145	4350	11	47,850
I	12W 30	107	3210	4	12,840
II	12W 30	107	3210	4	12,840
F-1	12W 30	107	3210	4	12,840
G	12W 30	107	3210	4	12,840
H	12W 30	107	3210	2	6,420

Plant	121,380# =	61 Tons	
Warehouse	12,840# =	6 Tons	153,480#
Office	19,260 =	10 Tons	=
		77 Tons	77 Tons
			Total

14 W Base Plates

Each = 18" x 18" x 1.5" = 486 C.I.
= 486 C.I. x 0.283#/C.I. = 138# (x 22 = 3036#)

12 W Base Plates

Each = 14" x 14" x 1.5" = 294 C.I.
= 294 C.I. x 0.283#/C.I. = 84# (x 18 = 1512#)

4548#

Base Plates

Plant 22 x 138# = 3036
8 x 84# = 672
3708#

Warehouse 6 x 84# = 504#

Office 4 x 84# = 336#

Figure 6.15 Column summary (05120).

Composite metal facing panels (07420) including insulation (16-gauge aluminum outside, 1½ in insulation, 18-gauge steel inside):

$$\text{Plant} = 30\text{ ft }(300\text{ ft} + 100\text{ ft} + 100\text{ ft} + 50\text{ ft} + 50\text{ ft}) = 30\text{ ft }(600\text{ ft}) = 18{,}000\text{ sf}$$

$$\text{Warehouse} = 30\text{ ft }(60\text{ ft} + 50\text{ ft})2 = 30\text{ ft }(220) = 6600\text{ sf}$$

Division 8, doors and windows

Metal panels to be 10 ft × 6 ft. Support off girt system 10 ft center to center, made up of 12-in channels (MC12–31 lb). Clip to girders at top elevation.

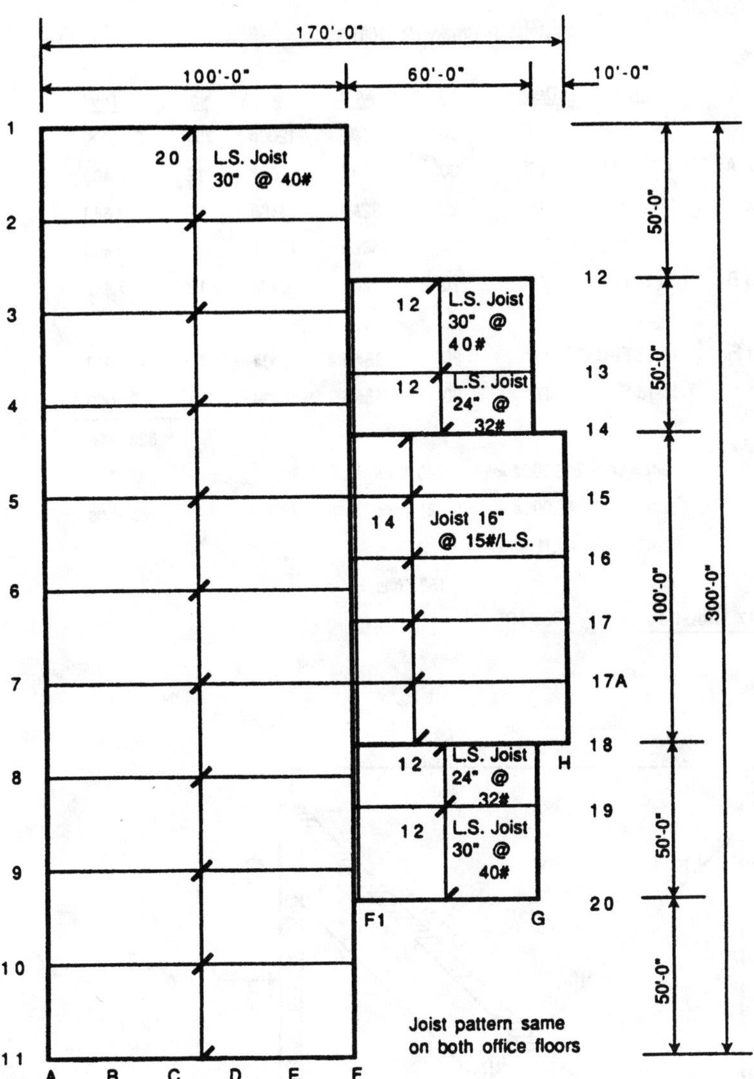

Figure 6.16 Joist schedule.

Girts:

Plant = (100 ft + 300 ft + 100 ft + 50 ft + 50 ft) 3 = (600 ft × 3)
= 1800 lf

Weight = 1800 lf × 31 lb/lf = 55,800 lb

Warehouse = (60 ft + 50 ft + 50 ft + 60 ft) 3 = (220 ft × 3) = 660 lf

Weight = (660 ft × 31 lb/lf) = 20,460 lb = 10.25 tons

Joist Summary (052100)

Place	Area	Depth	FT	#/FT	#	No.	Total
Plant	1-11/A-F	30"	30'	40#/1	1200#	200	240,000
Warehouse A	12/13/FI-G	30"	30'	40#/1	1200#	12	14,400
Warehouse A	13-14/FI-G	24"	20'	32#/1	640#	12	7,680
Warehouse B	18-19/FI-G	30"	30'	40#/1	1200#	12	14,400
Warehouse B	19-20/FI-G	24"	20'	32#/1	640#	12	7,680
Office							
1st Flr	14-18/FI-M	16"	20'	15#/1	300#	70	21,000
Roof	14-18/FI-M	16"	20'	15#/1	300#	70	21,000
							326,160#

Plant = 240,000# = 120 Tons
Office = 42,000# = 21 Tons
Warehouse = 44,160# = 22 Tons
163 Tons

163 Tons

Figure 6.17 Joist summary (05210).

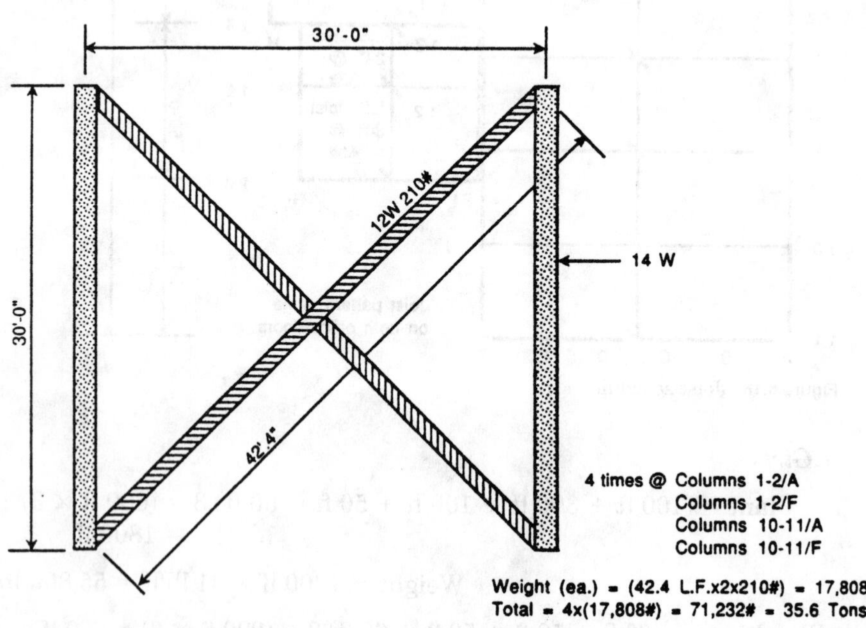

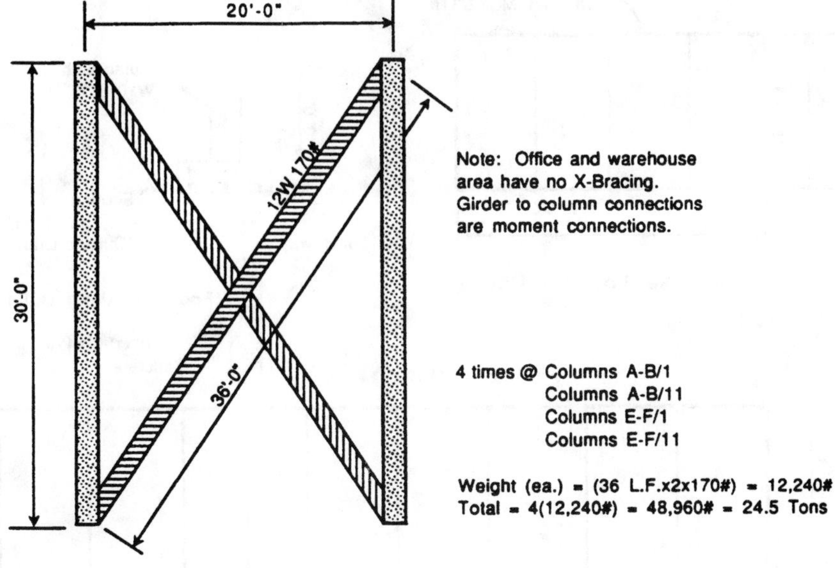

Figure 6.19 Cross bracing in plant (E - W).

Exterior doors (Division 8). From Fig. 6.22:

1. Vertical multileaf industrial-type doors, 16 ft × 16 ft (08365)
2. Steel personnel, 3 ft 0 in × 7 ft 0 in (08110)
3. Aluminum double swing, 6 ft 0 in × 7 ft 0 in (08410)

	Type 1	Type 2	Type 3
Plant	4	5	0
Warehouses	2	2	0
Office	0	1	1

Interior partitions, office. (See Figs. 6.23 and 6.24). From Figs. 6.10, 6.23, and 6.24:

Masonry (04220):

$$\text{Outside walls} = (70\text{ ft} + 70\text{ ft} + 100\text{ ft})\ 28\text{ ft} = 6{,}720\text{ sf}$$

$$\text{Stair wells} = (30\text{ ft} + 30\text{ ft})\ 28\text{ ft} = 1{,}680\text{ sf}$$

$$\text{Elevator shaft} = (10\text{ ft} + 10\text{ ft} + 10\text{ ft} + 10\text{ ft})\ 28\text{ ft} = 1{,}120\text{ sf}$$

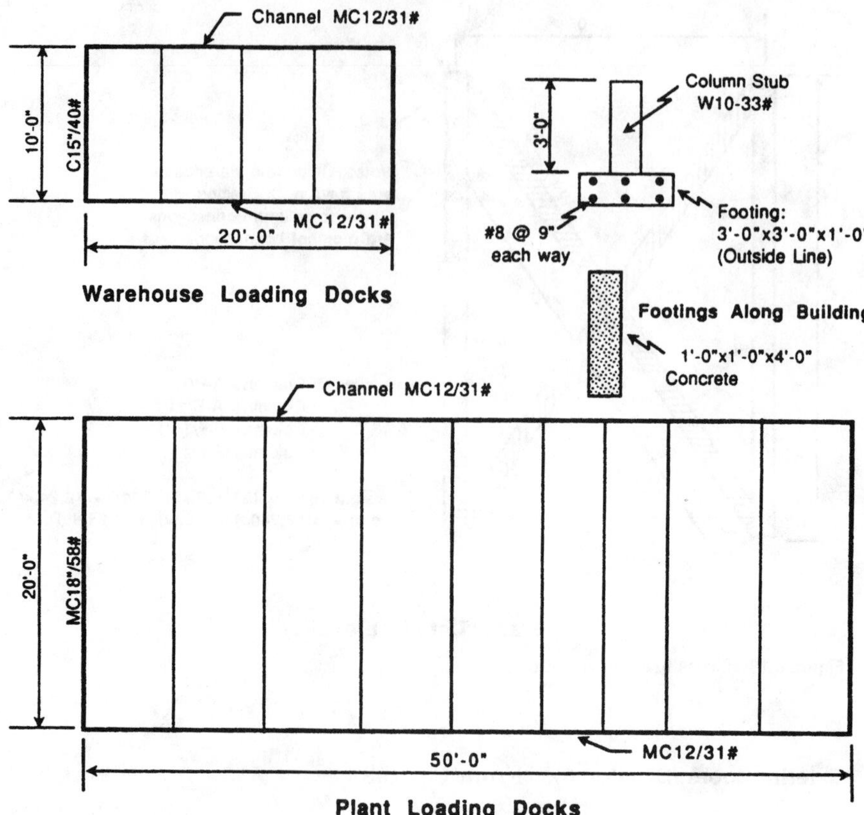

Figure 6.20 Loading docks.

First floor baths = (7 ft × 15 ft) 10 ft = 1,050 sf

Second floor baths = (7 ft × 15 ft) 10 ft = 1,050 sf

Boiler room and load center = (15 ft + 30 ft + 15 ft + 15 ft) 14 = 1,050 sf

Total = 12,670 sf

Drywall partitions (*09260*) (*metal stud*):

L.F. first floor (left to right; then top to bottom):

15 ft + 30 ft + 15 ft + 20 ft + 15 ft + 30 ft + 10 ft + 30 ft + 45 ft + 85 ft + 30 ft + 15 ft + 15 ft + 15 ft + 20 ft + 30 ft + 15 ft + 15 ft + 20 ft + 20 ft + 20 ft = 510 lf

Loadmaster Steel Section Properties

Description	Units	Standard Duty (2)	Heavy Duty (2)			Extra Duty (2)			
Gauge Number	—	28 ga.	25 ga.	24 ga.	22 ga.	25 ga.	24 ga.	22 ga.	20 ga.
Design Thickness	in.	0.0149	0.0205	0.0239	0.0299	0.0205	0.0239	0.0299	0.0359
Nominal Pitch	in.	2.5	3.75	3.75	3.75	5	5	5	5
Nominal Depth	in.	9/16	15/16	15/16	15/16	1 5/16	1 5/16	1 5/16	1 5/16
Weight Painted	PSF	.78	1.04	1.24	1.57	1.07	1.34	1.67	2.01
Weight Galvanized	PSF	.86	1.13	1.36	1.68	1.18	1.44	1.78	2.11
Minimum Yield Strength	PSI	80,000	80,000	80,000	80,000	80,000	80,000	80,000	80,000

Description	Units	Pyro Span (2) (3)				Super Span (2)	
Gauge Number	—	24 ga.	22 ga.	20 ga.	18 ga.	22 ga.	20 ga.
Design Thickness	in.	0.0239	0.0295	0.0358	0.0474	0.0299	0.0359
Nominal Pitch	in.	6	6	6	6	6	6
Nominal Depth	in.	1½	1½	1½	1½	2	2
Weight Painted	PSF	1.39	1.73	2.08	2.77	1.89	2.27
Weight Galvanized	PSF	1.47	1.81	2.16	2.85	2.01	2.38
Minimum Yield Strength	PSI	80,000	33,000	33,000	33,000	80,000	80,000

1. Physical properties were computed in accordance with specifications section of the "Light Gage Cold Formed Steel Design Manual" published by the American Iron and Steel Institute (AISI); and also conform to the specifications of the Steel Deck Institute.
2. Steel sections are roll formed from high tensile strength, cold steel, conforming to ASTM specification A-446 Grade E.
3. Steel Sections are roll formed from cold steel conforming to ASTM specification A-611 Grade C (minimum) for painted finish or ASTM specification A-446 for galvanized finish.
4. White Paint Finish — Steel is roller coat painted with flexible primer, then oven cured.
5. Galvanized Finish — Galvanized coating applied to the steel conforms to ASTM specification A-525 class G90, G60, G01 or Federal Specification QQ-S-775 Class d.

Figure 6.21 Roof dock information. (*From Sweet's Catalog.*)

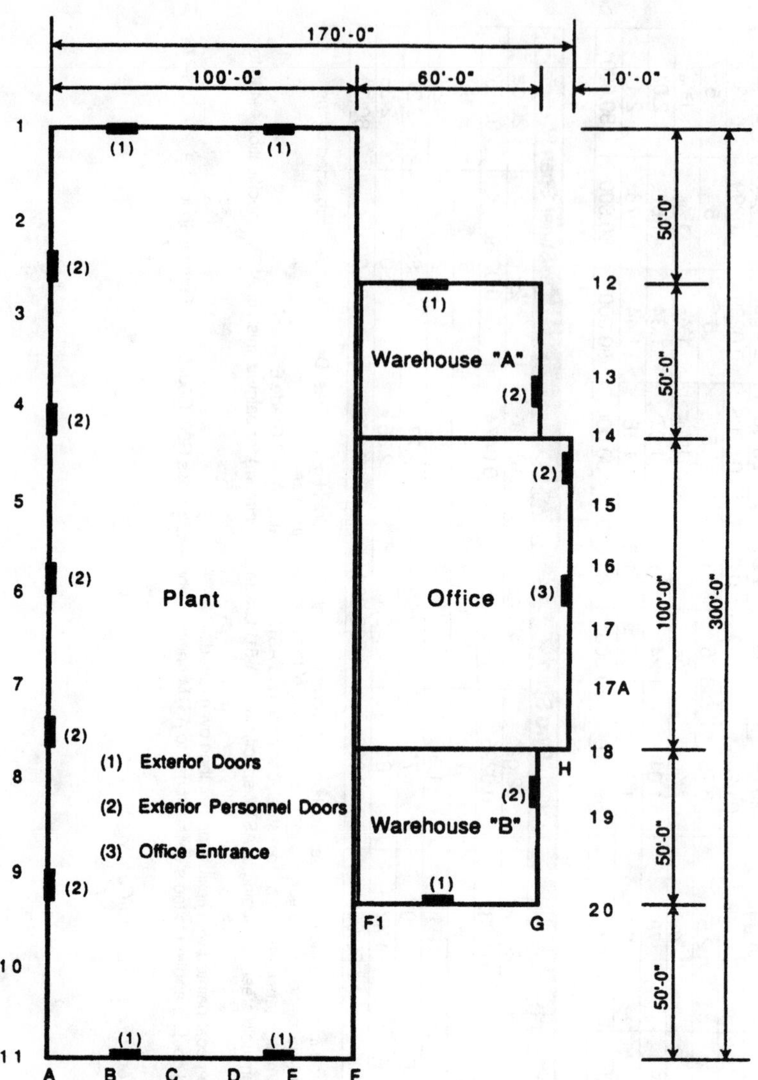

Figure 6.22 Exterior doors.

L.F. second floor:

$$20 \text{ ft} + 35 \text{ ft} + 15 \text{ ft} + 15 \text{ ft} + 30 \text{ ft} + 30 \text{ ft} + 10 \text{ ft} + 15 \text{ ft} + 30 \text{ ft} + 15 \text{ ft} + 30 \text{ ft} + 30 \text{ ft} + 15 \text{ ft} + 30 \text{ ft} + 15 \text{ ft} + 20 \text{ ft} + 20 \text{ ft} + 35 \text{ ft} = 380 \text{ lf}$$

$$\text{Partitions} = (510 \text{ ft} + 380 \text{ ft})(10 \text{ ft}) = 8900 \text{ sf}$$

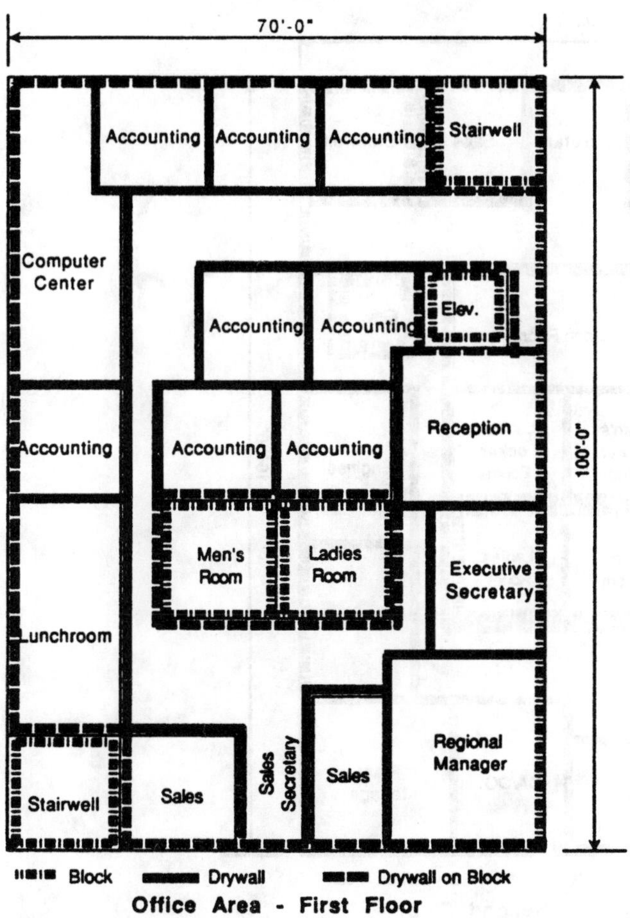

Figure 6.23 Office area partitions, first floor.

Drywall on block (*09260*):

L.F. first floor:

85 ft + 15 ft + 55 ft + 15 ft + 15 ft + 40 ft + 30 ft + 30 ft + 30 ft + 15 ft + 15 ft + 55 ft = 400 ft

+ (75% × 120 ft) = (400 ft + 90 ft) = 490 lf

L.F. second floor:

15 ft + 55 ft + 15 ft + 30 ft + 15 ft + 15 ft + 15 ft + 15 ft + 40 ft = (215 ft + 75% 120 ft) = 305 lf

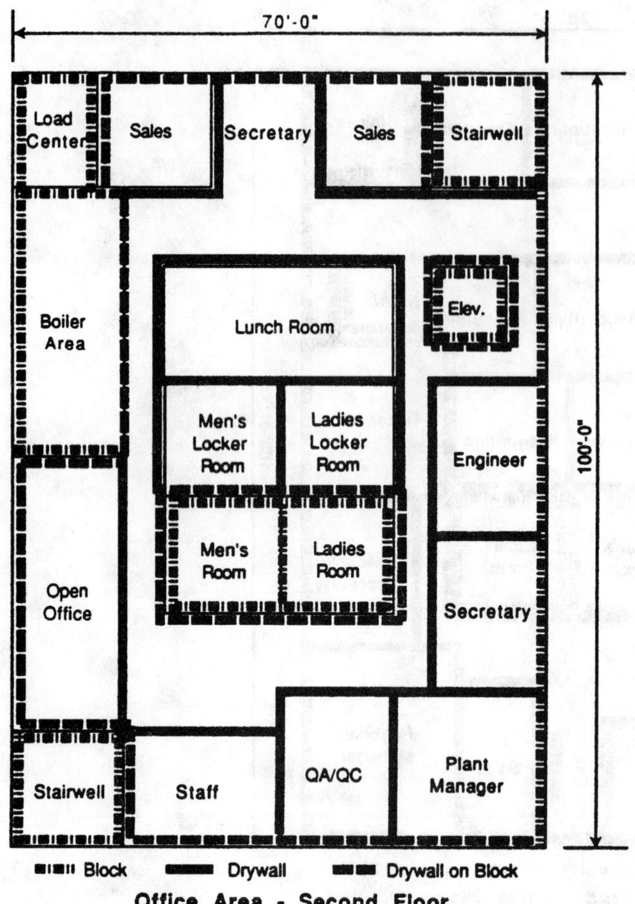

Figure 6.24 Office area partitions, second floor.

Drywall on block = (490 ft + 305 ft) × 10 ft = 7950 sf
Insulation on block = (6150 + 2700) sf (front wall) = 8850 sf

Interior doors, office. See Figs. 6.25 and 6.26.

Wood doors (08210)	*Fire doors (08110)*
First floor 20	First floor 2
Second floor 16	Second Floor 5
Total 36	Total 7

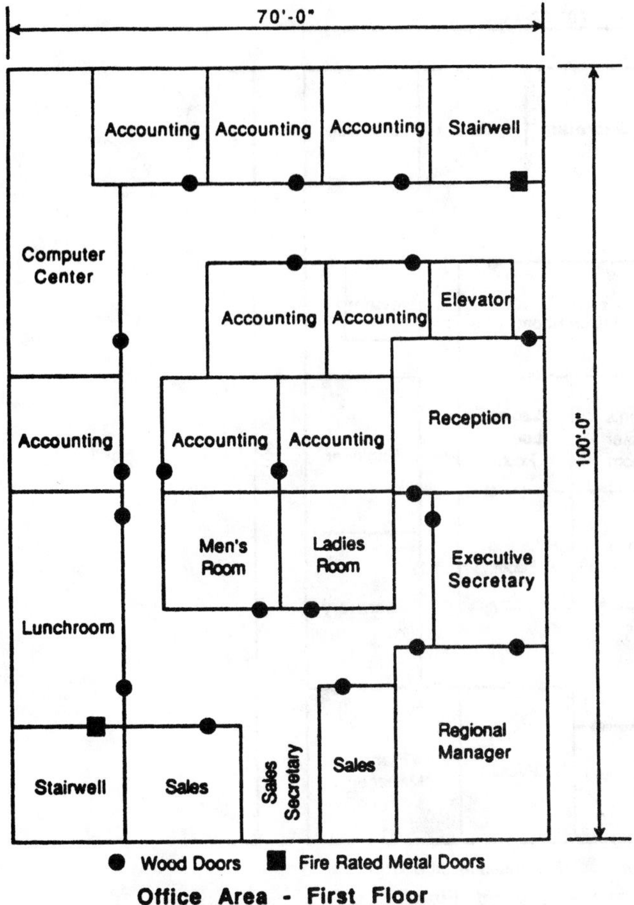

Figure 6.25 Interior doors, office area, first floor.

Flooring. See Figs. 6.27 and 6.28. Using the 5-ft grid, count spaces and multiply by 25 sf:

Ceramic tile (09310):

First floor, 18 × 25 sf = 450 sf

Second floor, 18 × 25 sf = 450 sf

Vinyl tile (09660):

First floor, 51 × 25 sf = 1275 sf

Second floor, 102 × 25 sf = 2550 sf

3825 sf

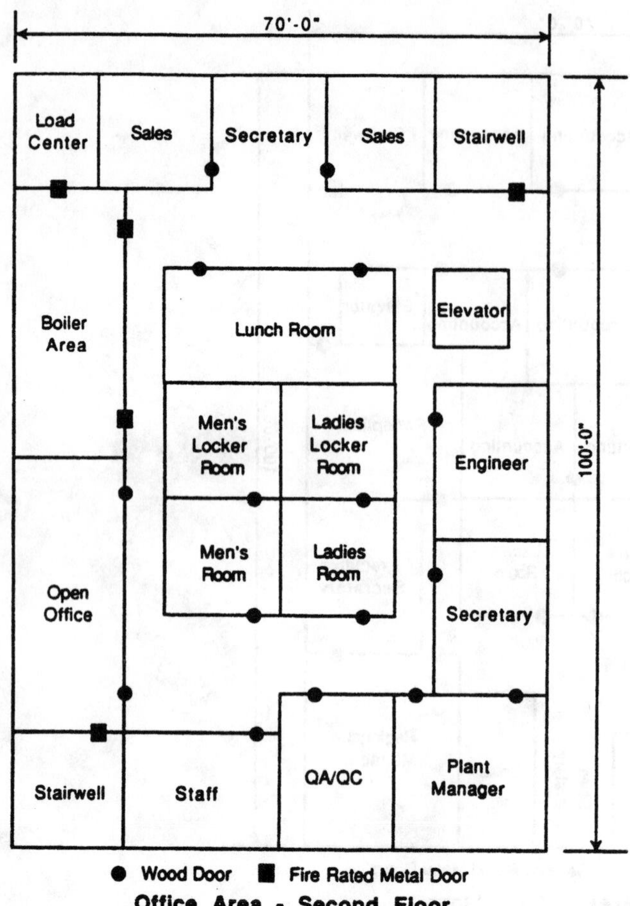

Figure 6.26 Interior doors, office area, second floor.

Carpet (*09680*)

First floor, 205 × 25 sf = 5125 sf

Second floor, 110 × 25 sf = 2750 sf

7875 sf = 875 sy

Acoustic hung ceiling (*09510*): See Figs. 6.29 and 6.30. Using 5-ft grid system, calculate overall square feet and deduct areas without ceiling (i.e., number of clear blocks times 25 sf):

First floor, (100 ft × 70 ft) − (22 × 25 sf) = (7000 sf − 550 sf) = 6,450 sf

Second floor, [7000 sf = (46 × 25 sf)] = (7000 sf − 1150 sf) = 5,850 sf

12,300 sf

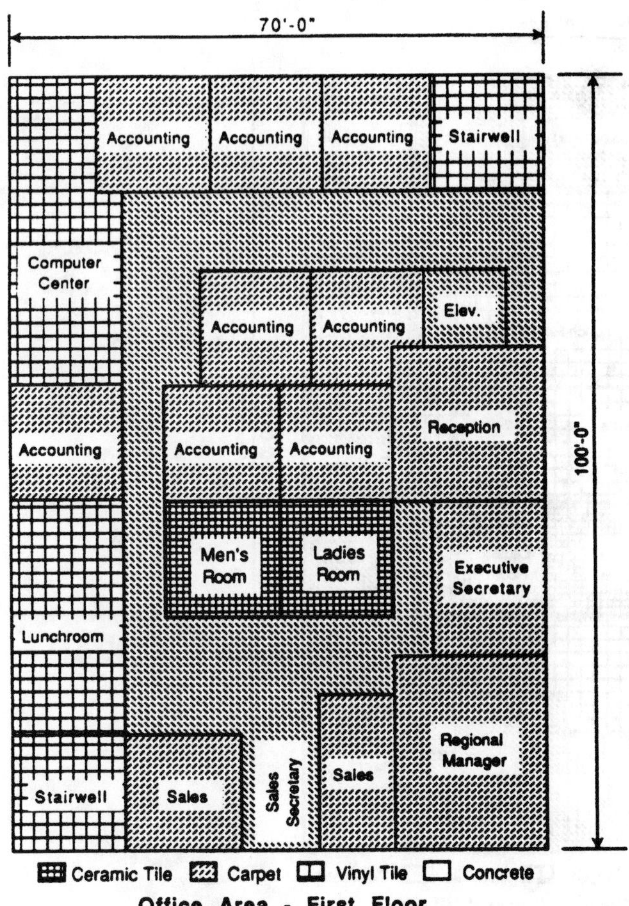

Figure 6.27 Flooring, office area, first floor.

Paint

Interior finishes (09920): In Fig. 6.36, the paint area (in sf) for each lineal foot (lf) of each structural shape is calculated. In Fig. 6.37, total sf is calculated, as follows:

	Primer	Two coats
Plant	47,632 sf	95,264 sf
Warehouse	15,160 sf	30,320 sf
Office	17,500 sf	35,000 sf

Assume prime coat and first finish coats are applied at the shop with final coat after field erection.

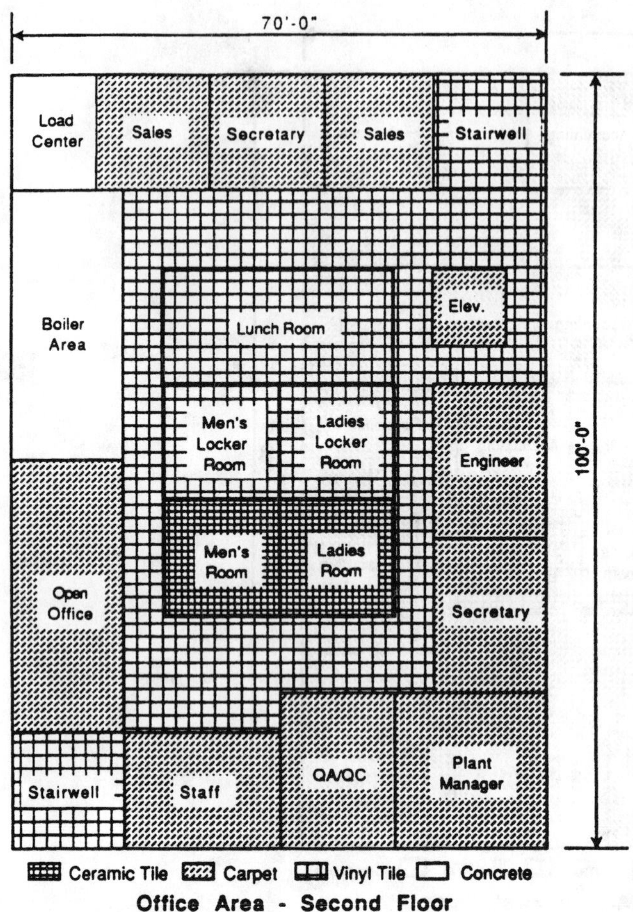

Figure 6.28 Flooring, office area, second floor.

Paint in the office interior is taken from Figs. 6.23 and 6.24, as follows:
Paint exposed masonry:

Plant (100 ft × 30 ft) = 3000 sf × 2 coats = 6,000 sf

Warehouse (60 ft × 60 ft) 28 ft = 3360 sf × 2 coats = 6,720 sf

Office (2 stairs) (15 ft + 15 ft + 15 ft + 15 ft)(28)(2) = 3,360 sf

(Boiler room) (15 ft + 15 ft + 30 ft + 30 ft) 14 ft = 1,260 sf

(Load center) (10 ft + 15 ft + 10 ft + 15 ft) 14 ft = 700 sf

Subtotal 5,320 sf

At 2 coats × 2

10,640 sf

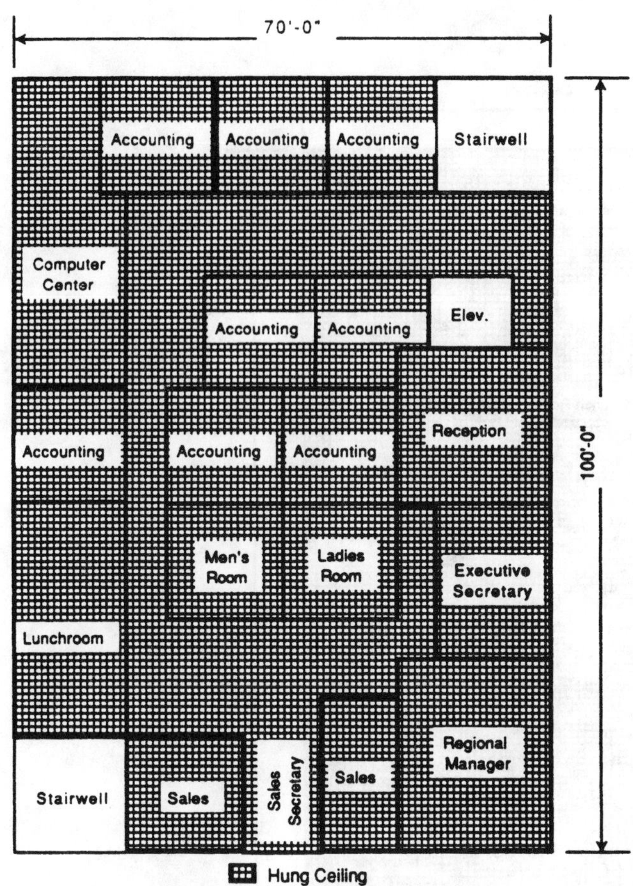

Figure 6.29 Acoustic hung ceiling, office area, first floor.

Paint drywall:

Partitions	2 × 8900 sf =	17,800 sf
	Board on block =	7,950 sf
	Subtotal paint	25,750 sf
	At 2 coats	× 2
		51,500 sf

Paint doors:

Plant:

Vertical multileaf	4 × 16 ft × 16 ft =	1,024 sf
Personnel	5 × 3 ft × 7 ft =	105 sf
	Subtotal paint	1,129 sf
	2 sides × 2 coats	× 4
	Subtotal	4,516 sf

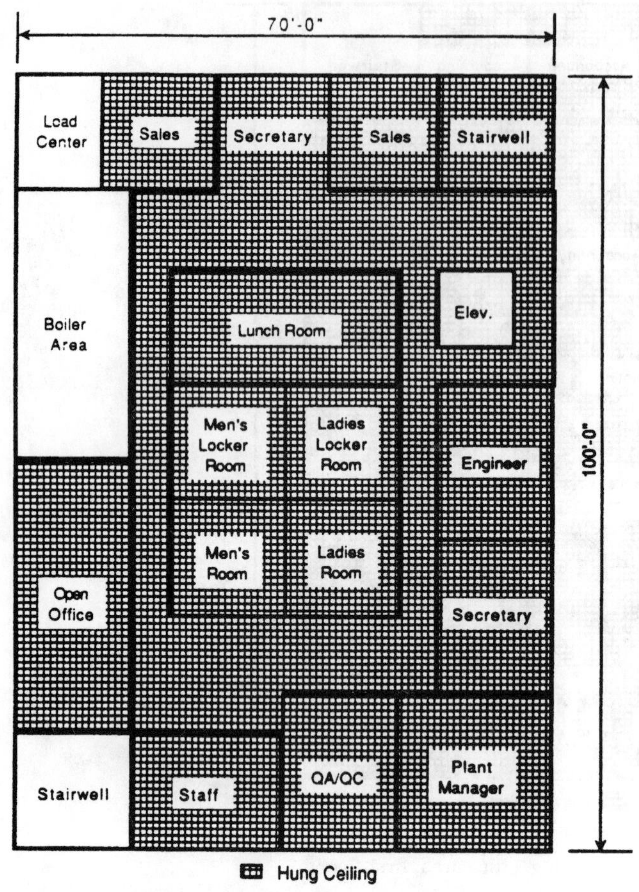

Figure 6.30 Acoustic hung ceiling, office area, second floor.

Warehouse:

$$\text{Vertical multileaf} = 2 \times 16 \text{ ft} \times 16 \text{ ft} = 512 \text{ sf}$$

$$\text{Personnel} = 2 \times 3 \text{ ft} \times 7 \text{ ft} = \underline{\ 42 \text{ sf}}$$

$$554 \text{ sf}$$

$$2 \text{ sides} \times 2 \text{ coats} \quad \underline{\times 4}$$

$$2216 \text{ sf}$$

Office:

$$(\textit{Exterior})\ \text{Personnel} = 1 \times 3\text{ ft} \times 7\text{ ft} = 21\text{ sf}$$

$$(\textit{Interior})\ (42 \times 3\text{ ft} \times 6\text{ ft }8\text{ in}) = \underline{840\text{ sf}}$$

$$\text{Subtotal}\quad 861\text{ sf}$$

$$2\text{ sides} \times 2\text{ coats}\quad \underline{\times 4}$$

$$3444\text{ sf}$$

Exterior rain spouts and gutters. See Fig. 6.31 for exterior rain spouts and gutters (07630):

	Rain spouts	Gutters
Plant	7 at 35 ft = 245 lf	400 lf
Warehouse	2 at 35 ft = 70 lf	100 lf
Office	2 at 35 ft = 70 lf	100 lf

Conveying systems (Division 14)

Crane rails (14605)

Channels = MC18 in × 58 lb × 600 lf

Weight = 58 lb × 600 lf = 34,800 lb

Erection: 7-person crew × 20 days

Bridge crane (14630). This makes use of a 50-ton top running bridge crane.

Equipment cost (1996 delivery) = $360,000

(See Fig. 6.32 for telephone record and Fig. 6.33 for typical crane configuration.)

Plus:

- Erection crew—7 workers, 4 days
- 100-ton crane—2 days
- Electrical service and hookup
- Power pickup use—$30,000
- Hookup crew 5 workers—10 days

Monorail hoists (14610). This makes use of two 5-ton electrical monorail hoists.

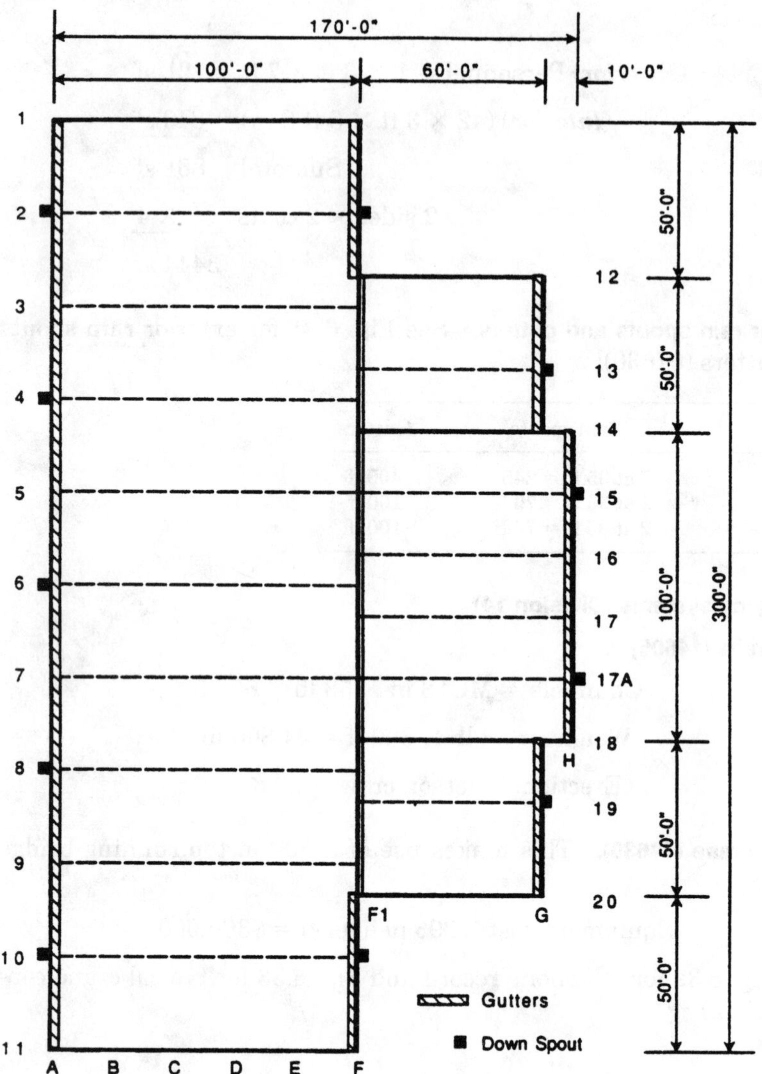

Figure 6.31 Exterior rain downspouts and gutters.

Equipment cost (1996 delivery) = $84,000 each

+ 20 ft of track and two switches (each)

Total $168,000

Plus:

- Erection crew—7 workers, 8 days
- Electrical hookup 5 workers—10 days

<table>
<tr><td colspan="2">TELEPHONE CONVERSATION RECORD</td><td>DATE
June 07, 1995</td></tr>
<tr><td colspan="3">SUBJECT OF CONVERSATION
BRIDGE CRANE</td></tr>
<tr><td colspan="3">INCOMING CALL</td></tr>
<tr><td>PERSON CALLING</td><td>ADDRESS</td><td>PHONE NUMBER AND EXTENSION</td></tr>
<tr><td>PERSON CALLED</td><td>OFFICE</td><td>PHONE NUMBER AND EXTENSION</td></tr>
<tr><td colspan="3">OUTGOING CALL</td></tr>
<tr><td>PERSON CALLING
Jim O'Brien</td><td>ADDRESS
OK - Pennsauken, NJ</td><td>PHONE NUMBER AND EXTENSION
609-665-2000</td></tr>
<tr><td>PERSON CALLED
Don Howard</td><td>OFFICE
Mannesmann Demag
Phoenixville</td><td>PHONE NUMBER AND EXTENSION
215-933-3355</td></tr>
<tr><td colspan="3">SUMMARY OF CONVERSATION:
• Suggest top running rather than under running
• 50 ton capacity includes dynamic impact factor of 15%
• Designed by harmonic: also torsional design reduce wheel loading
• Classification: CMA/HMI - Class C (Hoist Mfg. Institute)
• Support structure should have maximum deflection of 1/880
• Cost F.O.G. Phoenixville (1995) is $360,000</td></tr>
</table>

OKA FORM 42

Figure 6.32 Phone record.

(See Fig. 6.34 for telephone record and Fig. 6.35 for typical configuration.)

Elevator (14240). One passenger elevator, two-story hydraulic, 2500-lb capacity, 14-ft floor-to-floor, speed 150 fpm.

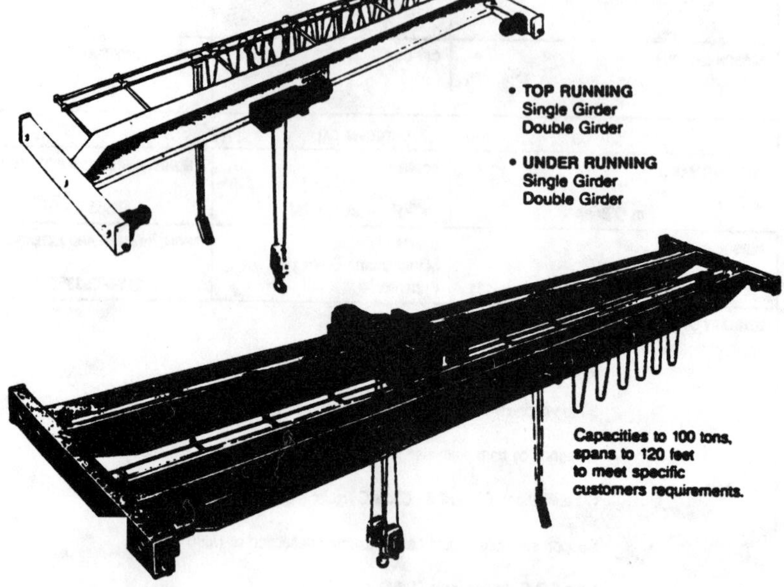

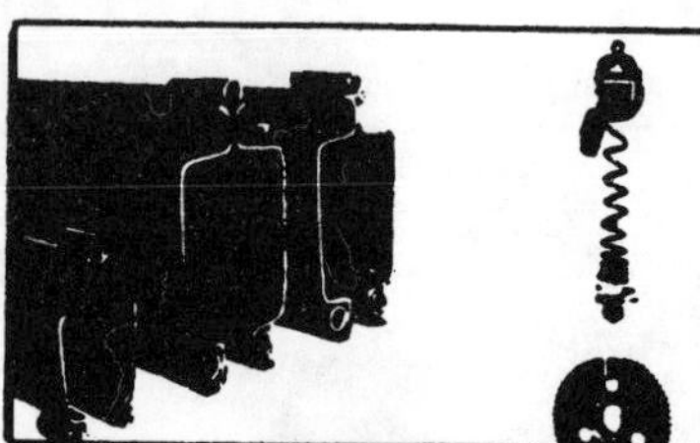

Figure 6.33 Bridge crane configuration.

TELEPHONE CONVERSATION RECORD		DATE June 07, 1995
SUBJECT OF CONVERSATION MONORAILS		
INCOMING CALL		
PERSON CALLING	ADDRESS	PHONE NUMBER AND EXTENSION
PERSON CALLED	OFFICE	PHONE NUMBER AND EXTENSION
OUTGOING CALL		
PERSON CALLING Jim O'Brien	ADDRESS OK - Pennsauken, NJ	PHONE NUMBER AND EXTENSION 609-665-2000
PERSON CALLED Steve	OFFICE Southern Service Monorail	PHONE NUMBER AND EXTENSION 404-448-7777

SUMMARY OF CONVERSATION:

Yale Patriot

5 Ton (including 200 feet of track and 2 switches)	$84,000 - $96,000
10 Ton (no switches, 200 feet of track)	$132,000 - $144,000
	(1995 Cost)

Figure 6.34 Phone record.

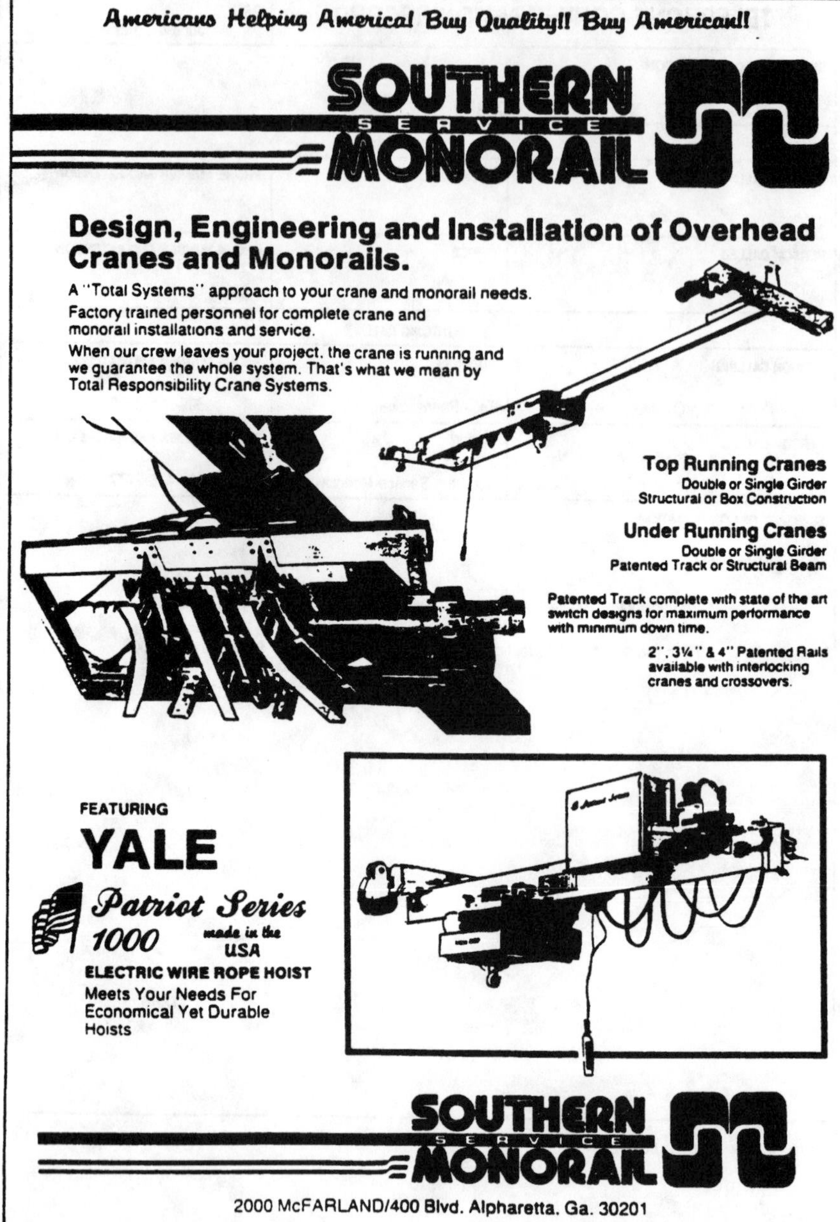

Figure 6.35 Monorail configuration.

Painting Areas for Structural Steel by Shape

SHAPE	FLANGE	HEIGHT	4 x FLANGE " + 2 x HEIGHT " = 12	SF/LF
Channels				
MC 12"-31#	4"	12"		3.3
MC 18"-58#	4"	18"		4.2
Columns				
12" W 170#	4"	12"		3.3
12" W 210#	5"	12"		3.7
14" W 145#	4"	14"		3.7
Beams				
18 W 58#	6"	18"		5.0
Girders				
G-45"	16"	45"		12.8
G-57"	18"	57"		15.5
Trusses				
Assume 2 SF/LF				

Figure 6.36 Painting areas for structural steel by shape.

Summary

Following is a summary of quantities. Note that preparation—first of the list of items in CSI order and then listing quantities for the various calculations—provides a built-in checklist.

Plant

Item	CSI	Quantity	Unit
Site preparation	02110	0.7	acre
Excavation			
Excavation, footprint	02220	3,345	cy
Pile caps	02220	74	cy
Grade beams	02220	223	cy
Structural backfill	02220	1672	cy
Standard backfill	02220	112	cy
Spread footings	02220	12	cy
Piles			
H-piles (14 in × 90 lb)	02360	1,760	lf

(Continued)

Plant (*Continued*)

Item	CSI	Quantity	Unit
Concrete			
Slab-on-grade	03310	1,111	cy
Pile caps	03310	74	cy
Spread footings	03310	12	cy
Grade beams	03310	111	cy
Loading docks	03310	56	cy
Reinforcement, slab-on-grade	03200	34,320	lb
Pile caps		4,699	lb
Spread footings		1,357	lb
Load docks		3,008	lb
Grade beams		6,144	lb
Cementitious roof deck	03500	333	cy
Waterstop (from Fig. 6.7)	03310	700	lf
Metals			
Structural steel	05120		
Plant, main steel beams and girders (Fig. 6.12)		202	tons
Plant, main steel columns (Fig. 615)		61	tons
Crane rails		17.4	tons
Plant cross bracing (Figs. 6.18 and 6.19)		60	tons
Girts, plant		28	tons
Loading docks		14.7	tons
Subtotal		383.1	tons
Column base plates (from Fig. 6.15)	05120	3708	lb
Joists (from Fig. 6.17)	05210	120	tons
Metal deck			
Roofing + loading docks	05310	31,000	sf
Thermal and Moisture Protection			
Roofing	07510	300	squares
Dampproofing, grade beam	07160	1,805	sf
Insulation, roof	07220	30,000	sf
Manufactured wall panels	07410	18,000	sf
Flashing	07620	800	lf
Insulation, grade beam	07210	1,805	sf
Exterior, rain spouts	07630	245	lf
Gutters	07630	400	lf
Doors			
Steel doors and frames	08110	5	each (3 ft × 7 ft)
Vertical lift doors	08365	4	each (20 ft × 16 ft)
Door hardware	08710	5	sets
Finishes			
Painting			
Exterior	09910	4,516	sf
Interior	09920		

Plant (*Continued*)

Item	CSI	Quantity	Unit
Finishes (*Cont.*)			
Structural steel			
Primer		47,632	sf
Finish		95,264	sf
Block wall		6,000	sf
Equipment			
Dock levelers	11161	4	each
Conveying Systems			
Bridge crane	14630	1	each
Mechanical			
Use schematic			
Electrical			
Use schematic			

Warehouses A and B

Item	CSI	Quantity	Unit
Site preparation	02110	0.14	acre
Excavation			
Excavation, footprint	02220	669	cy
Grade beams	02220	77	cy
Spread beams	02220	9	cy
Structural backfill	02220	334	cy
Backfill	02220	39	cy
Concrete			
Slab-on-grade	03310	167	cy
Spread footings	03310	9	cy
Grade beams	03310	40	cy
Loading beams	03310	11	cy
Reinforcement, slab-on-grade	03200	4,400	lb
Spread footings		1,018	lb
Load docks		602	lb
Grade beams		2,208	lb
Cementitious roof deck	03500	67	cy
Metals			
Structural steel	05120		
Warehouse, main steel			
Beams and girders (Fig. 6.12)		40	tons
Warehouse, main steel			
Columns (Fig. 6.15)		6	tons
Warehouse, loading docks		3.25	tons
Warehouse, girts		10.25	tons
Subtotal		59.5	tons

(Continued)

Warehouses A and B (*Continued*)

Item	CSI	Quantity	Unit
Metals (Cont.)			
Column base plates (Fig. 6.15)		504	lb
Joists (from Fig. 6.17)	05210	22	tons
Deck	05310		
Roofing + loading docks		6,400	sf
Thermal and Moisture Protection			
Roofing	07510	60	squares
Dampproofing-grade beam	07160	649	sf
Insulation, roof	07220	6,000	sf
Manufactured wall panels	07410	6,600	sf
Flashing	07620	440	lf
Insulation, grade beam	07210	649	sf
Rain spouts	07630	70	lf
Gutters	07630	100	lf
Doors and Windows			
Steel doors and frames	08110	2	each (3 ft × 7 ft)
Vertical lift doors	08365	2	each (16 ft × 16 ft)
Door hardware	08710	2	sets
Finishes			
Painting			
Exterior	09910	2,216	sf
Interior	09920		
Structural steel			
Primer		15,160	sf
Finish		30,320	sf
Block wall		3,600	sf
Equipment			
Dock levelers		4	each
Conveying equipment			
Monorail hoists 5 ton with 200 ft track and 2 switches	14610	2	each
Mechanical			
Use schematic			
Electrical			
Use schematic			

Office

Item	CSI	Quantity	Unit
Site preparation	02110	0.16	acre
Excavation			
Excavation, footprint	02220	764	cy
Grade beams	02220	45	cy

Office (*Continued*)

Item	CSI	Quantity	Unit
Excavation (*Cont.*)			
Spread beams	02220	6	cy
Structural backfill	02220	382	cy
Backfill	02220	22	cy
Concrete			
Stairs, forms	03310	125	sf
Concrete	03310	7	cy
Slab-on-grade	03310	130	cy
Spread footings	03310	6	cy
Grade beams	03310	23	cy
Second floor slab	03310	185	cy
Reinforcement, slab-on-grade	03200	5,170	lb
Spread footings		710	lb
Grade beams		1,248	lb
Second floor		10,528	lb
Waterstop (from Fig. 6.7)		50	lf
Cementitious roof deck	03500	79	cy
Unit Masonry			
Facade			
Scaffolding, 2-story	01525	36	csf
Backup wall masonry block	04220		
8-in thick, reinforced	04210	2,700	sf
Face brick common			
8-in face brick, 4 in thick		2,700	sf
Coping, precast, 14 in wide	04210	120	lf
Block partitions	04220	12,670	sf
Use 8-in block, reinforced			
Metals			
Structural steel	05120		
Office beams and girders (Fig. 6.12)		69	tons
Office columns (Fig. 6.15)		10	tons
Subtotal		79	tons
Column base plates (Fig. 6.15)		336	lb
Joists (from Fig. 6.17)	05210	21	tons
Metal deck roof	05310	7,000	sf
Handrails	05520		
Aluminum, 3-rail anodized		75	lf
Thermal and Moisture Protection			
Roofing	07510	70	squares
Dampproofing, grade beam	07160	367	sf
Insulation, roof	07220	7,000	sf
Insulation, building	07210	8,850	sf
Flashing	07620	140	lf
Insulation, grade beams	07210	367	sf
Rain downspouts	07630	70	lf
Gutters	07630	100	lf

(*Continued*)

Office (*Continued*)

Item	CSI	Quantity	Unit
Doors and Windows			
Fire doors (metal)(3 ft × 7 ft)	08110	8	each
Wood doors (3 ft × 2 ft 6 in) (Fig. 6.25)	08205	36	each
Entrance door (double 6 ft × 7 ft)	08410	1	each
Glass windows	08920	900	sf
Hardware	08710	45	sets
Finishes			
Acoustical suspension system	09130	12,300	sf
Drywall partitions (metal stud)	09260	8,900	sf
Drywall on block	09260	7,950	sf
Ceramic tile	09310	900	sf
Acoustic tile	09510	123,000	sf
Resilient tile	09660	3,825	sf
Carpet	09685	875	sy
Exterior painting	09910	42	sf
Interior	09920		
Structural steel			
Primer		17,500	sf
Finish		35,000	sf
Masonry		10,640	sf
Drywall		51,500	sf
Doors		3,444	sf
Specialties			
Metal toilet partitions	10160	12	each
Lockers	10505	20	each
Conveying Equipment			
Elevators	14240	1	each
Mechanical			
Use schematic			
Electrical			
Use schematic			

Structural Steel Painting Surface Areas

Plant	
G57 x 11 x 100' x 15.5 SF/LF =	17,050 SF
G45 x 20 x 30' x 12.8 SF/LF =	7,680 SF
14 W 145# x 22 x 30' x 3.7 SF/LF =	2,442 SF
Crane Rail MC 18-58# x 600' x 4.2 =	2,520 SF
Trusses 200 x 30' x 2.0 SF/LF =	12,000 SF
Firts 1800' x 3.3 SF/ =	5,940 SF
	47,632 SF
Warehouse	
G45 (240' + 200') x 12.8 SF/LF =	5,632 SF
12 W 170# 6 x 30' = 180' x 3.3 SF/LF =	594 SF
Trusses 12 x 100' = 1200' x 2.0 SF/LF =	2,400 SF
Girts 660' x 3.3 SF/1 =	6,534 SF
	15,160 SF
Office	
G45 400' x 12.8 SF/LF =	5,120 SF
G45 280' x 12.8 SF/LF =	3,584 SF
WF 18 560' x 5.0 SF/LF =	2,800 SF
12 W 170# 4 x 30' x 3.3. SF/LF =	396 SF
Trusses 28 x 100' x 2.0 SF/LF =	5,600 SF
	17,500 SF

Figure 6.37 Structural steel painting surface areas.

Chapter

7

Cost Databases

There are a number of well-recognized cost databases which can be used to add the cost dimension to the quantities taken off in the estimate. One of the more prominent is R.S. Means. Several databases by the R.S. Means Publishing Company published annually include:

Building Construction Cost Data

Building Construction Cost Data, Metric Version

Building Construction Cost Data, Western Version

Means Assemblies Cost Data

Means Concrete Cost Data

Means Electrical Change Order Cost Data

Means Electrical Cost Data

Means Facilities Cost Data

Means Heavy Construction Cost Data

Means Interior Cost Data

Means Labor Rates for the Construction Industry

Means Light Commercial Cost Data

Means Mechanical Cost Data

Means Open Shop Building Construction Cost Data

Means Plumbing Cost Data

Means Repair and Remodeling Cost Data

Means Residential Cost Data

Means Site Work and Landscape Cost Data

Means Square Foot Costs

The following description of the *Means Facilities Cost Data* is paraphrased from that document (with permission):

> The *Means Facilities Cost Data* (approximately 900 pages) is devoted specifically to the needs of professionals responsible for the maintenance, construction and renovation of commercial, industrial, municipal and institutional properties. This reference provides immediate access to cost associated with facilities construction, renovation and maintenance—with over 40,000 unit price line items.
>
> Since 1942, R. S. Means Company, Inc., has been actively engaged in construction cost publishing and consulting throughout North America. The company says their primary objective is to provide the construction industry professional—the contractor, the owner, the architect, the engineer, the facilities manager—with current and comprehensive construction cost data.
>
> Data are collected and organized into a format that is instantly accessible. The data are useful for all phases of construction cost determination—from the preliminary budget to the detailed unit price estimate.
>
> The Means organization is prepared to assist in the solution of construction problems through the services of its four major divisions; Construction and Cost Data Publishing, Computer Data and Software Services, Consulting Services, and Educational Seminars.

Development of Cost Data

The staff at R. S. Means Company, Inc., continuously monitors developments in the construction industry in order to ensure reliable, thorough, and up-to-date cost information. While *overall* construction costs may vary relative to general economic conditions, price fluctuations within the industry are dependent upon many other factors. Individual price variations may, in fact, be opposite to overall economic trends. Therefore, costs are monitored and updated and new items are added in response to industry changes.

All costs represent U.S. national averages and are given in U.S. dollars. The Means City Cost Indexes can be used to convert costs to a particular location. The City Cost Indexes for Canada can be used to convert U.S. national averages to local costs in Canadian dollars.

Material costs are determined by contacting manufacturers, dealers, distributors, and contractors throughout the United States. If current material costs are available for a specific location, adjustments can be made to reflect differences from the national average. Material costs do not include sales tax.

Labor costs are based on the average of wage rates from 30 major U.S. cities. Rates are determined from agreements or prevailing wages for construction trades for the current year. Rates are listed on the inside back cover of the book. If wage rates in your area vary from those used in the book, or if rate increases are expected within a given year, labor costs should be adjusted accordingly.

Labor costs reflect productivity based on actual working conditions. These figures include time spent during a normal work day on items other than actual installation such as material receiving and handling, mobilization, site movement, breaks, and cleanup. Productivity data are developed over an extended period so as not to be influenced by abnormal variations, and reflect a typical average.

Equipment costs as presented include not only rental costs but also operating costs. Equipment prices are obtained from industry sources throughout the country—contractors, suppliers, dealers, and manufacturers.

Factors Affecting Costs

Quality

The prices for materials and the workmanship upon which productivity is based are in line with U.S. government specifications and represent good sound construction.

Overtime

No allowance has been made for overtime. If premium time or work during other than normal working hours is anticipated, adjustments to labor costs should be made accordingly.

Productivity

The productivity, daily output, and worker-hours figures for each line item are based on working an 8-h day in daylight hours. For other than normal working hours, productivity may decrease.

Size of project

The size and type of construction project can have a significant impact on cost. Economy of scale can reduce costs for large projects. Conversely, costs may be higher for small projects owing to higher percentage overhead costs, small-quantity material purchases, and minimum labor and/or equipment charges. Costs in the book are intended for the size and type of project as described in the "How to Use This Book" pages. Costs for projects of a significantly different size or type should be adjusted accordingly.

Location

Material prices are for metropolitan areas. Beyond a 20-mile radius of large cities, extra trucking or other transportation charges will increase the material costs slightly. This material increase may be offset by lower wage rates. Both of these factors should be considered when preparing an estimate, especially if the job site is remote. Highly specialized subcontract items may require high travel and per diem expenses for mechanics.

Other factors affecting costs are season of year, contractor management, weather, local union restrictions, building code requirements, and the availability of adequate energy, skilled labor, and building materials. General business conditions influence the "in-place" cost of all items. Substitute materials and construction methods may have to be employed, and these may increase the installed cost and/or life cycle costs. Such factors are difficult to evaluate and cannot be predicted on the basis of the job's location in a particular section of the country. Thus there may be a significant but unavoidable cost variation where these factors are concerned.

CSI Masterformat

Unit price data in the book are organized according to the *MasterFormat*™ system of classification and numbering as developed by the Construction Specifications Institute (CSI) and Construction Specifications Canada. This system, widely accepted in the industry, is used extensively by architects and engineers for construction specifications, by contractors for estimating and record keeping, and by manufacturers and suppliers for categorization of construction materials and products. R. S. Means has organized unit price data in this system to help construction professionals categorize all aspects of the construction process.

How the Book Is Arranged

The book is divided into four sections: Unit Price, Assemblies, Reference, and an Appendix.

Unit price section

All cost data have been divided into the 16 divisions of the Construction Specifications Institute's (CSI) *MasterFormat*™ plus a sf (square foot) and cf (cubic foot) cost division (17). A listing of these divisions and an outline of their subdivisions is shown in the Table of Contents page at the beginning of the Unit Price section.

Numbering. Each unit price line item has been assigned a unique 10-digit code. A graphic explanation of the numbering system is shown on the "How to Use Unit Price" page.

Descriptions. Each line item number is followed by a description of the item. Subitems and additional sizes are indented beneath appropriate line items. The first line or two after the main (boldface) item often contain descriptive information that pertains to all line items beneath this boldface listing.

Crew. The "Crew" column designates the typical trade or crew to install the item. When an installation is done by one trade and requires no power equipment, that trade is listed. For example, "2 Carp" indicates that the installation is done with 2 carpenters. Where a composite crew is appropriate, a crew code designation is listed. For example, a "C-2" crew is made up of 1 supervisor, 4 carpenters, 1 laborer, plus power tools. All crews are listed at the beginning of the Unit Price section. Costs are shown both with bare labor rates and with the contractor's overhead and profit. For each, the total crew cost per 8-h day and the composite cost per worker-hour are listed (see Fig. 7.1).

Crew equipment cost. The power equipment required for each crew is included in the crew cost. The daily cost for crew equipment is based on dividing the weekly bare rental rate by 5 (working days per week) and then adding the hourly operating cost times 8 (hours per day). This "Crew Equipment Cost" is listed in Division 016 (see Fig. 7.2).

Daily output. To the right of every "Crew" code listing, a "Daily Output" figure is given. This is the number of units that the listed crew will install in a normal 8-h day.

Worker-hours. The column following "Daily Output" is "Worker-hours." This figure represents the worker-hours required to install

Crew C-2	Hourly	Daily	Hourly	Daily	Bare costs	Inc. O&P
1 Carpenter supervisor (out)	$23.40	$187.20	$37.60	$300.80	$20.97	$33.67
4 Carpenters	21.40	684.80	34.35	1099.20		
1 Building laborer	16.85	134.80	27.05	216.40		
Power tools		30.00		33.00	.62	.68
48 worker-hours, daily totals		$1036.80		$1649.40	$21.59	$34.35

Figure 7.1 Crew Composition and Cost, 1989 (R. S. Means Co., Inc., Kingston, MA).

016 | Material and Equipment

	016 400	Equipment Rental		Unit	Hourly Oper. Cost.	Rent per Day	Rent per Week	Rent per Month	Crew Equipment Cost	
406	0010	CONCRETE EQUIPMENT RENTAL								406
	0100	without operators								
	0200	Bucket, concrete lightweight, ½ C.Y.		Ea.	.11	20	59	180	12.70	
	0300	1 C.Y.			.15	28	78	235	16.80	
	0400	1-½ C.Y.	(13)		.18	30	90	270	19.45	
	0500	2 C.Y.			.22	35	105	315	22.75	
	0600	Cart, concrete, operator walking, 10 C.F.			1	60	180	540	44	
	0700	Operator riding, 18 C.F.			2.35	145	435	1,300	105.80	
	0800	Conveyer for concrete, portable, gas, 16" wide, 26' long			2.73	115	350	1,050	91.85	
	0900	46' long			2.90	140	425	1,275	108.20	
	1000	56' long			3.10	155	470	1,400	118.80	
	1100	Core drill, electric, 2-½ H.P., 1" to 8" bit diameter			.40	48	145	435	32.20	
	1200	Finisher, concrete floor, gas, riding trowel, 48" diameter			2.30	73	220	660	62.40	
	1300	Gas, manual, 3 blade, 36" trowel			.70	33	100	300	25.60	
	1400	4 blade, 48" trowel			1.10	43	130	390	34.80	
	1500	Float, hand-operated (Bull float) 48" wide			.12	6	18	55	4.55	
	1570	Curb builder , 14 H.P., gas, single screw			1.20	45	130	390	35.60	
	1590	Double screw			1.95	48	145	435	44.60	
	1600	Grinder, concrete and terrazzo, electric, floor			1.10	43	130	390	34.80	
	1700	Wall grinder			.39	23	70	210	17.10	
	1800	Mixer, powered, mortar and concrete, gas, 6 C.F., 18 H.P.			.90	43	130	390	33.20	
	1900	10 C.F., 25 H.P.			1.18	54	160	480	41.45	
	2000	16 C.F.			1.35	72	215	645	53.80	
	2100	Concrete, stationary, tilt drum, 2 C.Y.			6.70	295	885	2,650	230.60	
	2120	Pump, concrete, truck mounted, 4" line, 80' boom			8.70	740	2,225	6,675	514.80	
	2140	5" line, 110' boom			10.55	900	2,700	8,100	624.40	

11

Figure 7.2 Equipment costs, Division 016, 1991 (R. S. Means Co., Inc., Kingston, MA).

one "unit" of work. Unit worker-hours are calculated by dividing the total daily crew hours (as seen in the Crew Tables) by the Daily Output.

Unit. To the right of the "Worker-hour" column is the "Unit" column. The abbreviated designations indicate the unit upon which the price, production, and crew are based. The Appendix has a complete list of abbreviations.

Material. The first column under the "Bare Cost" heading lists the unit material cost for the line item. This figure is the "bare" material cost with no overhead and profit allowances included. Costs shown reflect national average material prices for January of the current year and include delivery to the jobsite.

Labor. The second "Bare Cost" column is the unit labor cost. This cost is derived by dividing the daily labor cost by the daily output. The wage rates used are listed on the inside back cover.

Equipment. The third "Bare Cost" column lists the unit equipment cost. This figure is the daily crew equipment cost divided by the daily output.

Total. The last "Bare Cost" column lists the total bare cost of the item. This is the arithmetic total of the three previous columns: "Material," "Labor," and "Equipment."

Total including overhead and profit. The figure in this column is the sum of three components: the bare material cost plus 10%; the bare labor cost plus overhead and profit; and the bare equipment cost plus 10%. A sample calculation is shown on the "How to Use Unit Price" page.

Division 17. This division contains square foot and cubic foot costs for 59 different building types. These figures include contractor's overhead and profit but do not include architectural fees or land costs (see Fig. 7.3).

Assemblies section

This section uses an "Assemblies" format grouping all the functional elements of a building into 12 "Uniformat" Construction Divisions (see Fig. 7.4).

171 | S.F., C.F. and % of Total Costs

	171 000	S.F. & C.F. Costs	UNIT	UNIT COSTS ¼	UNIT COSTS MEDIAN	UNIT COSTS ¾	% OF TOTAL ¼	% OF TOTAL MEDIAN	% OF TOTAL ¾	
390	1800	Equipment	S.F.	1.40	3.88	6.30	2.90%	6.30%	8.40%	390
	2720	Plumbing		1.98	3.12	6.25	4.90%	7.40%	11%	
	2730	Heating & ventilating		2.91	4	4.83	5.20%	7.20%	9.50%	
	2900	Electrical		2.71	4.49	6.30	7%	9%	11.10%	
	3100	Total: Mechanical & Electrical		6.40	11.35	16.30	15.70%	21.90%	27.80%	
400	0010	GARAGES, MUNICIPAL (repair)	S.F.	39	53.50	81				400
	0020	Total project costs	C.F.	2.44	3.40	4.55				
	0500	Masonry	S.F.	4.42	6.90	10.30	7%	10.60%	15.50%	
	1140	Roofing		2.56	3.51	5.95	6.50%	7.30%	10.10%	
	2720	Plumbing		2.06	3.87	6.35	4.10%	6.90%	8.70%	
	2730	Heating & ventilating		2.58	4.62	6.95	6.10%	7.80%	11.60%	
	2900	Electrical		2.97	4.65	6.80	5.60%	8.10%	10.50%	
	3100	Total: Mechanical & Electrical		7.80	15.50	21.90	17%	24.40%	32.40%	
410	0010	GARAGES, PARKING		17.75	22.50	37.35				410
	0020	Total project costs	C.F.	1.53	2.03	3.34				
	2720	Plumbing	S.F.	.32	.62	.90	2.10%	2.80%	3.80%	
	2900	Electrical		.75	1.09	1.76	4.20%	5.20%	6.50%	
	3100	Total: Mechanical & Electrical		1.29	1.74	2.60	7.10%	8.30%	9.50%	
	3200									
	9000	Per car, total cost	Car	5,675	7,725	10,900				
	9500	Total: Mechanical & Electrical	"	405	610	910				
430	0010	GYMNASIUMS	S.F.	46.15	62.60	80.70				430
	0020	Total project costs	C.F.	2.32	3.23	4.03				
	1800	Equipment	S.F.	1.21	2.18	3.52	2%	3.30%	6.70%	
	2720	Plumbing		2.87	3.91	4.85	4.80%	7.20%	8.30%	
	2770	Heating, ventilating, air conditioning		3.03	5.30	8.65	7.20%	9.70%	14%	
	2900	Electrical		3.92	4.78	6.60	6.20%	8.90%	10.80%	
	3100	Total: Mechanical & Electrical		8.45	13.10	16.95	16.60%	21.80%	27.40%	
	3500	See also division 114-801 & 114-805								

Figure 7.3 Square foot–cubic foot costs, 1991 (R. S. Means Co., Inc., Kinston, MA).

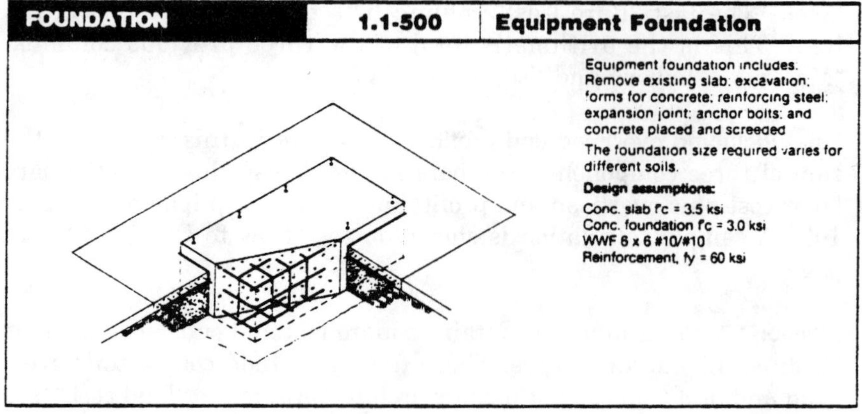
FOUNDATION | 1.1-500 | **Equipment Foundation**

Equipment foundation includes: Remove existing slab; excavation; forms for concrete; reinforcing steel; expansion joint; anchor bolts; and concrete placed and screeded

The foundation size required varies for different soils.

Design assumptions:
Conc. slab f'c = 3.5 ksi
Conc. foundation f'c = 3.0 ksi
WWF 6 x 6 #10/#10
Reinforcement, fy = 60 ksi

System Components	QUANTITY	UNIT	COST EACH MAT.	COST EACH INST.	COST EACH TOTAL
SYSTEM 01.1-500					
CONC. EQUIPMENT FOUNDATION, 4'X8'X30" DEEP, 3,000 PSI.					
Cut out and break up slab; 6"	45.000	S.F.	48.72	473.58	522.30
Remove concrete; hand	3.133	C.Y.		59.68	59.68
Excavate; hand	2.500	C.Y.		142.50	142.50
Forms, 1 use	60.000	SFCA	108	648	756
Reinforcing steel, #6 bar	318.000	Lb.	95.40	98.58	193.98
Concrete, 3000 psi, for foundation, incl. premium delv. chg.	3.000	C.Y.	232.29		232.29
Anchor bolts, 1" diam., with sleeves	8.000	Ea.	24.24	50.96	75.20
Place concrete foundation	3.000	C.Y.		56.83	56.83
Backfill by hand	.866	C.Y.		14.16	14.16
Compaction in 6" layers, hand tamp	.866	C.Y.		9.61	9.61
Premolded, bituminous fiber, ½"x6"	24.000	L.F.	12.72	18.24	30.96
Welded wire fabric: 6x6-#10/10	.130	C.S.F.	1.02	2.49	3.51
Concrete, 3500 psi, for slab, incl. premium delv. chg.	.241	C.Y.	19.55		19.55
Place conc. slab	.241	C.Y.		3.08	3.08
Finish slab; steel trowel	13.000	S.F.		7.28	7.28
Finish equip. foundation; screed	32.000	S.F.		9.60	9.60
TOTAL			541.94	1,594.59	2,136.53
COST per C.Y. of FOUNDATION			180.65	531.53	712.18

Figure 7.4 Assemblies Cost Data (R. S. Means Co., Inc., Kingston, MA).

At the top of each "Assembly" cost table is an illustration, a brief description, and the design criteria used to develop the cost. Each of the components and its contributing cost to the system is shown.

For a complete breakdown and explanation of a typical "Assemblies" page, see "How to Use Assemblies Cost Tables" at the beginning of the section.

Material. These cost figures include a standard 10% markup for "handling." They are national average material costs as of January of the current year and include delivery to the jobsite.

Installation. The installation costs include labor and equipment, plus a markup for the installing contractor's overhead and profit.

Reference section

Following the items in the "Unit Price" pages, there are frequently found large numbers in circles. These numbers refer the reader to data in this Reference Section (see Fig. 7.5). This material includes estimating procedures, alternate pricing methods, technical data, and cost derivations. This section also includes information on design and economy in construction.

Appendix

Included in this section are Historical and City Cost Indexes, a list of abbreviations, and a comprehensive index.

Historical cost index. This index provides annual data to adjust construction costs overtime.

City cost indexes. These indexes provide data to adjust the "national average" costs in this book to 162 major cities throughout the United States and Canada.

Abbreviations and index. A listing of the abbreviations used throughout the book, along with the terms they represent, is included. Following the abbreviations list is an index of all sections.

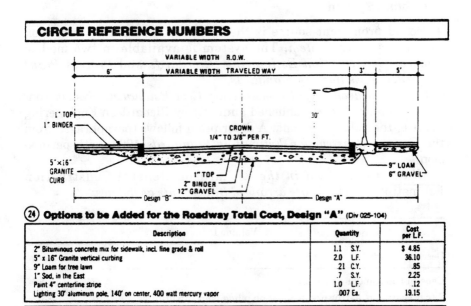

CIRCLE REFERENCE NUMBERS

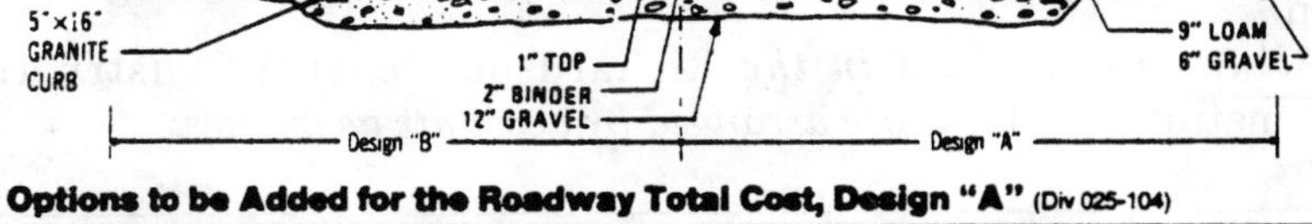

(24) Options to be Added for the Roadway Total Cost, Design "A" (Div 025-104)

Description	Quantity	Cost per L.F.
2" Bituminous concrete mix for sidewalk, incl. fine grade & roll	1.1 S.Y.	$ 4.85
5" x 16" Granite vertical curbing	2.0 L.F.	36.10
9" Loam for tree lawn	.21 C.Y.	.85
1" Sod, in the East	.7 S.Y.	2.25
Paint 4" centerline stripe	1.0 L.F.	.12
Lighting 30' aluminum pole, 140' on center, 400 watt mercury vapor	.007 Ea.	19.15

Figure 7.5 Circle Reference Numbers (R. S. Means Co., Inc., Kingston, MA).

Project size

The book is aimed primarily at industrial and commercial projects costing $5000 to $500,000. With reasonable exercise of judgment the figures can be used for any building project, but they do not apply to civil engineering structures such as bridges, dams, or highways.

Rounding of costs

In general, all unit prices in excess of $5 have been rounded to make them easier to use and still maintain adequate precision of the results. The rounding rules are as follows:

Price from $5.01 to $20.00 rounded to the nearest 5 cents

Price from $20.01 to $100.00 rounded to the nearest $1

Price from $100.01 to $1,000.00 rounded to the nearest $5

Price from $1,000.01 to $10,000.00 rounded to the nearest $25

Price from $10,000.01 to $50,000.00 rounded to the nearest $100

Price from $50,000.01 rounded to the nearest $500

The Richardson Rapid Construction Cost Estimating System

A second prominent source is the *Richardson Rapid Construction Cost Estimating System.* The system is available in two modes: *General Construction Estimating Standards* and *Process Plant Construction Estimating Standards.*

The *Richardson Rapid Construction Cost Estimating System* consists of three volumes published annually by Richardson Engineering Service, Inc., Mesa, Arizona. A description follows (paraphrased from the *General Construction Estimating Standards 1992,* with permission).

All data contained in the Richardson General Construction Estimating Standards *are arranged by accounts as follows:*

Volume 1	
Description	Account
General notes and instructions	1-0
Site preparation, earthwork, foundation piling, and site utilities	2-0
Concrete foundations, columns, slabs, walls, tiltup, precast, and pile caps	3-0

Volume 2	
Description	Account
Hollow and solid unit masonry	4-0
Structural steel, miscellaneous steel, and prefabricated metal buildings	5-0
Rough and finish carpentry	6-0
Thermal and moisture protection, metal siding, roofing, sheet-metal work, roof hatches and skylights, caulks, and sealants	7-0
Doors and windows	8-0
Finishes, steel studs, lath and plaster, fireproofing, acoustical ceilings, tile and terrazzo, hardwood and resilient flooring, painting and wall covering	9-0
Specialties and conveying systems	10-0

Volume 3	
Description	Account
Air conditioning, heating, ventilating, plumbing, process piping, and instrumentation	15-0
Electrical lighting, power, and instrument wiring	16-0

An alphabetical key word index and abbreviations appear at the back of each volume.

Richardson states:

> The purpose of Richardson Engineering Services Publications is to provide assistance in upgrading the estimating skills of others. Richardson is dedicated to the publication of authentic construction cost information. The information contained within these standards is based on data obtained from sources believed to be reliable. Averages have been weighted and adjusted for anticipated economic conditions to arrive at the forecasted data. It is noted that estimates can vary significantly on projects even when detailed plans and specifications are available. Costs can vary greatly depending on location, market conditions (material and labor), experience, work delays, etc. Richardson Engineering Services, Inc., or the editors, makes no warranty or guaranty as to the accuracy, correctness, or sufficiency of data contained within these standards and assumes no responsibility or liability in connection with their use.
>
> The format, estimating techniques, language, method of expression, worker-hours for performing the work, costs, and other information herein are unique to the Richardson Rapid System. The data have been developed by Richardson Engineering Services, Inc.

Additional services furnished without additional charge during the subscription period include:

1. *The Cost Trend Reporter,* updated and issued quarterly, containing: *Wage Rates,* for 16 construction trades in 160 cities; *Revision Sheets,*

for insertion in the standards to maintain them in an up-to-date condition; *Construction Cost Indexes,* for total construction cost fractionated into 17 different categories (1987 = 100 as basis for indexes); and *Material Price Indexes,* for all construction materials fractionated into 65 different categories (1982 = 100 as basis for indexes).

2. A master set of estimating forms specially designed for use with the Richardson Rapid System.

Basis for the data on the Standards, continued from Account 1-0 in Volume 1:

10. **Table of Contents, Account 4-0**

    Hollow metal unit masonry
    Lightweight concrete block
    Screen blocks
    Surface bonded concrete blocks
    Accessory materials
    Split face concrete blocks
    Fluted concrete blocks
    Slump concrete blocks
    Glazed concrete blocks
    "Thru the Wall" hollow clay units
    Glass block partitions

    Solid unit masonry
    Brick masonry units
    Accessory materials

    Natural stone masonry
    Random rubble
    Sawn ashlar
    Split ashlar
    Round rubble
    Indiana limestone
    Marble and granite

    Acid brick tile floors

    Brick floors and stairways

    Richardson Rapid System Estimates
    Concrete blocks and "Thru the Wall" hollow clay brick
    Veneer and brick-block composite walls

    Insulated block masonry

11. **Purpose of the Standards.** The purpose of the data in Work Accounts 4-1 through 4-100 is to present average material prices, worker-hours, and subcontract costs of items required for estimating masonry. The worker-hours shown in the accounts that follow are for direct labor only and are for work performed at the jobsite. The quantity of worker-hours is shown in parentheses ( ), to indicate that the figures shown are for worker-hours, *not* dollars.

TABLE 7.1 Composite Crew Rates (14)—Masonry

Crew	Basic hourly rate	H & W	Pensions	Vacation	Education and/or app. tr.	Total hourly rate	Percentage of 1 h	Cost per hour
Bricklayer supervisor	$19.30	$1.60	$1.60	$1.00	$0.50	$24.00	10	$2.40
Bricklayer	18.80	1.60	1.60	1.00	0.50	23.50	40	9.40
Bricklayer helper	14.35	1.60	1.80	0.50	0.10	18.35	50	9.18
Composite							100	$20.98

Average, per worker-hour; use $21.00.

12. **Composite Crew Rates.** The composite crew rate used in the work accounts and Rapid System Estimates has been determined. See Table 7.1.

13. **Takeoff Procedure.** When making a takeoff, it is essential that the notes in each work account be followed so correct quantities will be obtained and the variables receive proper consideration. A separate takeoff should be made and totaled for each work account. In the case of a *firm price bid estimate,* material prices, labor rates, and subcontract costs in effect at the project site should be substituted for those shown.

14. **To Complete an Estimate, Indirect Costs Must Be Considered.** When estimates of direct costs (material, labor, and subcontracts) have been made in accord with the data in the various work accounts that follow, indirect costs must be considered. Proceed as follows:

(a) *Preliminary or budget-type estimates*

1. Using the chart below, multiply the main account summary total *worker-hours* by the rate per worker-hour per total story height or project maximum height shown.

2. Add 10% to the total subcontract costs in the account to allow for the prime contractor's cost for administration.

3. Add applicable sales tax to the total material costs.

(*Special note:* This method for estimating job indirect costs is for *preliminary estimates only* and cannot be used unless the takeoffs have been made per the instructions in these standards.) See Tables 7.2 and 7.3.

TABLE 7.2 Dollars per Worker-hour for Job Indirect Costs Not Including Profit When Building Has a Total Story Height or Project Has a Maximum Height Not Exceeding

Total story height	1 story	2 stories	3 stories	4 stories	5 stories	6 stories
Project maximum height	20 ft 0 in	36 ft 0 in	52 ft 0 in	68 ft 0 in	84 ft 0 in	100 ft 0 in
Account 4-0	$10.92	$11.45	$11.97	$12.50	$13.01	$13.51

10. Table of Contents, Account 15-0

TABLE 7.3 Example, Main Account Summary, Account 4-0, Masonry

		Materials		Labor				Subcontract		
Description	Quantity	Unit price	Amount	Unit worker-hours	Total worker-hours	Rate	Amount	Unit price	Amount	Total
4-62 reinforced masonry wall No. 10	23,000 sf	—	$153,870	—	(6187)	—	$129,927	—	—	$283,797
Subtotal, direct costs	23,000 sf	—	153,870	—	(6187)	—	129,927	—	—	283,797
Job indirect costs*										
3-story rate per worker-hour	—	—	—	—	(6187)	11.97	74,058	—	—	74,058
Sales tax on materials	—	5%	7,694	—	—	—	—	—	—	7,694
Total Account 4-0	—	—	$161,564	—	(6187)	—	$203,985	—	—	$365,549

*Refer to Account 1-0, page 5, for definition of indirect costs and procedures for preparing estimates of total project costs. For firm price estimates, follow instructions given in Account 1-0, part (b), page 10.

Steam trap assemblies

Total installed costs, preliminary estimates

Air conditioning, heating, and ventilating systems, total installed costs

Plumbing fixtures, total installed costs

Fire protection systems, total installed costs

Service piping

CPVC plastic piping system

A-120 steel piping systems

Copper piping systems

Red brass piping systems

Cast-iron drain, waste, and vent piping

A-120 steel drain, waste, and vent piping systems

Polypropylene, ABS, and PVC drain, waste, and vent piping systems

Grooved end A-53 carbon steel piping system

Process and Steam Piping

A-106 carbon steel piping systems

A-53 carbon steel piping systems—field fabricated

Flanged carbon steel plastic-lined piping systems

PVDF (Kynar) plastic piping systems

Fiberglass reinforced piping systems (RTRP)

Self-contained regulating valves and backflow preventers

Instruments

Liquid level gauges and pressure gauges

Thermometers, thermocouples, and thermowells and orifice plates

Safety relief valves

Piping specialties

Expansion joints and flexible pump connectors

Steam tracing

Bolt and gasket sets

11. **Purpose of the Standards.** The purpose of the data in Work Accounts 15-1 through 15-105 is to present average material prices, worker-hours, and subcontract costs of items for estimating mechanical work.

    The worker-hours shown in the accounts that follow are for direct labor only and are for work performed at the jobsite. The quantity of worker-hours is shown in parentheses ( ), to indicate that the figures shown are for worker-hours, *not* dollars.

12. **Composite Crew Rates.** See Table 7.4.

13. **Takeoff Procedures.** When making a takeoff, it is essential that the notes in each work account be followed so correct quanti-

TABLE 7.4 Composite Crew Rates (15)—Mechanical

Crew	Basic hourly rate	H & W	Pensions	Vacation	Education and/or app. tr.	Total hourly rate	Percentage of 1 h	Cost per hour
Plumber and pipefitter supervisor	$22.50	$1.80	$1.80	$1.00	$0.50	$27.60	10	$ 2.76
Plumber and pipefitter	22.00	1.80	1.80	1.00	0.50	27.10	70	18.97
Equipment operator	19.20	1.80	1.80	1.00	0.30	24.10	10	2.41
Laborer	14.35	1.60	1.80	0.50	0.10	18.35	10	1.835
Composite crew rate							100	$25.975

Average, per worker-hour; use $26.00.

ties will be obtained and the variables receive proper consideration. A separate takeoff should be made and totaled for each work account. In the case of a *firm price bid estimate,* material prices, labor rates, and subcontract costs in effect at the project site should be substituted for those shown.

10. **Table of Contents, Account 16-0**

Panel enclosures and junction boxes

Bus duct, trolley duct

Feeder or plug-in duct and accessories

Trolley duct and accessories

Power line poles and power line transmission wire

Electric heat tracing system

Richardson Rapid System estimates

Circuit feeders

Branch circuits with convenience outlets

Branch circuits with toggle switches

Branch circuits for lighting

Motor feeders

Flexible conduit connections at electrical motors

Total installed costs, preliminary estimates

Electrical systems for air conditioning, heating, and ventilation

Total installed costs

Electrical wiring and fixtures, total installed costs

Radiant heat cable, total installed costs

Transformers, switchgear, switchboards

Metal clad switchgear

Switchboards

Transformers

Panelboards, switches, circuit breakers, motor starters

Circuit breaker distribution panelboards

Panelboards, load centers, subpanels, general-purpose

Panelboards, load centers, subpanels, hazardous areas

Safety switches

Circuit breakers

Motor starters

Meter panels

Motor control centers, control stations

Motor control centers

Control stations, pushbuttons, and pilot lights

Ground fault protection systems

Convenience outlets and toggle switches

Lighting fixtures

Lighting fixtures, general-purpose

Lighting fixtures, vaportight and explosion-proof

Outdoor floodlights and floodlight poles

Grounding systems

10. **Purpose of the Standards.** The purpose of the data in Work Accounts 16-1 through 16-103 is to present average material prices, worker-hours, and subcontract costs of items for estimating electrical work. The worker-hours shown in the accounts that follow are for direct labor only and are for work performed at the jobsite. The quantity of worker-hours is shown in brackets ( ), to indicate that the figures shown are for worker-hours, *not* dollars.
12. **Composite Crew Rates.** See Table 7.5.
13. **Takeoff Procedure.** When making a takeoff, it is essential that the notes in each work account be followed so correct quantities will be obtained and the variables receive proper consideration. A separate takeoff should be made and totaled for each work account. In the case of a *firm price bid estimate,* material prices, labor rates, and subcontract costs in effect at the project site should be substituted for those shown.

Note that the Richardson Account numbers 1.0 through 16.0 follow the basic CSI divisions. However, the CSI subaccount codes are not used within the account.

Richardson shows pictures of items and, sometimes, subassemblies. Other publishers offer estimating databases and/or estimating procedures, including (but not limited to) the following.

Computerized Databases

Both the R.S. Means and Richardson databases are available in computer accessible format, as are other databases such as the Lee Saylor database.

TABLE 7.5 Composite Crew Rates (16)—Electrical

Crew	Basic hourly rate	H & W	Pensions	Vacation	Education and/or app. tr.	Total hourly rate	Percentage of 1 h	Cost per hour
Electrician supervisor	$21.90	$1.80	$1.80	$1.00	$0.50	$27.00	10	$ 2.700
Electrician	21.40	1.80	1.80	1.00	0.50	26.50	70	18.550
Equipment operator	19.20	1.80	1.80	1.00	0.30	24.10	10	2.410
Laborer	14.35	1.60	1.80	0.50	0.10	18.35	10	1.835
Composite crew rate							100	$25.495

Average, per worker-hour; use $25.50.

In specialized areas, such as LRV (light rail vehicle) and heavy rail (railroads and subways), estimating consultants may have to develop their own database.

Most of the major cost estimating systems utilize one (or more) of the recognized databases.

Other Databases

Saylor publications, Inc., publishes these databases annually:

Commercial Square Foot Building Costs
Current Construction Costs
Remodeling and Repair Construction Costs
Residential Construction Costs

Craftsman Book Company publishes these databases annually:

Building Cost Manual
Electrical Construction Estimator (disk included)
National Construction Estimator (estimate writer disk included)
National Plumbing and HVAC Estimating (disk included)
Painting Cost Guide

Chapter

8

Pricing and Costing the Takeoff

When the material takeoff is complete for the project, the results should be summarized into the CSI format. To expedite completion of the estimate, this summary can occur as takeoff for each division is *complete*. Pricing of a division of work should not proceed until the takeoff is complete for that division.

The base price of an item is calculated by locating the cost per unit in a cost database (such as R.S. Means), multiplying that by the number of units, and extending that to the item cost.

The results should be listed on a PC format which permits identification of the following: CSI number, description, number of units, labor cost, material cost, subcontractor overhead, and profit and total cost. This format should be a Lotus spreadsheet, dBase, or similar software that permits reorganizing and totaling by category.

At the design development stage, the description of the work may not facilitate a breakdown by labor, material, and overhead and profit (OH&P). Where this occurs there are two choices: first, only the lump-sum price can be inserted, or, second, the breakdown can be assumed by the nature of the item. A typical breakdown would include 15% OH&P. Then, the total of labor and material would be (price)/1.15 = 87% (price). If labor and material were equal, each would be 87%/2 = 43.5%.

If the item were material-intensive (i.e., such as the WHATCO bridge crane), the purchase cost (i.e., material cost) might be 90% of the net and labor 10%. This would result in the following:

Labor	8.7%
Material	78.3%
Subtotal	87.0%
OH&P	13.0%
Total	100.0%

Other sources of cost data are material suppliers, equipment vendors, and subcontractors. The sources can be identified in either Sweet's Catalogs or Thomas Register.

WHATCO Project

Table 8.1 is a summary of the WHATCO quantities in CSI order. They are priced from the 1995 Means Cost Book (as assumed by

TABLE 8.1 Summary of Quantities and Costs

Item	CSI No.	R. S. Means I.D. No.	Unit	No.	R. S. Means 1995	Cost
		Plant				
Site preparation (clear)	02110	021 104 2000	acre	0.7	4,640	3,248
Excavation, footprint	02220	022 246 4220	cy	3,345	5.36	17,929
Excavation, pile caps and spread footing	02220	022 250 0100	cy	86	70	6,020
Excavation, grade beams	02220	022 254 0050	cy	223	5.95	1,327
Borrow	02220	022 216 0020	cy	1,672	3.10	5,183
Deliver borrow	02220	022 216 3800	cy	1,672	9.34	15,616
Spread	02220	022 216 2500	cy	1,672	5.22	8,728
Compaction	02220	022 226 5040	cy	1,672	0.88	1,471
Fill	02220	022 262 0170	cy	112	2.64	296
Piles (14 in × 89#)	02360	023 608 1100	lf	1,760	41	72,160
Concrete reinf., footings, pile caps	03200	032 107 0500	lb	6,056	0.78	4,724
Concrete reinf., SOG, load deck	03200	032 107 0750	lb	37,328	0.60	22,397
Concrete reinf., grade beam	03200	032 107 0100	lb	6,144	0.67	4,116
Slab-on-grade, 12 in	03310	033 130 4900	sf	30,000	3.85	115,500
Pile caps and spread footers	03310	033 130 3850	cy	86	160	13,760
Grade beams concrete	03310	033 130 4260	cy	111	486	53,946
Loading dock concrete	03310	033 130 3150	sf	4,000	2.07	8,280
Insulating lightweight fill	03500	035 212 0250	sf	30,000	1.14	34,200
Structural steel, heavy	05120	051 255 4900	tons	323	2,240	723,520
Structural steel, light	05120	051 255 4600	tons	06.1	2,560	153,856
Base plates	05120	051 255 4300	lb	3,708	1.05	3,893
Joists	05210	052 110 0030	tons	120	1,504	180,480
Metal deck	05310	053 104 0440	sf	31,000	5.12	158,720

Item	CSI No.	R.S. Means I.D. No.	Unit	No.	R.S. Means 1995	Cost
		Plant (*Continued*)				
Sandblast	05120	051 255 6130	sf	47,632	0.81	41,440
Primer	05120	051 255 6520	sf	47,632	0.54	25,721
Intermediate coat	05120	051 255 6630	sf	47,632	0.78	37,153
Top coat (elevated)	05120	051 255 7030	sf	47,632	1.19	56,682
Dampproof grade beam	07160	071 602 0300	sf	1,805	0.45	812
Insulation grade beam	07210	072 203 1740	sf	1,805	1.28	2,310
Insulation roof	07220	072 203 0100	sf	30,000	1.04	31,200
MFG wall panels	07410	074 202 0400	sf	18,000	14.53	259,740
Roofing	07510	075 102 0200	square	300	170	48,000
Flashing	07620	076 201 5800	lf	800	4.44	3,552
Ext. downspouts	07630	076 201 2500	lf	245	9.14	2,239
Ext. gutters	07630	076 210 3400	lf	400	13.31	5,324
Steel doors and frames	08110	081 103 0100	each	5	209	1,045
Vertical lift doors	08365	083 654 0010	each	4	21,376	85,504
Door hardware	08710	087 125 3110	each	5	429	2,145
Exterior paint	09910	099 106 2400	sf	4,516	0.54	2,439
Interior paint, block	09920	099 106 2400	sf	6,000	0.54	3,240
Dock levelers	11161	111 601	each	4	7,296	29,184
Bridge crane	14630	Quote	each	1	NA	418,264
		Warehouses				
Site preparation (clear)	02110	021 104 2000	acre	0.14	4,640	650
Excavation, footprint	02220	022 246 4220	cy	669	5.36	3,586
Excavation, spread footing	02220	022 250 0100	cy	9	70	630
Excavation, grade beam	02220	022 254 0050	cy	77	5.95	458
Borrow	02220	022 216 0020	cy	334	3.10	1,035
Deliver borrow	02220	022 216 3800	cy	334	9.34	3,120
Spread	02220	022 216 2500	cy	334	5.22	1,743
Compaction	02220	022 226 5040	cy	334	0.88	294
Fill	02220	022 262 0170	cy	39	2.64	103
Concrete reinf., footings	03200	032 107 0500	lb	1,018	0.78	794
Concrete reinf., SOG, load deck	03200	032 107 0750	lb	5,002	0.60	3,001
Concrete reinf., grade beam	03200	032 107 0100	lb	2,208	0.67	1,479
Slab-on-grade, 9 in	03310	033 130 4840	sf	6,000	2.79	16,740
Spread footers	03310	033 130 3850	cy	9	160	1,440
Grade beams	03310	033 130 4260	cy	40	486	19,440
Loading dock concrete	03310	033 130 3150	sf	400	2.07	8.28

(*Continued*)

TABLE 8.1 Summary of Quantities and Costs (*Continued*)

Item	CSI No.	R. S. Means I.D. No.	Unit	No.	R. S. Means 1995	Cost
		Warehouses (*Continued*)				
Insulating lightweight fill	03500	035 212 0250	sf	6,000	1.14	6,840
Structural steel, heavy	05120	051 255 4900	tons	46	2,240	103,040
Structural steel, light	05120	051 255 4600	tons	13.5	2,560	34,560
Base plates	05120	051 255 4300	lb	504	1.05	529
Joists	05210	052 110 0030	tons	22	1,504	33,088
Metal deck	05310	053 104 0440	sf	6,400	5.12	32,768
Sandblast	05120	051 255 6130	sf	15,160	0.87	13,189
Primer	05120	051 255 6520	sf	15,160	0.54	8,186
Intermediate coat	05120	051 255 6630	sf	15,160	0.78	11,825
Top coat (elevated)	05120	051 255 7030	sf	15,160	1.19	18,040
Dampproof grade beam	07160	071 602 0300	sf	649	0.45	292
Insulation, grade beam	07210	072 203 1740	sf	649	1.28	831
Insulation, roof	07220	072 203 0100	sf	6,000	1.04	6,240
MFG wall panels	07410	074 202 0400	sf	6,000	14.53	95,898
Roofing	07510	075 102 0200	square	60	160	9,600
Flashing	07620	076 201 5800	lf	440	4.44	1,954
Ext. downspouts	07630	076 201 2500	lf	70	9.14	640
Ext. gutters	07630	076 210 3400	lf	100	13.31	1,331
Steel doors and frames	08110	081 103 0100	each	2	209	418
Vertical lift doors	08365	083 654 0010	each	2	21,400	42,800
Door hardware	08710	087 125 3110	each	2	429	858
Exterior paint	09910	099 106 2400	sf	2,216	0.54	1,197
Interior paint, block	09920	099 106 2400	sf	3,600	0.54	1,944
Dock levelers	11161	111	each	4	7,300	29,200
Monorails	14620	Quote	each	2	NA	201,036
		Office				
Site preparation	02111	021 140 2000	acre	0.16	4,640	742
Excavation, footprint	02220	022 246 4220	cy	764	5.36	4,095
Excavation, spread footing	02220	022 250 0100	cy	6	70	420
Excavation, grade beam	02220	022 254 0050	cy	45	5.95	268
Borrow	02220	022 216 0020	cy	382	3.10	1,184
Deliver borrow	02220	022 216 3800	cy	382	9.34	3,568
Spread	02220	022 216 2500	cy	382	5.22	1,994
Compaction	02220	022 226 5040	cy	382	0.88	336
Fill	02220	022 262 0170	cy	22	2.64	58

Item	CSI No.	R. S. Means I.D. No.	Unit	No.	1995	R. S. Means Cost
		Office (*Continued*)				
Concrete reinf., footings	03200	032 107 0500	lb	710	0.78	554
Concrete reinf., SOG, load deck	03200	032 107 0750	lb	5,170	0.60	3,102
Concrete reinf., grade beam	03200	032 107 0100	lb	1.248	0.67	836
Concrete reinf., second floor	03200	032 107 0750	lb	10,528	0.60	6,317
Slab-on-grade, 69 in	03310	033 130 4820	sf	7,000	1.92	13,440
Spread footings	03310	033 130 3850	cy	6	160	960
Grade beams	03310	033 130 4260	cy	23	486	11,178
Insulating lightweight fill	03500	035 212 0250	sf	1.14	7,980	
Scaffold, 2-story	01525	015 254 4100	csf	36	82	2,952
Coping, precast, 14 in wide	04210	042 116 0150	lf	120	35	4,200
Face brick, common, 8 in long, 4 in thick	04210	042 184 0800	sf	2,700	9.98	26,946
Backup wall, masonry block, 8 in thick, reinforced	04220	042 216 1150	sf	2,700	6.11	16,497
Block partitions, 8-in block, reinforced	04220	042 216 1150	sf	12,670	6.11	77,414
Structural steel, heavy	05120	051 255 4900	tons	79	2,240	176,960
Base plates	05120	051 255 4300	lb	336	1.05	353
Joists	05210	052 110 0030	tons	21	1,504	31,584
Metal deck	05310	053 104 0400	sf	7,000	5.12	35,840
Handrails, aluminum 3-rail, anodized	05520	055 203 0220	lf	75	42	3,150
Sandblast	05120	051 255 6130	sf	17,500	0.87	15,225
Primer	05120	051 255 6520	sf	17,500	0.54	9,450
Intermediate coat	05120	051 255 6630	sf	17,500	0.78	13,650
Top coat (elevated)	05120	051 255 7030	sf	17,500	1.19	20,825
Dampproof grade beam	07160	071 602 0300	sf	367	0.45	165
Insulate grade beam	07210	072 203 1740	sf	367	1.28	470
Insulation, building	07220	072 203 0100	sf	8,850	1.28	11,328
Insulation, roof	07220	072 203 0100	sf	7,000	1.04	7,280
Roofing	07510	075 102 0200	square	70	160	11,200
Flashing	07620	076 201 5800	lf	140	4.44	622
Ext. downspouts	07630	076 201 2500	lf	70	9.14	640
Ext. gutters	07630	076 201 3400	lf	100	13.31	1,331
Steel doors (3 ft × 7 ft)	08110	081 103 0100	each	8	209	1,672

(Continued)

TABLE 8.1 Summary of Quantities and Costs (*Continued*)

Item	CSI No.	R. S. Means I.D. No.	Unit	No.	R. S. Means 1995	Cost
		Office (*Continued*)				
Wood doors (3 ft × 6 ft 8 in)	08205	082 082 1640	each	36	314	11,304
Entrance doors	08410	084 105 1000	sf	42	21.82	916
Hardware	08710	087 125 0030	each	45	262	11,790
Glass windows	08920	089 204 5100	sf	900	23.17	20,853
Acoustical suspension system	09130	091 304 0050	sf	12,300	0.92	11,316
Drywall partitions, metal stud, 2 sides	09260	092 608 0350	sf	17,800	1.05	18,690
Metal studs	09260	092 612 2200	sf	8,900	1.19	10,591
Drywall on block (drywall only)	09260	092 608 0150	sf	7,950	0.59	46,905
Ceramic tile	09310	093 102 0010	sf	14.85	13,365	
Acoustic tile	09510	095 106 0810	sf	12,300	2.46	30,258
Resilient tile	09660	096 601 0100	sf	3,825	1.98	7,574
Carpet	09685	096 852 1100	sy	875	38	33,250
Exterior paint	09910	099 106 2400	sf	42	0.54	23
Interior paint, masonry	09920	099 106 2400	sf	10,640	0.54	5,746
Interior paint, drywall	09920	099 106 2400	sf	51,500	0.54	27,810
Interior paint, doors	09920	099 204 2500	sf	3,444	0.91	3,134
Metal toilet partitions	10160	101 602 0400	each	12	1,146	13,752
Lockers	10505	105 054 0300	each	20	84	1,680
Elevator	14240	142 011 1000	each	1	51,200	51,200

author). Figure 8.1 shows how the R. S. Means identification number is developed. The first three digits are always the first three digits of the CSI number. The balance may be the same or similar. The R.S. Means identification number is very specific and allows retrieval of specific cost data from the R.S. Means database. Figures 8.2 and 8.3 calculate the costs of the installed bridge crane and monorails (respectively). These values have been inserted in Table 8.1.

Figure 8.4 has schematic costs from Chap. 3. The design development phase does not define the mechanical (Division 15) and electrical (Division 16) areas. For the design development estimate, an *average* of the Composer Gold and I.C.E. estimates will be used (from Fig. 8.4):

Line Number Determination
Each line item is identified by a unique ten-digit number.

MASTERFORMAT
Division

061 104 0200

Subdivision

MASTERFORMAT
Mediumscope

061 100
061 **104** 0200

Major Classification

061 104 **0200**

Individual Line Number

Description
The meaning of this line shows wood let-in bracing with 1" x 6" boards will be installed by an F-1 crew at a rate of .035 man-hours per linear foot. Stud spacing is 24" on center.

061 | Rough Carpentry

	061 100 | Wood Framing		CREW	DAILY OUTPUT	MAN-HOURS	UNIT	BARE COSTS MAT.	LABOR	EQUIP.	TOTAL	TOTAL INCL O&P	
102	0011	BLOCKING										102
	2600	Miscellaneous, to wood construction										
	2620	2" x 4"	F-1	.17	47.080	M.B.F.	400	1,025	47	1,472	2,175	
	2660	2" x 8"	"	.27	29.630	"	405	650	30	1,085	1,550	
	2720	To steel construction										
	2740	2" x 4"	F-1	.14	57.140	M.B.F.	400	1,250	57	1,707	2,550	
	2780	2" x 8"	"	.21	38.100	"	405	840	38	1,283	1,850	
104	0010	BRACING Let-in, with 1" x 6" boards, studs @ 16" O.C.	F-1	150	.053	L.F.	.28	1.17	.05	1.48	2.27	104
	0200	Studs @ 24" O.C.		230	.035	"	.26	.77	.03	1.06	1.58	
106	0010	BRIDGING Wood, for joists 16" O.C., 1" x 3"		130	.062	Pr.	.31	1.35	.06	1.72	2.62	106
	0100	2" x 3" bridging		130	.062		.57	1.35	.06	1.98	2.91	
	0300	Steel, galvanized, 18 ga., for 2" x 10" joists at 12" O.C.	1 Carp	130	.062		.78	1.35		2.13	3.07	
	0400	24" O.C.		140	.057		1.19	1.26		2.45	3.36	
	0900	Compression type, 16" O.C., 2" x 8" joists		200	.040		1.15	.88		2.03	2.70	
	1000	2" x 12" joists		200	.040		1.23	.88		2.11	2.79	
108	0010	FRAMING, LIGHT Average for all light framing	F-2	1.05	15.240	M.B.F.	410	335	15.25	760.25	1,025	108
	2300	Joists, fir, 2" x 4"		.85	18.820		390	415	18.80	823.80	1,125	
	3300	Mud sills, redwood, construction grade, 2" x 4"		.80	28.670		1,000	585	27	1,612	2,100	
	3400	Nailers, treated, 2" x 4" to 2" x 8" wood construction		.75	21.330		540	470	21	1,031	1,375	
	4300	Post, columns & girts, 4" x 4"		.52	30.770		775	675	31	1,481	2,000	
	5400	Roof cants, split 4" x 4"		6.50	2.460	C.L.F.	100	54	2.46	156.46	200	
	5500	Split 6" x 6"		6	2.670	"	220	59	2.67	281.67	340	
	6100	Rough bucks, treated, for doors or windows, 2" x 6"		400	.040	L.F.	.55	.88	.04	1.47	2.09	
	6500	Sills, 4" x 4", treated		.78	20.510	M.B.F.	840	450	21	1,311	1,875	
	6600	Sleepers on concrete, treated, 1" x 2"		2,350	.007	L.F.	.09	.15	.01	.25	.35	
	7300	Stair stringers, fir, 2" x 10"		.30	53.330	M.B.F.	510	1,175	53	1,738	2,525	
110	0010	FRAMING, BEAMS & GIRDERS										110
	1000	Single, 2" x 6"	F-2	700	.023	L.F.	.39	.50	.02	.91	1.28	
	1140	3" x 12" (86)		450	.036		1.80	.78	.04	2.62	3.30	
	1160	3" x 14"		400	.040		2.23	.88	.04	3.15	3.93	
	1180	4" x 8"	F-3	1,000	.040		1.65	.89	.38	2.92	3.68	
	1200	4" x 10"	"	950	.042		2.12	.94	.40	3.46	4.29	

Figure 8.1 Unit Price Line Numbers (R. S. Means Co., Inc., Kingston, MA).

Bridge Crane		
Crane Cost (1995) (Fig. 6.32/6.33)		$360,000
Erection Crew (1995)		
• Foreman	1 @ 32 MH @ $40.92	1,309
• Crew	6 @ 32 MH @ $39.42	7,569
Crane		
• Equipment	2 days @ $1,538	3,076
• Operating Engineer	16 MH @ $35.37	566
• Oiler	16 MH @ $29.50	472
Electrical Material	$30,000	30,000
• Foreman	1 @ 80 MH @ $39.38	3,150
• Crew	4 @ 80 MH @ $37.88	12,122
	TOTAL	$418,264

Figure 8.2 Bridge crane, installed.

Monorail Hoists		
Two 5 Ton Electrical Hoists (1995) (Fig. 6.34/6.35) Includes 20 feet of track and 2 switches (each)	Each $84,000 x 2	$168,000
Erection Crew (1995)		
• Foreman	1 @ 64 MH @ $40.92	2,619
• Crew	6 @ 64 MH @ $39.42	15,137
Electrical Hookup (1995)		
• Foreman	1 @ 80 MH @ $39.38	3,158
• Crew	4 @ 80 MH @ $37.88	12,122
	TOTAL	$201,036

Figure 8.3 Monorails, installed.

	Mechanical	Electrical
Plant	$419,910	$348,754
Warehouse	101,515	42,062
Office	71,542	40,892

In summary, the manual costing of the WHATCO project follows: Base construction costs/(1995) are:

Schematic Costs (From Chapter 3)

Composer Gold (includes General Conditions)		ICE (includes 10% Contingency and Escalation to 1995)	
To computed results, add:			
Overhead	8%	General Conditions	8%
Profit	7%	Home Office OH	5%
Bond	1.5%	Bond	1.5%
Escalation (to 1995)	15%	Profit	7%
Contingency	10%		21.5%
	41.5%		

	Base	With Markup	Base	With Markup	Average With Markups
Mechanical					
Plant	429,100	607,177	191,475	232,642	419,910
Warehouse	79,400	112,351	74,880	90,979	101,510
Office	33,800	47,827	78,400	95,256	71,542
Electrical					
Plant	325,500	460,583	195,000	236,925	348,754
Warehouse	30,600	43,299	33,600	40,824	42,062
Office	26,200	37,073	36,800	44,712	40,892

Figure 8.4 Schematic costs.

	Plant		Warehouse		Office
	$2,665,384		$ 705,489		$ 846,943
GC/HOE/bond/profit at 21.5%	573,058		151,680		182,093
Contingency at 5%	133,269		35,274		42,347
Subtotal Divisions 1 to 14	$3,371,711		$ 892,443		$1,071,383
Divisions 15, 16 (10% contingency)	768,664		143,572		112,434
	4,140,375	+	1,036,015	+	1,183,817
Total (not including site work) $6,360,207					

Fred Seidell III, CCC, of Building Systems Design using Composer Gold made a design development estimate of the WHATCO project.

The results follow (pp. 224–269):

	No. of pages (full size)
Title page	1
Basis of estimate	1
Table of contents	1
Summary of reports	5
Detailed estimate	17

Kevitt Adler, president of MC^2 (Management Computer Controls, Inc.) provided the following design development estimate using their I.C.E. program. The estimate includes (pp. 270–317):

	No. of pages (full sizd)
Summary	1
Section 01—Plant Summary	1
Section 01—Plant Estimate	4
Section 02—Warehouse Summary	1
Section 02—Warehouse Estimate	3
Section 03—Office Summary	1
Section 03—Office Estimate	5
Sort by Item Code No.	1
Quantity Survey Report	16

Phillip R. Waier, Cheif Editor of the R. S. Means Facilities Cost Data, prepared a design development estimate.

TABLE 8.2

	Manual	Composer Gold	R. S. Means	I.C.E.
Plant	$4,140,375	$3,884,963	$4,215,381	$3,917,622
Warehouse	1,036,015	1,030,734	982,078	1,028,742
Office	1,183,817	1,185,643	1,249,303	1,108,215
Total	$6,360,207	$6,101,340	$6,446,762	$6,054,579

Summary

The design development estimates are compared in Table 8.2. Comparing the Composer Gold (B.S.D.), I.C.E. (MC^2), and R. S. Means estimates to the manual, the variances are shown in Table 8.3. As noted in the schematic estimate summary, the design development estimates are in a much tighter range.

Please note tha the use of the author's manual estimate as a baseline (i.e., 100%) does not suggest that it is the most accurate estimate.

TABLE 8.3

	Manual	Composer Gold	R.S. Means	I.C.E.
Plant	100%	94%	102%	95%
Warehouse	100%	99.5%	95%	99%
Office	100%	100.2%	105.5%	94%
Total	100%	100.6%	101.4%	95%

O'Brien - Kreitzburg & Associates, Inc.
PROJECT WHATCO: Whatco Manufacturing-Whse-Office - Demonstration Project
Design Development Estimate TITLE PAGE 1

Whatco Manufacturing-Whse-Office
Demonstration Project

Designed By: James J. O'Brien, P.E.
Estimated By: Building Systems Design

Prepared By: Fred M. Seidell III, C.C.C.
Senior Technical Consultant

Date: 01/04/93
Est Construction Time: 360 Days

O'Brien · Kreitzburg & Associates, Inc.
PROJECT WHATCO: Whatco Manufacturing-Whse-Office - Demonstration Project
PROJECT NOTES Design Development Estimate TITLE PAGE 2

BASIS OF ESTIMATE

This estimate was produced from Design Development information received from James J O'Brien, P.E. on December 16th, 1992.

Sketch Plans & Designs	: Dated	"NOT DATED"
Original Specifications	: Dated	"NOT DATED"
Sitework Budgeting	: Dated	October 92 information

The estimate has been compliled using the 1993 R.S.Means Facilities Data and Davis Bacon wage rates for Philadelphia, Pennsylvania.

BASIS FOR PRICING

Pricing shown reflects probable construction costs obtainable in the Southeastern Pennsylvania area on the date of this statement of costs. This estimate is a determination of fair market value for the construction of this project. It is not a prediction of low bid. Pricing assumes competitive bidding for every portion of the construction work for all subcontractors, as well as the general contractor; that is to mean 6 to 7 bids. If less bids are received, bid results can be expected to be higher.

Length of construction is assumed to be 12 months. Any costs for excessive overtime to meet stringent milestone dates are not included in this estimate.

Bid date is assumed to be April 1995. A value of 5% per annual escalation is added to the cost for the construction which is assumed to be completed March 1996.

The General Contractor's Overhead is set at 8% and his Profit margins are set at 7% on all of his work, and 5% on all of his Subcontractors. Bond is set at 1.5%. The Subcontractors have been properly teired for this project and this is shown in the Indirect Cost Summary Report.

An Owners'10% Design and Pricing Contingency has been included.

PROJECT DESCRIPTION

This project consists of : The construction of a Plant Office Warehouse on Property in Philadelphia, Pennsylvania.

STATEMENT OF PROBABLE COST

Building Systems Design has no control over the cost of labor and materials, the general contractor's or any subcontractor's method of determining prices, or competitive bidding and market conditions. This opinion of probable cost of construction is made on the basis of experience, qualifications, and best judgement of Building Systems Design estimators familar with the construction industry. We cannot and do not guarantee that proposals, bids or actual construction costs will not vary from this or subsequent cost estimates.

O'Brien - Kreitzburg & Associates, Inc.

PROJECT WHATCO: Whatco Manufacturing-Whse-Office - Demonstration Project

TABLE OF CONTENTS Design Development Estimate CONTENTS PAGE 1

SUMMARY REPORTS SUMMARY PAGE

DETAILED ESTIMATE DETAIL PAGE

No Backup Reports...

* * * END TABLE OF CONTENTS * * *

O'Brien - Kreitzburg & Associates, Inc.

PROJECT WHATCO: Whatco Manufacturing-Whse-Office - Demonstration Project

Design Development Estimate

SUMMARY PAGE 1

** PROJECT DIRECT SUMMARY - LEVEL 1 (Rounded to 100's) **

		QUANTITY UOM	Labor	Equipmnt	Material	SubContr	TOTAL COST	UNIT COST
10	Plant Building	30000.00 SF	549,300	112,800	1,426,700	577,000	2,665,900	88.86
20	Warehouse Buildings "A" & "B"	6000.00 SF	158,700	23,500	417,500	107,700	707,300	117.89
30	Office Building	7000.00 SF	294,700	26,300	408,300	84,300	813,600	116.23
40	Site Support	6.00 ACR	20,300	13,000	31,000	140,800	205,100	34183.48
	Whatco Manufacturing-Whse-Office	1.00 EA	1,023,000	175,700	2,283,500	909,800	4,391,900	4391932
	Overhead @ 8.0 %						351,400	
	SUBTOTAL						4,743,300	
	Profit @ 7.0 %						238,100	
	SUBTOTAL						4,981,300	
	Bond @ 1.5 %						74,700	
	TOTAL INCL INDIRECTS						5,056,100	
	Escalation @ 5.0 % / Annum						758,400	
	SUBTOTAL						5,814,500	
	Contingency @ 10.0 %						581,400	
	TOTAL INCL OWNER COSTS						6,395,900	

O'Brien - Kreitzburg & Associates, Inc.

PROJECT WHATCO: Whatco Manufacturing-Whse-Office - Demonstration Project

Design Development Estimate — SUMMARY PAGE 3

** PROJECT DIRECT SUMMARY - LEVEL 2 (Rounded to 100's) **

	QUANTITY	UOM	Labor	Equipmnt	Material	SubContr	TOTAL COST	UNIT COST
Site Support	6.00	ACR	20,300	13,000	31,000	140,800	205,100	34183.48
Whatco Manufacturing-Whse-Office	1.00	EA	1,023,000	175,700	2,283,500	909,800	4,391,900	4391932
Overhead @ 8.0 %							351,400	
SUBTOTAL							4,743,300	
Profit @ 7.0 %							238,100	
SUBTOTAL							4,981,300	
Bond @ 1.5 %							74,700	
TOTAL INCL INDIRECTS							5,056,100	
Escalation @ 5.0 % / Annum							758,400	
SUBTOTAL							5,814,500	
Contingency @ 10.0 %							581,400	
TOTAL INCL OWNER COSTS							6,395,900	

O'Brien - Kreitzburg & Associates, Inc.

PROJECT WHATCO: Whatco Manufacturing-Whse-Office - Demonstration Project

Design Development Estimate

SUMMARY PAGE 4

** DIVISION OWNER SUMMARY (Rounded to 100's) **

	CONTRACT COST	Escalatn	Contincy	TOTAL COST
01 General Requirements	3,000	400	300	3,700
02 Site Work	330,900	49,600	38,100	418,600
03 Concrete	378,600	56,800	43,500	478,900
04 Masonry	133,900	20,100	15,400	169,300
05 Metals	1,513,600	227,000	174,100	1,914,700
07 Moisture-Thermal Control	504,500	75,700	58,000	638,200
08 Doors, Windows, & Glass	201,600	30,200	23,200	255,000
09 Finishes	163,600	24,500	18,800	207,000
10 Specialties	13,600	2,000	1,600	17,200
11 Equipment	58,900	8,800	6,800	74,500
14 Conveying Systems	783,000	117,400	90,000	990,500
15 Mechanical	512,000	76,800	58,900	647,600
16 Electrical	459,200	68,900	52,800	580,800
	5,056,100	758,400	581,400	6,395,900

O'Brien - Kreitzburg & Associates, Inc.

PROJECT WHATCO: Whatco Manufacturing-Whse-Office - Demonstration Project

Design Development Estimate

** CONTRACTOR INDIRECT SUMMARY (Rounded to 100's) **

	TOTAL DIRECT	Overhead	Profit	Bond	TOTAL COST	UNIT COST
GM General Markup						
AT Acoustical Contractor	31,700	1,900	2,700	0	36,300	36323.45
CA Carpet & Resilent Contractor	27,200	1,600	1,700	0	30,600	30595.45
CW Concrete Subcontractor						
WP Water Proofing Subcontractor	1,400	100	200	0	1,700	1686.22
Subtotal Subcontract Work	1,400	100	200	0	1,700	1686.22
Indirect on Subcontracts	1,700	100	100	0	1,900	1930.39
Indirect on Own Work	289,000	23,100	18,700	0	330,900	330896.50
CW Concrete Subcontractor	290,700	23,300	18,800	0	332,800	332826.89
EL Electrical Subcontractor	363,200	0	0	0	363,200	363160.18
EV Elevator & Coveying Subcontractr	51,300	5,100	4,500	0	60,900	60941.34
GL Glazing Subcontractr	18,200	1,500	1,400	0	21,000	21042.87
GW Gypsum Subcontractr	20,200	1,200	1,300	0	22,700	22720.74
HT Hard Tile Subcontractr	5,200	300	300	0	5,900	5876.11
MA Masonry Subcontractr	107,300	8,600	7,000	0	122,900	122861.10
ME Mechanical Subcontractr	549,800	0	0	0	549,800	549794.00
MR Membrane Roofing Subcontractr	112,700	13,500	10,100	0	136,300	136342.04
MW Metal Wall Panel Subcontractr	249,800	22,500	21,800	0	294,100	294069.27

PS Painting Subcontractor	173,700	9,600	11,900	0	195,200	195220.38
SD Speciality Subcontractor						
CR Crane & Monorail Supplier	619,300	0	0	0	619,300	619300.25
Subtotal Subcontract Work	619,300	0	0	0	619,300	619300.25
Indirect on Subcontracts	619,300	0	0	0	619,300	619300.25
Indirect on Own Work	144,400	14,400	15,900	0	174,800	174750.68
SD Speciality Subcontractor	763,700	14,400	15,900	0	794,100	794050.93
SS Structural Steel Subcontractor	1,018,800	73,400	74,300	0	1,166,400	1166413
SW Sitework Subcontractor						
PL Piling Subcontractor	57,000	5,100	5,000	0	67,100	67110.46
Subtotal Subcontract Work	57,000	5,100	5,000	0	67,100	67110.46
Indirect on Subcontracts	67,100	6,700	7,400	0	81,200	81203.65
Indirect on Own Work	113,300	11,300	12,500	0	137,000	137040.21
SW Sitework Subcontractor	180,400	18,000	19,800	0	218,200	218243.87
Subtotal Subcontract Work	3,964,100	194,900	191,500	0	4,350,500	4350481
Indirect on Subcontracts	4,350,500	348,000	234,900	74,000	5,007,400	5007448
Indirect on Own Work	41,500	3,300	3,100	700	48,600	48618.97
GM General Markup	4,391,900	351,400	238,100	74,700	5,056,100	5056067

O'Brien - Kreitzburg & Associates, Inc.

PROJECT WHATCO: Whatco Manufacturing-Whse-Office - Demonstration Project

DETAILED ESTIMATE — Design Development Estimate — DETAIL PAGE 1

10. Plant Building

10.01. Substructure	QUANTY UOM	Labor	Equipmnt	Material	SubContr	TOTAL COST

10. Plant Building

The Plant Building is 300' X 100' rectangle area of 30,000 Gross Square Feet.

SUBSTRUCTURE AND STRUCTURE:

One Story Steel Frame Structure With Heavy Wide Flange Columns On Concrete Foundations Supported by Driven Piles. This system supports The Roof Structure and the overhead Bridge Crane. The Exterior Walls Are Supported By Grade Beams And Spread Footings. The Slab On Grade is a 12 inch Thicknesses. A Monorail Crane Runway Structure, and A Bridge Crane Runway Structure are also provided for.

ROOFING

The Roofing is a Built-up Roofing Over Metal Decking. Rigid Insulation Covers The Entire Roof Area and Roof sloping is accomplished with the insulation.
Gutters And Downspouts Are Used To Channel Rain Water Off The Roof Area.

EXTERIOR WALL

Consists Of Structural Steel and Girt System. Insulated Metal Panels Are Used As A Veneer. The Soffit And Fascia Consists Of Insulated Metal Panels On Metal Studs. Hollow Metal Doors Are Used To Enter And Exit The Facility.

INTERIOR

There are no Interior Partitions.

SPECIALITIES

Includes Dock Levelers, And a Bridge Crane.

MECHANICAL

The Mechanical is a Subcontractor Price.

ELECTRICAL

The Electrical is a Subcontractor Price.

O'Brien : Kreitzburg & Associates, Inc.
PROJECT WHATCO: Whatco Manufacturing-Whse-Office - Demonstration Project
DETAILED ESTIMATE
Design Development Estimate
10. Plant Building

10.01. Substructure	QUANTY UOM	Labor	Equipmnt	Material	SubContr	TOTAL COST
10.01. Substructure						
RSM SW Site Preparation		1954.19	1497.99	0.00	0.00	3452.17
	0.70 ACR	1,655	1,269	0	0	2,924
RSM SW Borrow; Loaded At Pit, No Haul		0.00	0.00	2.59	0.00	2.59
	1672.00 TON	0	0	5,240	0	5,240
RSM SW Spread W/200Hp Dozer, Bank Yards		1.33	2.68	0.00	0.00	4.01
	1672.00 CY	2,698	5,421	0	0	8,119
RSM SW Borrow Delivery Charge, Min 12		2.55	4.62	0.00	0.00	7.17
	1672.00 CY	5,162	9,340	0	0	14,502
RSM SW Compaction, Riding Vibrating		0.29	0.24	0.00	0.00	0.53
	1672.00 CY	578	492	0	0	1,070
RSM SW Excavating Bulk - Footprint		0.67	3.04	0.00	0.00	3.72
	3345.00 CY	2,724	12,322	0	0	15,046
RSM SW Excavating, Structural, Hand,		54.03	0.00	0.00	0.00	54.03
	86.00 CY	5,622	0	0	0	5,622

Description	Quantity					
RSM SW Excav Grade Beam, Common Earth,		3.20	1.34	0.00	0.00	4.54
	223.00 CY	863	361	0	0	1,224
RSM SW Spread Fill W/ Ldr, Crwlr, 300'		0.62	1.27	0.00	0.00	1.89
	112.00 CY	84	173	0	0	256
RSM PL Piles Steel No Mobil/Demobilize		4.30	2.46	21.00	0.00	27.76
	1760.00 VLF	10,782	6,167	52,646	0	69,596
RSM PL Mobilization, Rule Of Thumb		5184.18	2965.18	0.00	0.00	8149.36
	1.00 EA	7,384	4,224	0	0	11,608
RSM CW Reinforcing - Footings,Pile Caps		0.32	0.00	0.23	0.00	0.55
	6056.00 LB	2,230	0	1,595	0	3,825
RSM CW Reinforcing - Grade Beam		0.19	0.00	0.25	0.00	0.44
	6144.00 LB	1,320	0	1,758	0	3,079
RSM WP Bituminous Dampproof Grade Beam		0.34	0.00	0.15	0.00	0.49
	1805.00 SF	860	0	375	0	1,235
RSM CW Insulation Grade Beam,		0.41	0.00	0.20	0.00	0.61
	1805.00 SF	849	0	413	0	1,262
Substructure		42,814	39,768	62,027	0	144,609

O'Brien - Kreitzburg & Associates, Inc.
PROJECT WHATCO: Whatco Manufacturing-Whse-Office - Demonstration Project
DETAILED ESTIMATE — Design Development Estimate — DETAIL PAGE 3
10. Plant Building

10.02. Structural Frame	QUANTY UOM	Labor	Equipmnt	Material	SubContr	TOTAL COST
10.02. Structural Frame						
RSM CW Reinforcing In Place Slab On		0.32	0.00	0.25	0.00	0.57
	37328 LB	13,747	0	10,683	0	24,431
RSM CW Conc. In Place, Load Deck		0.62	0.20	0.70	0.00	1.52
	4000.00 SF	2,837	925	3,205	0	6,968
RSM CW Concrete In Place,Spread Footing		36.40	5.44	64.00	0.00	105.83
	86.00 CY	3,583	535	6,301	0	10,419
RSM CW Concrete In Place, Grade Beam		184.51	9.33	138.00	0.00	331.84
	111.00 CY	23,446	1,185	17,536	0	42,168
RSM CW Conc In Place, Ground Slab, 12"		0.58	0.19	1.94	0.00	2.71
	30000 SF	19,868	6,481	66,627	0	92,976
RSM CW Insul Lightweight Fill 3" Thick		0.22	0.07	0.75	0.00	1.04
	30000 SF	7,418	2,421	25,758	0	35,598
RSM SS Structural Steel Base Plates		0.34	0.00	0.49	0.00	0.83
	3708.00 LB	1,434	0	2,080	0	3,514

RSM SS Structural Steel Beams Light		205.65	92.90	1300.00	0.00	1598.55
	6.10 TON	1,436	649	9,079	0	11,164
RSM SS Structural Steel Beams Heavy		188.08	84.96	1125.00	0.00	1398.04
	323.00 TON	69,551	31,419	416,027	0	516,997
RSM PS Sand Blast		0.43	0.06	0.14	0.00	0.63
	47632 SF	22,965	3,270	7,493	0	33,727
RSM PS Paints & Protect Coats, Alkyds,		0.14	0.01	0.06	0.00	0.21
	47632 SF	7,546	407	3,211	0	11,164
RSM PS Paints & Protect Coats, Epoxy,		0.18	0.01	0.14	0.00	0.33
	47632 SF	9,703	524	7,493	0	17,720
RSM PS Top Coat (Elevated)		0.28	0.02	0.18	0.00	0.48
	47632 SF	15,092	813	9,633	0	25,539
RSM SS Open Web Joists, H Series,		257.55	95.11	750.00	0.00	1102.66
	120.00 TON	35,385	13,066	103,041	0	151,492
RSM SS Metal Deck Galvanized 2" 20-20Ga		0.94	0.05	2.40	0.00	3.39
	31000 SF	33,234	1,785	85,180	0	120,200
Structural Frame		267,246	63,482	773,347	0	1,104,076

O'Brien - Kreitzburg & Associates, Inc.

PROJECT WHATCO: Whatco Manufacturing-Whse-Office - Demonstration Project

DETAILED ESTIMATE

Design Development Estimate

10. Plant Building

10.03. Roofing	QUANTY UOM	Labor	Equipmnt	Material	SubContr	TOTAL COST
10.03. Roofing						
RSM MR Roof Deck Insulation, Fiberboard		0.36	0.00	0.60	0.00	0.96
	30000 SF	12,951	0	21,773	0	34,724
RSM MR Built Up Roofing		91.56	6.75	33.00	0.00	131.31
	300.00 SQ	33,226	2,448	11,975	0	47,649
RSM MR Downspout, Galvanized, 3" X 4"		2.16	0.00	1.25	0.00	3.41
	800.00 LF	2,092	0	1,210	0	3,302
RSM MR Flashing		2.11	0.00	3.15	0.00	5.26
	800.00 LF	2,046	0	3,048	0	5,095
RSM MR Gutter, 5" X 6"		5.22	0.00	2.50	0.00	7.72
	400.00 LF	2,528	0	1,210	0	3,737
Roofing		52,844	2,448	39,215	0	94,507

10.04. Exterior Closure

Item	Quantity					
RSM MW MFG Wall Panels-Galv. Steel Int.		5.43	0.13	4.85	0.00	10.41
	18000 SF	115,096	2,687	102,770	0	220,552
RSM GM Steel Doors & Frame 3' X 7'		32.87	0.97	146.00	0.00	179.84
	5.00 EA	164	5	730	0	899
RSM SD Vert Lift Drs,To 30'X18',Mtr Op		1949.79	276.67	14800.00	0.00	17026.46
	4.00 EA	9,437	1,339	71,632	0	82,408
RSM GM Door Hardware		31.04	0.00	263.00	0.00	294.04
	5.00 EA	155	0	1,315	0	1,470
RSM PS Painting Exterior Siding		0.29	0.00	0.14	0.00	0.43
	4516.00 SF	1,450	0	710	0	2,160
		----------	----------	----------	----------	----------
Exterior Closure		126,302	4,031	177,157	0	307,489

10.06. Interior

Item	Quantity					
RSM PS Interior Painting		0.29	0.00	0.14	0.00	0.43
	6000.00 SF	1,926	0	944	0	2,870
		----------	----------	----------	----------	----------
Interior		1,926	0	944	0	2,870

O'Brien - Kreitzburg & Associates, Inc.

PROJECT WHATCO: Whatco Manufacturing-Whse-Office - Demonstration Project

DETAILED ESTIMATE — Design Development Estimate — DETAIL PAGE 5

10. Plant Building

10.07. Specialties	QUANTY UOM	Labor	Equipmnt	Material	SubContr	TOTAL COST
10.07. Specialties						
RSM SD Dock Platform Levelers, 7'X 8',		553.94	68.00	2900.00	0.00	3521.94
	4.00 EA	2,681	329	14,036	0	17,046
Specialties		2,681	329	14,036	0	17,046
10.10. Mechanical Systems						
USR ME Mechanical		0.00	0.00	0.00	315000.00	315000.00
	1.00 EA	0	0	0	315,000	315,000
Mechanical Systems		0	0	0	315,000	315,000

10.11. Electrical

USR EL Electrical		0.00	0.00	0.00	262000.00	262000.00
	1.00 EA	0	0	0	262,000	262,000
Electrical		0	0	0	262,000	262,000

10.13. Equipment & Conveying

B RSM CR Crane, Bridge No Equip		55488.00	2776.25	360000.00	0.00	418264.25
	1.00 EA	55,488	2,776	360,000	0	418,264

*** MATERIAL BACKUP FOR 14605 QUOTE Crane, Bridge No Equip ***

ID	Name	City/State	Date	Telephone	Quote
BRIDGEMAN1	Bridge Crane Quote Company	Annapolis Maryland	12/28/92	(212)456-1234	360000

Equipment & Conveying	55,488	2,776	360,000	0	418,264
Plant Building	549,300	112,834	1,426,727	577,000	2,665,861

O'Brien - Kreitzburg & Associates, Inc.
PROJECT WHATCO: Whatco Manufacturing-Whse-Office - Demonstration Project
DETAILED ESTIMATE Design Development Estimate DETAIL PAGE 6
20. Warehouse Buildings "A" & "B"

20.01. Substructure	QUANTY UOM	Labor	Equipmnt	Material	SubContr	TOTAL COST

20. Warehouse Buildings "A" & "B"
This represents two 50' X 60' warehouses flanking the office complex on both sides. The Gross Floor Area is 6000 S. F.

SUBSTRUCTURE and SUPERSTRUCTURE

These Two Buildings Are Constructed Of Steel Framed Roofs Supported By Heavy Steel Frames founded on Spread Footings and Grade Beams.

The Slab On Grade Is Designed Using a 9" Thicknesses.

ROOFING

The Roof Is A Metal Deck on a Heavy Steel Frame. Rigid Insulation and a built-up bitiumnous roof system is also included.
There are Gutters And Downspouts.

EXTERIOR WALL

The Exterior is To Be covered with Insulated Metal Siding (PrePainted). The Exterior Doors Are Both Hollow Metal Doors And 2 Overhead Doors With Motors.

SPECILATIES

The Only Specialty Items Are Dock Levelers and Monorails for owner supplied Lifting devices.

MECHANICAL

The Mechanical Price is a Subcontractor value.

ELECTRICAL

The Electrical Price is a Subcontractor value.

This information was infered from the information received.

O'Brien - Kreitzburg & Associates, Inc.
PROJECT WHATCO: Whatco Manufacturing-Whse-Office - Demonstration Project
DETAILED ESTIMATE Design Development Estimate DETAIL PAGE 7
20. Warehouse Buildings "A" & "B"

20.01. Substructure	QUANTY UOM	Labor	Equipmnt	Material	SubContr	TOTAL COST
20.01. Substructure						
RSM SW Site Preparation		1954.19	1497.99	0.00	0.00	3452.17
	0.14 ACR	331	254	0	0	585
RSM SW Borrow; Loaded At Pit, No Haul		0.00	0.00	2.59	0.00	2.59
	334.00 CY	0	0	1,047	0	1,047
RSM SW Spread W/200Hp Dozer, Bank Yards		1.33	2.68	0.00	0.00	4.01
	334.00 CY	539	1,083	0	0	1,622
RSM SW Borrow Delivery Charge, Min 12		2.55	4.62	0.00	0.00	7.17
	334.00 CY	1,031	1,866	0	0	2,897
RSM SW Compaction, Riding Vibrating		0.29	0.24	0.00	0.00	0.53
	334.00 CY	116	98	0	0	214
RSM SW Excavating Bulk - Footprint		0.67	3.04	0.00	0.00	3.72
	669.00 CY	545	2,464	0	0	3,009
RSM SW Excavating, Structural, Hand,		54.03	0.00	0.00	0.00	54.03
	9.00 CY	588	0	0	0	588

Item	Quantity					
RSM SW Excav Grade Beam, Common Earth,		3.20	1.34	0.00	0.00	4.54
	77.00 CY	298	125	0	0	423
RSM SW Spread Fill W/ Ldr, Crwlr, 300'		0.62	1.27	0.00	0.00	1.89
	39.00 CY	29	60	0	0	89
RSM CW Reinforcing - Footings		0.32	0.00	0.23	0.00	0.55
	1018.00 LB	375	0	268	0	643
RSM CW Reinforcing - Grade Beam		0.19	0.00	0.25	0.00	0.44
	2208.00 LB	474	0	632	0	1,106
RSM WP Bituminous Dampproof Grade Beam		0.34	0.00	0.15	0.00	0.49
	649.00 SF	309	0	135	0	444
RSM CW Insulation Grade Beam,		0.41	0.00	0.20	0.00	0.61
	649.00 SF	305	0	149	0	454
		----------	--------	----------	----------	---------
Substructure		4,941	5,950	2,230	0	13,121

20.02. Structural Frame

Item	Quantity					
RSM CW Reinforcing In Place Slab On		0.32	0.00	0.25	0.00	0.57
	5002.00 LB	1,842	0	1,432	0	3,274
RSM CW Reinforcing In Place, Load Dock		0.62	0.20	0.70	0.00	1.52
	5002.00 SF	3,547	1,157	4,008	0	8,713

O'Brien - Kreitzburg & Associates, Inc.

PROJECT WHATCO: Whatco Manufacturing-Whse-Office - Demonstration Project

DETAILED ESTIMATE

Design Development Estimate

20. Warehouse Buildings "A" & "B"

DETAIL PAGE 8

20.02. Structural Frame	QUANTY UOM	Labor	Equipmnt	Material	SubContr	TOTAL COST
RSM CW Concrete In Place,Spread Footing		36.40	5.44	64.00	0.00	105.83
	9.00 CY	375	56	659	0	1,090
RSM CW Concrete In Place, Grade Beam		184.51	9.33	138.00	0.00	331.84
	40.00 CY	8,449	427	6,319	0	15,196
M RSM CW Conc In Place,Grnd Slab, 9"Thick.		0.51	0.17	1.46	0.00	2.13
	6000.00 SF	3,489	1,138	10,028	0	14,655
RSM CW Insul Lightweight Fill 3" Thick		0.22	0.07	0.75	0.00	1.04
	6000.00 SF	1,484	484	5,152	0	7,120
RSM SS Structural Steel Base Plates		0.34	0.00	0.49	0.00	0.83
	504.00 LB	195	0	283	0	478
RSM SS Structural Steel Beams Light		205.65	92.90	1300.00	0.00	1598.55
	13.50 TON	3,179	1,436	20,093	0	24,707
RSM SS Structural Steel Beams Heavy		188.08	84.96	1125.00	0.00	1398.04
	46.00 TON	9,905	4,475	59,248	0	73,628

RSM PS Sand Blast		0.43	0.06	0.14	0.00	0.63
	15160 SF	7,309	1,041	2,385	0	10,734
RSM PS Paints & Protect Coats, Alkyds,		0.14	0.01	0.06	0.00	0.21
	15160 SF	2,402	129	1,022	0	3,553
RSM PS Paints & Protect Coats, Epoxy,		0.18	0.01	0.14	0.00	0.33
	15160 SF	3,088	167	2,385	0	5,640
RSM PS Top Coat (Elevated)		0.28	0.02	0.18	0.00	0.48
	15160 SF	4,803	259	3,066	0	8,128
RSM SS Open Web Joists, H Series,		257.55	95.11	750.00	0.00	1102.66
	22.00 TON	6,487	2,396	18,891	0	27,773
RSM SS Metal Deck Galvanized 2" 20-20Ga		0.94	0.05	2.40	0.00	3.39
	6400.00 SF	6,861	369	17,586	0	24,815
		----------	----------	----------	----------	----------
Structural Frame		63,415	13,533	152,557	0	229,505
20.03. Roofing						
RSM MR Roof Deck Insulation, Fiberboard		0.36	0.00	0.60	0.00	0.96
	6000.00 SF	2,590	0	4,355	0	6,945
RSM MR Built Up Roofing		91.56	6.75	33.00	0.00	131.31
	60.00 SQ	6,645	490	2,395	0	9,530

O'Brien - Kreitzburg & Associates, Inc.

PROJECT WHATCO: Whatco Manufacturing-Whse-Office - Demonstration Project

DETAILED ESTIMATE — Design Development Estimate — DETAIL PAGE 9

20. Warehouse Buildings "A" & "B"

20.03. Roofing	QUANTY UOM	Labor	Equipmnt	Material	SubContr	TOTAL COST
RSM MR Downspout, Galvanized, 3" X 4"		2.16	0.00	1.25	0.00	3.41
	70.00 LF	183	0	106	0	289
RSM MR Flashing		2.11	0.00	3.15	0.00	5.26
	440.00 LF	1,125	0	1,677	0	2,802
RSM MR Gutter, 5" X 6"		5.22	0.00	2.50	0.00	7.72
	100.00 LF	632	0	302	0	934
Roofing		11,176	490	8,834	0	20,500
20.04. Exterior Closure						
RSM MW MFG Wall Panels-18 Ga Galvanized		5.43	0.13	4.85	0.00	10.41
	6000.00 SF	38,365	896	34,257	0	73,517
RSM GM Steel Doors & Frames 3' X 7'		32.87	0.97	146.00	0.00	179.84
	2.00 EA	66	2	292	0	360

Item	Quantity					
RSM SD Vert Lift Drs, 30'X 18', Mtr Op		1949.79	276.67	14800.00	0.00	17026.46
	2.00 EA	4,719	670	35,816	0	41,204
RSM GM Door Hardware		31.04	0.00	263.00	0.00	294.04
	2.00 EA	62	0	526	0	588
RSM PS Painting Exterior		0.29	0.00	0.14	0.00	0.43
	2216.00 SF	711	0	349	0	1,060
		----------	--------	----------	----------	----------
Exterior Closure		43,923	1,567	71,239	0	116,729
20.06. Interior						
RSM PS Interior Painting		0.29	0.00	0.14	0.00	0.43
	3600.00 SF	1,156	0	566	0	1,722
		----------	--------	----------	----------	----------
Interior		1,156	0	566	0	1,722
20.07. Specialties						
RSM SD Dock Platform Levelers, 7'X 8',		553.94	68.00	2900.00	0.00	3521.94
	4.00 EA	2,681	329	14,036	0	17,046
		----------	--------	----------	----------	----------
Specialties		2,681	329	14,036	0	17,046

O'Brien - Kreitzburg & Associates, Inc.
PROJECT WHATCO: Whatco Manufacturing-Whse-Office - Demonstration Project
DETAILED ESTIMATE — Design Development Estimate — DETAIL PAGE 10
20. Warehouse Buildings "A" & "B"

20.10. Mechanical Systems	QUANTY UOM	Labor	Equipment	Material	SubContr	TOTAL COST
20.10. Mechanical Systems						
USR ME Mechanical		0.00	0.00	0.00	76134.00	76134.00
	1.00 EA	0	0	0	76,134	76,134
Mechanical Systems		0	0	0	76,134	76,134
20.11. Electrical						
USR EL Electrical		0.00	0.00	0.00	31547.00	31547.00
	1.00 EA	0	0	0	31,547	31,547
Electrical		0	0	0	31,547	31,547

O'Brien - Kreitzburg & Associates, Inc.

PROJECT WHATCO: Whatco Manufacturing-Whse-Office - Demonstration Project

DETAILED ESTIMATE | Design Development Estimate | DETAIL PAGE 11

30. Office Building

30.01. Substructure	QUANTY	UOM	Labor	Equipmnt	Material	SubContr	TOTAL COST

30. Office Building

This Building is an office building is a 70 ft x 100 ft square yeilding a gross Square Footage of 7000 sf.

STRUCTURE and SUBSTRUCTURE

The Foundation For This Facility Consists Of Spread Footings And Continous Grade Beams. A Floor Slab Is Poured Monolithically. The Floor Slab Is 6 and 9 inches thick Of 3000Psi Concrete, over a Vapor Barrier The Structural Frame Consists Of Structural Steel Columns and Roof Framing.

ROOFING

Roofing is a Built Up Roof on a Mineral Fiber Insulation Board set on a composite metal deck.

DOORS and WINDOWS

Entrance Exterior Doors Are Double Plate Glass and Hollow Metal Door Units. Exterior Windows Are Non-Operable, With 1 in Insulated Glass.

O'Brien - Kreitzburg & Associates, Inc.

PROJECT WHATCO: Whatco Manufacturing-Whse-Office - Demonstration Project

DETAILED ESTIMATE — Design Development Estimate — DETAIL PAGE 12

30. Office Building

30.01. Substructure	QUANTY UOM	Labor	Equipmnt	Material	SubContr	TOTAL COST
30.01. Substructure						
RSM SW Site Preparation		1954.19	1497.99	0.00	0.00	3452.17
	0.16 ACR	378	290	0	0	668
RSM SW Borrow; Loaded At Pit, No Haul		0.00	0.00	2.59	0.00	2.59
	382.00 CY	0	0	1,197	0	1,197
RSM SW Spread W/200Hp Dozer, Bank Yards		1.33	2.68	0.00	0.00	4.01
	382.00 CY	617	1,238	0	0	1,855
RSM SW Borrow Delivery Charge, Min 12		2.55	4.62	0.00	0.00	7.17
	382.00 CY	1,179	2,134	0	0	3,313
RSM SW Compaction, Riding Vibrating		0.29	0.24	0.00	0.00	0.53
	382.00 CY	132	112	0	0	245
RSM SW Excavating Bulk - Footprint		0.67	3.04	0.00	0.00	3.72
	764.00 CY	622	2,814	0	0	3,437
RSM SW Excavating, Structural, Hand,		54.03	0.00	0.00	0.00	54.03
	6.00 CY	392	0	0	0	392

Description	Quantity					
RSM SW Excav Grade Beam, Common Earth,		3.20	1.34	0.00	0.00	4.54
	45.00 CY	174	73	0	0	247
RSM SW Spread Fill W/ Ldr, Crwlr, 300'		0.62	1.27	0.00	0.00	1.89
	22.00 CY	16	34	0	0	50
RSM CW Reinforcing - Footings		0.32	0.00	0.23	0.00	0.55
	710.00 LB	261	0	187	0	448
RSM CW Reinforcing - Grade Beam		0.19	0.00	0.25	0.00	0.44
	1248.00 LB	268	0	357	0	625
RSM WP Bituminous Dampproof Grade Beam		0.34	0.00	0.15	0.00	0.49
	367.00 SF	175	0	76	0	251
RSM CW Insulation Grade Beam,		0.41	0.00	0.20	0.00	0.61
	367.00 SF	173	0	84	0	257
		-----------	---------	----------	----------	-------
Substructure		4,389	6,696	1,902	0	12,986
30.02. Structural Frame						
RSM CW Reinforcing In Place Slab On		0.32	0.00	0.25	0.00	0.57
	5170.00 LB	1,904	0	1,480	0	3,384
RSM CW Concrete Reinforcing 2nd Floor		0.62	0.20	0.70	0.00	1.52
	10528 SF	7,466	2,436	8,437	0	18,339

O'Brien - Kreitzburg & Associates, Inc.

PROJECT WHATCO: Whatco Manufacturing-Whse-Office - Demonstration Project

DETAILED ESTIMATE — Design Development Estimate — DETAIL PAGE 13

30. Office Building

30.02. Structural Frame	QUANTY UOM	Labor	Equipmnt	Material	SubContr	TOTAL COST
RSM CW Concrete In Place,Spread Footing		36.40	5.44	64.00	0.00	105.83
	6.00 CY	250	37	440	0	727
RSM CW Concrete In Place, Grade Beam		184.51	9.33	138.00	0.00	331.84
	23.00 CY	4,858	246	3,634	0	8,737
M RSM CW Conc In Place,Grnd Slab, 9"Thick		0.51	0.17	1.46	0.00	2.13
	7000.00 SF	4,070	1,328	11,700	0	17,098
RSM CW Insul Lightweight Fill 3" Thick		0.22	0.07	0.75	0.00	1.04
	7000.00 SF	1,731	565	6,010	0	8,306
RSM SS Structural Steel Base Plates		0.34	0.00	0.49	0.00	0.83
	336.00 LB	130	0	188	0	318
RSM SS Structural Steel Beams Heavy		188.08	84.96	1125.00	0.00	1398.04
	79.00 TON	17,011	7,685	101,753	0	126,448
RSM PS Sand Blast		0.43	0.06	0.14	0.00	0.63
	17500 SF	8,437	1,201	2,753	0	12,391

RSM PS Paints & Protect Coats, Alkyds,		0.14	0.01	0.06	0.00	0.21
	17500 SF	2,772	149	1,180	0	4,102
RSM PS Paints & Protect Coats, Epoxy,		0.18	0.01	0.14	0.00	0.33
	17500 SF	3,565	193	2,753	0	6,510
RSM PS Top Coat (Elevated)		0.28	0.02	0.18	0.00	0.48
	17500 SF	5,545	299	3,539	0	9,383
RSM SS Open Web Joists, H Series,		257.55	95.11	750.00	0.00	1102.66
	21.00 TON	6,192	2,287	18,032	0	26,511
RSM SS Metal Deck Galvanized 2" 20-20Ga		0.94	0.05	2.40	0.00	3.39
	7000.00 SF	7,505	403	19,234	0	27,142
		-----------	---------	----------	----------	-----------
Structural Frame		71,437	16,828	181,132	0	269,397
30.03. Roofing						
RSM SS Metal Deck		0.99	0.05	2.50	0.00	3.55
	7000.00 SF	7,969	428	20,036	0	28,432
RSM MR Roof Deck Insulation, Fiberboard		0.36	0.00	0.60	0.00	0.96
	7000.00 SF	3,022	0	5,080	0	8,102
RSM MR Built Up Roofing,Asphalt Base		91.56	6.75	33.00	0.00	131.31
	70.00 SQ	7,753	571	2,794	0	11,118

O'Brien - Kreitzburg & Associates, Inc.

PROJECT WHATCO: Whatco Manufacturing-Whse-Office - Demonstration Project

DETAILED ESTIMATE — Design Development Estimate — DETAIL PAGE 14

30. Office Building

30.03. Roofing	QUANTY UOM	Labor	Equipmnt	Material	SubContr	TOTAL COST
RSM MR Downspout, Steel, Galvanized,		2.16	0.00	1.25	0.00	3.41
	70.00 LF	183	0	106	0	289
RSM MR Flashing		2.11	0.00	3.15	0.00	5.26
	140.00 LF	358	0	533	0	892
RSM MR Gutter, Aluminum, 5" X 6"		5.22	0.00	2.50	0.00	7.72
	100.00 LF	632	0	302	0	934
Roofing		19,916	999	28,852	0	49,767
30.04. Exterior Closure						
RSM MA Scaffolding,Steel Tubular;Rented		47.30	0.00	15.15	0.00	62.45
	36.00 CSF	1,949	0	624	0	2,574
RSM MA Coping Precast Concr Stock 14"		6.39	0.00	12.50	0.00	18.89
	120.00 LF	878	0	1,717	0	2,596
RSM MA Walls, Common 8"X2-2/3"X4" 4"		6.02	0.00	1.79	0.00	7.81
	2700.00 SF	18,613	0	5,533	0	24,145

O'Brien - Kreitzburg & Associates, Inc.
PROJECT WHATCO: Whatco Manufacturing-Whse-Office - Demonstration Project
DETAILED ESTIMATE — Design Development Estimate — DETAIL PAGE 15
30. Office Building

30.04. Exterior Closure	QUANTY UOM	Labor	Equipmnt	Material	SubContr	TOTAL COST
Exterior Closure		97,683	97	72,690	0	170,470
30.06. Interior						
RSM AT Acoustical Suspension System		0.59	0.00	0.15	0.00	0.74
	12300 SF	8,370	0	2,112	0	10,482
RSM GW Gypsum On Block		0.28	0.00	0.15	0.00	0.43
	7950.00 SF	2,496	0	1,340	0	3,836
RSM GW Drywall, Gypsum, 5/8" Thick		0.28	0.00	0.19	0.00	0.47
	17800 SF	5,588	0	3,800	0	9,388
RSM GW Metal Studs 25 Ga. N.L.B.,		0.65	0.00	0.30	0.00	0.95
	8900.00 SF	6,497	0	3,000	0	9,497
RSM HT Ceramic Tile		2.85	0.00	2.96	0.00	5.81
	900.00 SF	2,883	0	2,993	0	5,876

Item	Quantity					Total
RSM AT Acoustical Tile Sus 2'X 4'X 5/8"		0.74	0.00	1.10	0.00	1.84
	12300 SF	10,352	0	15,489	0	25,842
RSM CA Resilient Tile		0.70	0.00	0.85	0.00	1.55
	3825.00 SF	3,002	0	3,653	0	6,655
RSM CA Carpet		4.90	0.00	19.45	0.00	24.35
	875.00 SY	4,819	0	19,122	0	23,941
RSM PS Exterior Paint		0.29	0.00	0.14	0.00	0.43
	42.00 SF	13	0	7	0	20
RSM PS Interior Painting		0.29	0.00	0.14	0.00	0.43
	10640 SF	3,415	0	1,674	0	5,089
RSM PS Interior Paint, Doors		0.60	0.00	0.11	0.00	0.71
	3444.00 SF	2,325	0	426	0	2,751
RSM PS Wall & Ceiling, Conc Drywall Or		0.43	0.00	0.11	0.00	0.54
	51500 SF	24,592	0	6,365	0	30,957
Interior		74,352	0	59,981	0	134,333

30.07. Specialties

Item	Quantity					Total
RSM SS Alum Pipe Railing, 3 Rail,		9.98	0.54	22.00	0.00	32.52
	75.00 LF	857	46	1,889	0	2,792

O'Brien - Kreitzburg & Associates, Inc.
PROJECT WHATCO: Whatco Manufacturing-Whse-Office - Demonstration Project
DETAILED ESTIMATE Design Development Estimate DETAIL PAGE 16
30. Office Building

30.07. Specialties	QUANTY UOM	Labor	Equipmnt	Material	SubContr	TOTAL COST
RSM GM Partitions, Toilet, Ceiling Hung		139.68	0.00	700.00	0.00	839.68
	12.00 EA	1,676	0	8,400	0	10,076
RSM GM Lockers		12.06	0.00	62.50	0.00	74.56
	20.00 OPN	241	0	1,250	0	1,491
RSM SD Dock Platform Levelers, Hinged		553.94	68.00	2900.00	0.00	3521.94
	4.00 EA	2,681	329	14,036	0	17,046
Specialties		5,455	375	25,575	0	31,405
30.10. Mechanical Systems						
USR ME Mechanical		0.00	0.00	0.00	53660.00	53660.00
	1.00 EA	0	0	0	53,660	53,660
Mechanical Systems		0	0	0	53,660	53,660

30.11. Electrical						
USR EL Electrical		0.00	0.00	0.00	30670.00	30670.00
	1.00 EA	0	0	0	30,670	30,670
		----------	--------	----------	----------	----------
Electrical		0	0	0	30,670	30,670
30.13. Equipment & Conveying						
RSM EV Freight Elev,2 Story,Hydraulic		18086.31	1111.11	32100.00	0.00	51297.42
	1.00 EA	21,487	1,320	38,135	0	60,941
		----------	--------	----------	----------	----------
Equipment & Conveying		21,487	1,320	38,135	0	60,941
		----------	--------	----------	----------	----------
Office Building		294,719	26,315	408,266	84,330	813,630

O'Brien - Kreitzburg & Associates, Inc.

PROJECT WHATCO: Whatco Manufacturing-Whse-Office - Demonstration Project

DETAILED ESTIMATE | Design Development Estimate | DETAIL PAGE 17

40. Site Support

40.10. Site Preparation	QUANTY UOM	Labor	Equipmnt	Material	SubContr	TOTAL COST
40. Site Support						
40.10. Site Preparation						
RSM SW Clear Light Trees To 6"Dia Cut &		1360.11	1042.60	0.00	0.00	2402.71
	6.00 ACR	9,874	7,569	0	0	17,444
RSM SW Grub Stumps & Remove Upto 6"Dia		346.11	750.10	0.00	0.00	1096.21
	6.00 ACR	2,513	5,446	0	0	7,958
Site Preparation		12,387	13,015	0	0	25,402

40.20. Site Electrical

Item	Quantity					Total
B RSM EL Electric & Telephone Sitework		39.72	0.00	155.00	0.00	194.72
	200.00 LF	7,943	0	31,000	0	38,943
RSM SW Pole Line Construction		0.00	0.00	0.00	9.85	9.85
	3000.00 LF	0	0	0	35,756	35,756
Site Electrical		7,943	0	31,000	35,756	74,699

40.30. Site Water

Item	Quantity					Total
M RSM ME Water Storage Tank Elevated		0.00	0.00	0.00	0.42	0.42
	250000 GAL	0	0	0	105,000	105,000
Site Water		0	0	0	105,000	105,000
Site Support		20,330	13,015	31,000	140,756	205,101
Whatco Manufacturing-Whse-Office		1,023,027	175,683	2,283,455	909,767	4,391,932

FILE ID - 256A

GENERAL CONTRACTOR SUMMARY OF ESTIMATE

PAGE - 2
DATE -
TIME -

PROJECT JOB NO -
PROJECT NAME - WHATCO PROJECT
PROJECT SIZE - 50,000 SQFT

** SECTION, ITEM CODE SEQUENCE **

MANAGEMENT COMPUTER CONTROLS, INC.
2881 DIRECTORS COVE
MEMPHIS, TENNESSEE 38131

DESCRIPTION	SECTION QUANTITY	UM	LABOR	MATERIAL	EQUIPMENT	SUBCONTRACT	TAX & INS INCLUDED COST/UNIT	TOTAL
SECTION 01 PLANT	30,000	SQFT	720,649	1,206,066	74,158	1,615,447	130.587	3,917,622
SECTION 02 WAREHOUSE	6,000	SQFT	196,168	367,831	13,529	365,932	171.457	1,028,742
SECTION 03 OFFICE	14,000	SQFT	323,085	248,480	17,387	408,311	79.158	1,108,215
***** ESTIMATE TOTAL			1,239,902	1,822,377	105,074	2,389,690	121.092	6,054,579

FILE ID - 256A
PROJECT JOB NO -
PROJECT NAME - WHATCO PROJECT
PROJECT SIZE - 30,000 SQFT
SECTION - 01 PLANT

E S T I M A T E   D E T A I L   R E C A P
** SECTION, ITEM CODE SEQUENCE **

PAGE - 1
DATE -
TIME -

MANAGEMENT COMPUTER CONTROLS, INC.
2881 DIRECTORS COVE
MEMPHIS, TENNESSEE 38131

DESCRIPTION	LABOR	MATERIAL	EQUIPMENT	SUBCONTRACT	TAX & INS INCLUDED COST/SQFT	TOTAL
GENERAL REQUIREMENTS				660,230	22.008	660,230
CONTINGENCY				186,553	6.218	186,553
CLEARING OF SITE	1,213		1,323		.096	2,875
EXCAV, GRADING & BACKFILL	35,096	8,295	19,047		2.432	72,963
PILING & CAISSONS	6,760	67,352	3,249		2.827	84,809
CONCRETE FINISHING	15,522	519			.681	20,421
FORM WORK	22,462	5,292			1.149	34,479
REINFORCING STEEL	18,123	12,467			1.223	36,694
CAST-IN-PLACE CONCRETE	22,925	74,802			3.677	110,320
CEMENTITIOUS DECKS	13,920	16,530			1.190	35,700
STRUCTURAL METALS	362,172	395,652	50,539		31.415	942,456
WATERPROOF & DAMPPROOF	4,820	949			.240	7,188
INSULATION	1,345	597			.079	2,368
PREFORMED ROOFING & SIDING	78,660	119,070			7.653	229,590
ROOFING,SHEETMETAL&ACCESS	57,250	37,069			3.780	113,387
METAL DOORS & FRAMES	457	1,158			.061	1,837
SPECIAL DOORS	5,292	52,920			2.135	64,060
HARDWARE	348	1,378			.065	1,937
PAINTING & WALL COVERING	6,711	1,886			.354	10,632
EQUIPMENT	3,337	13,230			.620	18,593
CONVEYING SYSTEMS	64,236	396,900			17.062	511,866
HVAC SYSTEMS & EQUIPMENT				419,910	13.997	419,910
ELEC SYSTEMS & EQUIPMENT				348,754	11.625	348,754
***** SECTION TOTAL	720,649	1,206,066	74,158	1,615,447	130.587	3,917,622

FILE ID - 256A
PROJECT JOB NO -
PROJECT NAME - WHATCO PROJECT
PROJECT SIZE - 30,000 SQFT
SECTION - 01 PLANT

E S T I M A T E   D E T A I L   R E P O R T
** SECTION, ITEM CODE SEQUENCE **

PAGE - 1
DATE -
TIME -

MANAGEMENT COMPUTER CONTROLS, INC.
2881 DIRECTORS COVE
MEMPHIS, TENNESSEE 38131

REF NO.	S D SC	ELEM	ITEM CODE	DESCRIPTION	UNIT MEAS	QUANTITY	LAB UNIT PRICE	MAT/EQP/ SUB UNIT	TOTAL LABOR	TOTAL MAT/ EQUIP/SUB	TAX & INS INCLUDED TOT UNIT PRICE	TAX & INS INCLUDED TOTAL PRICE
GENERAL REQUIREMENTS												
				OTHER GENERAL REQUIREMENTS								
1	01		185.901	OH/PROFIT/GC/BOND @ 21.5%	LS	1		660,230 S		660,230 S	660,230	660,230
				** TOTAL OTHER GENERAL REQUIREMENTS						660,230 S		660,230
				*** TOTAL GENERAL REQUIREMENTS						660,230 S		660,230
CONTINGENCY												
4	01		199.901	CONTINGENCY @ 5%	LS	1		186,553 S		186,553 S	186,553	186,553
				*** TOTAL CONTINGENCY						186,553 S		186,553
CLEARING OF SITE												
7	01		211.900	SITE PREP (CLEAR)	ACRE	1	1212.750	1323.000 E	1,213	1,323 E	2875.000	2,875
				*** TOTAL CLEARING OF SITE					1,213	1,323 E		2,875
EXCAV, GRADING & BACKFILL												
				SITE GRADING								
8	01		221.011	EXCAV-LOAD-HAUL SURP 3-4MILES	CUYD	3,345	3.027	3.638 E	10,125	12,169 E	7.513	25,131
				** TOTAL SITE GRADING					10,125	12,169 E		25,131
				EXCAVATION & BACKFILL								
13	01	0112	222.302	EXCAVATE PILE CAP	CUYD	84	8.875	1.268 E	746	107 E	12.631	1,061
14	01	0111	222.308	EXCAVATE GRADE BEAM	CUYD	289	8.875	1.103 E	2,565	319 E	12.464	3,602
18	01	0111	222.328	BACKFILL GRADE BEAM	CUYD	196	18.529		3,632		23.719	4,649
23	01	0112	222.332	FINE GRADE PILE CAP	SQFT	900	.529		476		.677	609
24	01	0211	222.334	FINE GRADE SLAB ON GRADE	SQFT	30,000	.370		11,100		.474	14,220
28	01		222.900	STRUCTURAL FILL	CUYD	1,672	3.859	4.961 M 3.859 E	6,452	8,295 M 6,452 E	14.169	23,691
				** TOTAL EXCAVATION & BACKFILL					24,971	8,295 M 6,878 E		47,832

				*** TOTAL EXCAV, GRADING & BACKFILL					35,096	8,295 M 19,047 E		72,963
PILING & CAISSONS												
33	01	0121	231.046	14X14-89LB H-SEC STL PILE	LNFT	1,760	3.841	38.268 M 1.846 E	6,760	67,352 M 3,249 E	48.187	84,809
				*** TOTAL PILING & CAISSONS					6,760	67,352 M 3,249 E		84,809
CONCRETE FINISHING												
				CONCRETE FINISH								
34	01	0211	301.019	TROWEL CEMENT FINISH	SQFT	30,000	.386		11,580		.494	14,820
37	01	0312	301.020	POINT & PATCH	SQFT	800	.128	.011 M	102	9 M	.176	141
39	01	0211	301.050	PROTECT & CURE	SQFT	30,000	.128	.017 M	3,840	510 M	.182	5,460
				** TOTAL CONCRETE FINISH					15,522	519 M		20,421
				*** TOTAL CONCRETE FINISHING					15,522	519 M		20,421
FORM WORK												
				FORM WORK								
42	01	0111	310.025	GRADE BEAM FORMS	SQFT	4,800	4.235	.937 M	20,328	4,498 M	6.435	30,888
46	01	0211	310.105	SLAB ON GRADE EDGE FORMS	SQFT	800	2.668	.992 M	2,134	794 M	4.489	3,591
				** TOTAL FORM WORK					22,462	5,292 M		34,479
				*** TOTAL FORM WORK					22,462	5,292 M		34,479
REINFORCING STEEL												
				RE-BARS								
47	01	0211	321.009	RE-STEEL @ SLAB ON GRADE	CWT	387	35.412	24.255 M	13,704	9,387 M	71.584	27,703
53	01	0112	321.120	RE-STEEL @ PILE CAP	CWT	63	34.798	24.255 M	2,192	1,528 M	70.794	4,460
54	01	0111	321.200	RE-STEEL @ GRADE BEAM	CWT	64	34.798	24.255 M	2,227	1,552 M	70.797	4,531
				** TOTAL RE-BARS					18,123	12,467 M		36,694

FILE ID - 256A
PROJECT JOB NO -
PROJECT NAME - WHATCO PROJECT
PROJECT SIZE - 30,000 SQFT
SECTION - 01 PLANT

E S T I M A T E   D E T A I L   R E P O R T

** SECTION, ITEM CODE SEQUENCE **

PAGE - 2
DATE -
TIME -

MANAGEMENT COMPUTER CONTROLS, INC.
2881 DIRECTORS COVE
MEMPHIS, TENNESSEE 38131

REF NO.	S D SC	ELEM	ITEM CODE	DESCRIPTION	UNIT MEAS	QUANTITY	LAB UNIT PRICE	MAT/EQP/ SUB UNIT	TOTAL LABOR	TOTAL MAT/ EQUIP/SUB	TAX & INS INCLUDED TOT UNIT PRICE	TOTAL PRICE
REINFORCING STEEL												
				RE-BARS								
				*** TOTAL REINFORCING STEEL					18,123	12,467 M		36,694
CAST-IN-PLACE CONCRETE												
				CAST-IN-PLACE CONCRETE								
59	01	0112	333.010	CONCRETE @ PILE CAP	CUYD	88	12.156	51.652 M	1,070	4,545 M	71.466	6,289
60	01	0111	333.025	CONCRETE @ GRADE BEAM	CUYD	117	14.761	51.652 M	1,727	6,043 M	74.812	8,753
63	01	0211	333.035	CONCRETE @ SLAB ON GRADE	CUYD	1,145	12.381	51.652 M	14,176	59,142 M	71.761	82,166
67	01		333.910	LOADING DOCK SLAB	SQFT	4,000	1.488	1.268 M	5,952	5,072 M	3.278	13,112
				** TOTAL CAST-IN-PLACE CONCRETE					22,925	74,802 M		110,320
				*** TOTAL CAST-IN-PLACE CONCRETE					22,925	74,802 M		110,320
CEMENTITIOUS DECKS												
69	01	0321	350.100	LTWT CONC ROOF FILL	SQFT	30,000	.464	.551 M	13,920	16,530 M	1.190	35,700
				*** TOTAL CEMENTITIOUS DECKS					13,920	16,530 M		35,700
STRUCTURAL METALS												
				STRUCTURAL METALS								
76	01		501.011	STRUCTURAL STEEL FRAME	CWT	6,460	35.756	38.588 M 5.513 E	230,984	249,278 M 35,614 E	93.053	601,122
79	01		501.051	MISC. SUPPORT FRAMING	CWT	122	71.513	49.613 M 5.513 E	8,725	6,053 M 673 E	150.762	18,393
81	01		501.900	BASE PLATE	LBS	3,708	.551	.496 M	2,043	1,839 M	1.242	4,605
84	01		510.901	SANDBLAST	SQFT	47,632	.276	.132 M .044 E	13,146	6,287 M 2,096 E	.540	25,721
87	01		510.902	PRIMER	SQFT	47,632	.088	.055 M	4,192	2,620 M	.184	8,764

							.011 E		524 E		
90	01		510.903 INTERMEDIATE COAT	SQFT	47,632	.121	.132 M	5,763	6,287 M	.309	14,718
							.011 E		524 E		
93	01		510.904 TOP COAT (ELEVATED)	SQFT	47,632	.276	.165 M	13,146	7,859 M	.543	25,864
							.011 E		524 E		
			** TOTAL STRUCTURAL METALS					277,999	280,223 M		699,187
									39,955 E		
		METAL JOIST									
96	01	0312	521.011 STEEL JOIST	CWT	2,400	29.447	36.383 M	70,673	87,319 M	81.487	195,569
							4.410 E		10,584 E		
			** TOTAL METAL JOIST					70,673	87,319 M		195,569
									10,584 E		
		METAL DECKING									
99	01	0312	532.010 1-1/2" METAL DECK	SQFT	30,000	.450	.937 M	13,500	28,110 M	1.590	47,700
			** TOTAL METAL DECKING					13,500	28,110 M		47,700
			*** TOTAL STRUCTURAL METALS					362,172	395,652 M		942,456
									50,539 E		
WATERPROOF & DAMPPROOF											
104	01		700.110 DAMPPROOFING	SQFT	1,805	.460	.110 M	830	199 M	.708	1,278
107	01	0211	710.000 VAPOR BARRIER @ SLAB	SQFT	30,000	.133	.025 M	3,990	750 M	.197	5,910
			*** TOTAL WATERPROOF & DAMPPROOF					4,820	949 M		7,188
INSULATION											
110	01	0111	720.030 2" FOUNDATION INSULATION	SQFT	1,805	.745	.331 M	1,345	597 M	1.312	2,368
			*** TOTAL INSULATION					1,345	597 M		2,368

PREFORMED ROOFING & SIDING

FILE ID - 256A
PROJECT JOB NO -
PROJECT NAME - WHATCO PROJECT
PROJECT SIZE - 30,000 SQFT
SECTION - 01 PLANT

E S T I M A T E   D E T A I L   R E P O R T
** SECTION, ITEM CODE SEQUENCE **

PAGE - 3
DATE -
TIME -

MANAGEMENT COMPUTER CONTROLS, INC.
2881 DIRECTORS COVE
MEMPHIS, TENNESSEE 38131

REF NO.	S D SC	ELEM	ITEM CODE	DESCRIPTION	UNIT MEAS	QUANTITY	LAB UNIT PRICE	MAT/EQP/ SUB UNIT	TOTAL LABOR	TOTAL MAT/ EQUIP/SUB	TAX & INS INCLUDED TOT UNIT PRICE	TOTAL PRICE
PREFORMED ROOFING & SIDING												
114	01	0411	740.010	INSULATED METAL SIDING	SQFT	18,000	4.370	6.615 M	78,660	119,070 M	12.755	229,590
				*** TOTAL PREFORMED ROOFING & SIDING					78,660	119,070 M		229,590
ROOFING,SHEETMETAL&ACCESS												
				ROOFING & ROOF INSULATION								
116	01	0501	751.002	3 PLY TAR& GRAVEL ROOFING	SQS	300	59.515	50.081 M	17,855	15,024 M	130.393	39,118
119	01	0503	755.003	2" FIBER BD ROOF INSULATION	SQFT	30,000	1.037	.659 M	31,110	19,770 M	2.040	61,200
				** TOTAL ROOFING & ROOF INSULATION					48,965	34,794 M		100,318
				FLASHING & SHEETMETAL								
122	01	0504	762.040	24 GA GALV IRON SHEETMETAL	SQFT	800	7.513	1.243 M	6,010	994 M	10.964	8,771
125	01	0504	762.064	6" GALV DOWNSPOUT	LNFT	245	3.807	2.143 M	933	525 M	7.192	1,762
128	01	0504	762.069	6" GALV GUTTER	LNFT	400	3.356	1.889 M	1,342	756 M	6.340	2,536
				** TOTAL FLASHING & SHEETMETAL					8,285	2,275 M		13,069
				*** TOTAL ROOFING,SHEETMETAL&ACCESS					57,250	37,069 M		113,387
METAL DOORS & FRAMES												
131	01	0616	805.100	HM FRAME	EACH	5	51.543	66.150 M	258	331 M	137.600	688
134	01	0616	805.200	HM DOOR	EACH	5	39.762	165.375 M	199	827 M	229.800	1,149
				*** TOTAL METAL DOORS & FRAMES					457	1,158 M		1,837
SPECIAL DOORS												
138	01		830.900	VERTICAL LIFT DOOR	EACH	4	1323.000	13,230 M	5,292	52,920 M	16,015	64,060
				*** TOTAL SPECIAL DOORS					5,292	52,920 M		64,060
HARDWARE												
141	01	0616	871.027	FINISH HARDWARE ALLOWANCE	OPNG	5	69.583	275.625 M	348	1,378 M	387.400	1,937
				*** TOTAL HARDWARE					348	1,378 M		1,937

PAINTING & WALL COVERING										
		PAINTING								
151	01	0423 990.008 PAINT EXTERIOR DOOR	SIDE	10	29.607	2.205 M	296	22 M	40.300	403
154	01	0616 990.032 PAINT DOOR FRAME	EACH	5	23.398	2.756 M	117	14 M	33.000	165
158	01	0621 990.081 PAINT MAS-CONC 3 CTS	SQS	60	69.082	19.845 M	4,145	1,191 M	109.917	6,595
161	01	0411 992.000 EXTERIOR PAINTING	SQS	46	46.797	14.333 M	2,153	659 M	75.413	3,469
		** TOTAL PAINTING					6,711	1,886 M		10,632
		*** TOTAL PAINTING & WALL COVERING					6,711	1,886 M		10,632
EQUIPMENT										
		LOADING DOCK EQUIPMENT								
166	01	1116 1116.000 DOCK LEVELER	EACH	4	834.314	3307.500 M	3,337	13,230 M	4648.250	18,593
		** TOTAL LOADING DOCK EQUIPMENT					3,337	13,230 M		18,593
		*** TOTAL EQUIPMENT					3,337	13,230 M		18,593
CONVEYING SYSTEMS										
		HOISTS & CRANES								
169	01	1430.900 BRIDGE CRANE	EACH	1	64,236	396,900 M	64,236	396,900 M	511,866	511,866
		** TOTAL HOISTS & CRANES					64,236	396,900 M		511,866
		*** TOTAL CONVEYING SYSTEMS					64,236	396,900 M		511,866
HVAC SYSTEMS & EQUIPMENT										
		HVAC SYSTEMS & EQUIP								
171	01	1599.901 MECHANICAL - PLANT	LS	1		419,910 S		419,910 S	419,910	419,910
		** TOTAL HVAC SYSTEMS & EQUIP						419,910 S		419,910
		*** TOTAL HVAC SYSTEMS & EQUIPMENT						419,910 S		419,910

FILE ID - 256A
PROJECT JOB NO -
PROJECT NAME - WHATCO PROJECT
PROJECT SIZE - 30,000 SQFT
SECTION - 01 PLANT

E S T I M A T E   D E T A I L   R E P O R T
** SECTION, ITEM CODE SEQUENCE **

PAGE - 4
DATE -
TIME -

MANAGEMENT COMPUTER CONTROLS, INC.
2881 DIRECTORS COVE
MEMPHIS, TENNESSEE 38131

REF NO.	S D	SC	ELEM	ITEM CODE	DESCRIPTION	UNIT MEAS	QUANTITY	LAB UNIT PRICE	MAT/EQP/ SUB UNIT	TOTAL LABOR	TOTAL MAT/ EQUIP/SUB	TAX & INS INCLUDED TOT UNIT PRICE	TAX & INS INCLUDED TOTAL PRICE
					ELEC SYSTEMS & EQUIPMENT								
					ELECTRICAL SYSTEMS								
174		01		1699.901	ELECTRICAL - PLANT	LS	1		348,754 S		348,754 S	348,754	348,754
					** TOTAL ELECTRICAL SYSTEMS						348,754 S		348,754
					*** TOTAL ELEC SYSTEMS & EQUIPMENT						348,754 S		348,754
					**** TOTAL PLANT					720,649	1206,066 M 74,158 E 1615,447 S		3,917,622

FILE ID - 256A
PROJECT JOB NO -
PROJECT NAME - WHATCO PROJECT
PROJECT SIZE - 6,000 SQFT
SECTION - 02 WAREHOUSE

ESTIMATE DETAIL RECAP

** SECTION, ITEM CODE SEQUENCE **

PAGE - 1
DATE -
TIME -

MANAGEMENT COMPUTER CONTROLS, INC.
2881 DIRECTORS COVE
MEMPHIS, TENNESSEE 38131

DESCRIPTION	LABOR	MATERIAL	EQUIPMENT	SUBCONTRACT	TAX & INS INCLUDED COST/SQFT	TOTAL
GENERAL REQUIREMENTS				173,372	28.895	173,372
CONTINGENCY				48,988	8.165	48,988
EXCAV, GRADING & BACKFILL	8,284	1,657	3,860		2.710	16,259
CONCRETE FINISHING	3,128	106			.686	4,116
FORM WORK	9,546	2,248			2.442	14,654
REINFORCING STEEL	3,024	2,086			1.022	6,129
CAST-IN-PLACE CONCRETE	3,437	11,974			2.894	17,362
CEMENTITIOUS DECKS	2,784	3,306			1.190	7,140
STRUCTURAL METALS	79,676	78,115	9,669		32.704	196,226
WATERPROOF & DAMPPROOF	1,097	221			.274	1,641
INSULATION	484	215			.142	851
PREFORMED ROOFING & SIDING	26,220	39,690			12.755	76,530
ROOFING,SHEETMETAL&ACCESS	13,701	7,845			4.338	26,025
METAL DOORS & FRAMES	183	463			.123	735
SPECIAL DOORS	2,646	26,460			5.338	32,030
HARDWARE	139	551			.129	775
PAINTING & WALL COVERING	3,728	1,059			.987	5,919
EQUIPMENT	1,669	6,615			1.550	9,297
CONVEYING SYSTEMS	36,422	185,220			41.187	247,121
HVAC SYSTEMS & EQUIPMENT				101,510	16.918	101,510
ELEC SYSTEMS & EQUIPMENT				42,062	7.010	42,062
***** SECTION TOTAL	196,168	367,831	13,529	365,932	171.457	1,028,742

FILE ID - 256A
PROJECT JOB NO -
PROJECT NAME - WHATCO PROJECT
PROJECT SIZE - 6,000 SQFT
SECTION - 02 WAREHOUSE

E S T I M A T E   D E T A I L   R E P O R T

** SECTION, ITEM CODE SEQUENCE **

PAGE - 1
DATE -
TIME -

MANAGEMENT COMPUTER CONTROLS, INC.
2881 DIRECTORS COVE
MEMPHIS, TENNESSEE 38131

REF NO.	S D	SC ELEM	ITEM CODE	DESCRIPTION	UNIT MEAS	QUANTITY	LAB UNIT PRICE	MAT/EQP/ SUB UNIT	TOTAL LABOR	TOTAL MAT/ EQUIP/SUB	TAX & INS INCLUDED TOT UNIT PRICE	TAX & INS INCLUDED TOTAL PRICE
GENERAL REQUIREMENTS												
				OTHER GENERAL REQUIREMENTS								
2	02		185.902	OH/PROFIT/GC/BOND @ 21.5%	LS	1		173,372 S		173,372 S	173,372	173,372
				** TOTAL OTHER GENERAL REQUIREMENTS						173,372 S		173,372
				*** TOTAL GENERAL REQUIREMENTS						173,372 S		173,372
CONTINGENCY												
5	02		199.902	CONTINGENCY @ 5%	LS	1		48,988 S		48,988 S	48,988	48,988
				*** TOTAL CONTINGENCY						48,988 S		48,988
EXCAV, GRADING & BACKFILL												
				SITE GRADING								
9	02		221.011	EXCAV-LOAD-HAUL SURP 3-4MILES	CUYD	669	3.027	3.638 E	2,025	2,434 E	7.513	5,026
				** TOTAL SITE GRADING					2,025	2,434 E		5,026
				EXCAVATION & BACKFILL								
11	02	0112	222.301	EXCAVATE COLUMN FOOTING	CUYD	9	8.875	1.268 E	80	11 E	12.556	113
15	02	0111	222.308	EXCAVATE GRADE BEAM	CUYD	114	8.875	1.103 E	1,012	126 E	12.465	1,421
19	02	0111	222.328	BACKFILL GRADE BEAM	CUYD	84	18.529		1,556		23.714	1,992
21	02	0112	222.331	FINE GRADE COLUMN FOOTING	SQFT	192	.529		102		.677	130
25	02	0211	222.334	FINE GRADE SLAB ON GRADE	SQFT	6,000	.370		2,220		.474	2,844
30	02		222.900	STRUCTURAL FILL	CUYD	334	3.859	4.961 M 3.859 E	1,289	1,657 M 1,289 E	14.171	4,733
				** TOTAL EXCAVATION & BACKFILL					6,259	1,657 M 1,426 E		11,233
				*** TOTAL EXCAV, GRADING & BACKFILL					8,284	1,657 M 3,860 E		16,259

CONCRETE FINISHING											
		CONCRETE FINISH									
35	02 0211	301.019	TROWEL CEMENT FINISH	SQFT	6,000	.386		2,316		.494	2,964
38	02 0312	301.020	POINT & PATCH	SQFT	340	.128	.011 M	44	4 M	.176	60
40	02 0211	301.050	PROTECT & CURE	SQFT	6,000	.128	.017 M	768	102 M	.182	1,092
			** TOTAL CONCRETE FINISH					3,128	106 M		4,116
			*** TOTAL CONCRETE FINISHING					3,128	106 M		4,116
FORM WORK											
		FORM WORK									
43	02 0111	310.025	GRADE BEAM FORMS	SQFT	2,040	4.235	.937 M	8,639	1,911 M	6.435	13,128
45	02 0211	310.100	SLAB ON GRADE EDGE FORMS	LNFT	340	2.668	.992 M	907	337 M	4.488	1,526
			** TOTAL FORM WORK					9,546	2,248 M		14,654
			*** TOTAL FORM WORK					9,546	2,248 M		14,654
REINFORCING STEEL											
		RE-BARS									
48	02 0211	321.009	RE-STEEL @ SLAB ON GRADE	CWT	52	35.412	24.255 M	1,841	1,261 M	71.577	3,722
51	02 0112	321.110	RE-STEEL @ COLUMN FOOTING	CWT	11	34.798	24.255 M	383	267 M	70.818	779
55	02 0111	321.200	RE-STEEL @ GRADE BEAM	CWT	23	34.798	24.255 M	800	558 M	70.783	1,628
			** TOTAL RE-BARS					3,024	2,086 M		6,129
			*** TOTAL REINFORCING STEEL					3,024	2,086 M		6,129
CAST-IN-PLACE CONCRETE											
		CAST-IN-PLACE CONCRETE									
57	02 0112	333.005	CONCRETE @ COLUMN FTG	CUYD	10	12.156	51.652 M	122	517 M	71.500	715
61	02 0111	333.025	CONCRETE @ GRADE BEAM	CUYD	40	14.761	51.652 M	590	2,066 M	74.825	2,993
64	02 0211	333.035	CONCRETE @ SLAB ON GRADE	CUYD	172	12.381	51.652 M	2,130	8,884 M	71.762	12,343
68	02	333.910	LOADING DOCK SLAB	SQFT	400	1.488	1.268 M	595	507 M	3.278	1,311
			** TOTAL CAST-IN-PLACE CONCRETE					3,437	11,974 M		17,362

FILE ID - 256A
PROJECT JOB NO -
PROJECT NAME - WHATCO PROJECT
PROJECT SIZE - 6,000 SQFT
SECTION - 02 WAREHOUSE

E S T I M A T E   D E T A I L   R E P O R T

** SECTION, ITEM CODE SEQUENCE **

PAGE - 2
DATE -
TIME -

MANAGEMENT COMPUTER CONTROLS, INC.
2881 DIRECTORS COVE
MEMPHIS, TENNESSEE 38131

REF NO.	S D	SC ELEM	ITEM CODE	DESCRIPTION	UNIT MEAS	QUANTITY	LAB UNIT PRICE	MAT/EQP/ SUB UNIT	TOTAL LABOR	TOTAL MAT/ EQUIP/SUB	TAX & INS INCLUDED TOT UNIT PRICE	TAX & INS INCLUDED TOTAL PRICE
CAST-IN-PLACE CONCRETE												
				CAST-IN-PLACE CONCRETE								
				*** TOTAL CAST-IN-PLACE CONCRETE					3,437	11,974 M		17,362
CEMENTITIOUS DECKS												
70	02	0321	350.100	LTWT CONC ROOF FILL	SQFT	6,000	.464	.551 M	2,784	3,306 M	1.190	7,140
				*** TOTAL CEMENTITIOUS DECKS					2,784	3,306 M		7,140
STRUCTURAL METALS												
				STRUCTURAL METALS								
77	02		501.011	STRUCTURAL STEEL FRAME	CWT	920	35.756	38.588 M	32,896	35,501 M	93.053	85,609
								5.513 E		5,072 E		
80	02		501.051	MISC. SUPPORT FRAMING	CWT	270	71.513	49.613 M	19,309	13,396 M	150.759	40,705
								5.513 E		1,489 E		
82	02		501.900	BASE PLATE	LBS	504	.551	.496 M	278	250 M	1.242	626
85	02		510.901	SANDBLAST	SQFT	15,160	.276	.132 M	4,184	2,001 M	.540	8,186
								.044 E		667 E		
88	02		510.902	PRIMER	SQFT	15,160	.088	.055 M	1,334	834 M	.184	2,790
								.011 E		167 E		
91	02		510.903	INTERMEDIATE COAT	SQFT	15,160	.121	.132 M	1,834	2,001 M	.309	4,685
								.011 E		167 E		
94	02		510.904	TOP COAT (ELEVATED)	SQFT	15,160	.276	.165 M	4,184	2,501 M	.543	8,232
								.011 E		167 E		
				** TOTAL STRUCTURAL METALS					64,019	56,484 M		150,833
										7,729 E		

			METAL JOIST								
97	02	0312	521.011 STEEL JOIST	CWT	440	29.447	36.383 M 4.410 E	12,957	16,009 M 1,940 E	81.484	35,853
			** TOTAL METAL JOIST					12,957	16,009 M 1,940 E		35,853
			METAL DECKING								
100	02	0312	532.010 1-1/2" METAL DECK	SQFT	6,000	.450	.937 M	2,700	5,622 M	1.590	9,540
			** TOTAL METAL DECKING					2,700	5,622 M		9,540
			*** TOTAL STRUCTURAL METALS					79,676	78,115 M 9,669 E		196,226
WATERPROOF & DAMPPROOF											
105	02		700.110 DAMPPROOFING	SQFT	649	.460	.110 M	299	71 M	.707	459
108	02	0211	710.000 VAPOR BARRIER @ SLAB	SQFT	6,000	.133	.025 M	798	150 M	.197	1,182
			*** TOTAL WATERPROOF & DAMPPROOF					1,097	221 M		1,641
INSULATION											
111	02	0111	720.030 2" FOUNDATION INSULATION	SQFT	649	.745	.331 M	484	215 M	1.311	851
			*** TOTAL INSULATION					484	215 M		851
PREFORMED ROOFING & SIDING											
115	02	0411	740.010 INSULATED METAL SIDING	SQFT	6,000	4.370	6.615 M	26,220	39,690 M	12.755	76,530
			*** TOTAL PREFORMED ROOFING & SIDING					26,220	39,690 M		76,530
ROOFING,SHEETMETAL&ACCESS											
			ROOFING & ROOF INSULATION								
117	02	0501	751.002 3 PLY TAR& GRAVEL ROOFING	SQS	60	59.515	50.081 M	3,571	3,005 M	130.400	7,824
120	02	0503	755.003 2" FIBER BD ROOF INSULATION	SQFT	6,000	1.037	.659 M	6,222	3,954 M	2.040	12,240
			** TOTAL ROOFING & ROOF INSULATION					9,793	6,959 M		20,064
			FLASHING & SHEETMETAL								
123	02	0504	762.040 24 GA GALV IRON SHEETMETAL	SQFT	440	7.513	1.243 M	3,306	547 M	10.961	4,823
126	02	0504	762.064 6" GALV DOWNSPOUT	LNFT	70	3.807	2.143 M	266	150 M	7.186	503
129	02	0504	762.069 6" GALV GUTTER	LNFT	100	3.356	1.889 M	336	189 M	6.350	635
			** TOTAL FLASHING & SHEETMETAL					3,908	886 M		5,961

FILE ID - 256A
PROJECT JOB NO -
PROJECT NAME - WHATCO PROJECT
PROJECT SIZE - 6,000 SQFT
SECTION - 02 WAREHOUSE

E S T I M A T E   D E T A I L   R E P O R T

** SECTION, ITEM CODE SEQUENCE **

PAGE - 3
DATE -
TIME -

MANAGEMENT COMPUTER CONTROLS, INC.
2881 DIRECTORS COVE
MEMPHIS, TENNESSEE 38131

REF NO.	S D	SC	ELEM	ITEM CODE	DESCRIPTION	UNIT MEAS	QUANTITY	LAB UNIT PRICE	MAT/EQP/ SUB UNIT	TOTAL LABOR	TOTAL MAT/ EQUIP/SUB	TAX & INS INCLUDED TOT UNIT PRICE	TOTAL PRICE
ROOFING,SHEETMETAL&ACCESS													
					FLASHING & SHEETMETAL								
					*** TOTAL ROOFING,SHEETMETAL&ACCESS					13,701	7,845 M		26,025
METAL DOORS & FRAMES													
132		02	0616	805.100	HM FRAME	EACH	2	51.543	66.150 M	103	132 M	137.500	275
135		02	0616	805.200	HM DOOR	EACH	2	39.762	165.375 M	80	331 M	230.000	460
					*** TOTAL METAL DOORS & FRAMES					183	463 M		735
SPECIAL DOORS													
139		02		830.900	VERTICAL LIFT DOOR	EACH	2	1323.000	13,230 M	2,646	26,460 M	16,015	32,030
					*** TOTAL SPECIAL DOORS					2,646	26,460 M		32,030
HARDWARE													
142		02	0616	871.027	FINISH HARDWARE ALLOWANCE	OPNG	2	69.583	275.625 M	139	551 M	387.500	775
					*** TOTAL HARDWARE					139	551 M		775
PAINTING & WALL COVERING													
					PAINTING								
152		02	0423	990.008	PAINT EXTERIOR DOOR	SIDE	4	29.607	2.205 M	118	9 M	40.500	162
155		02	0616	990.032	PAINT DOOR FRAME	EACH	2	23.398	2.756 M	47	6 M	33.000	66
159		02	0621	990.081	PAINT MAS-CONC 3 CTS	SQS	36	69.082	19.845 M	2,487	714 M	109.889	3,956
162		02	0411	992.000	EXTERIOR PAINTING	SQS	23	46.797	14.333 M	1,076	330 M	75.435	1,735
					** TOTAL PAINTING					3,728	1,059 M		5,919
					*** TOTAL PAINTING & WALL COVERING					3,728	1,059 M		5,919

```
EQUIPMENT
         LOADING DOCK EQUIPMENT
 167  02 1116 1116.000 DOCK LEVELER                EACH      2  834.314 3307.500 M  1,669    6,615 M4648.500     9,297
                  ** TOTAL LOADING DOCK EQUIPMENT                                   1,669    6,615 M               9,297

                 *** TOTAL EQUIPMENT                                                1,669    6,615 M               9,297

CONVEYING SYSTEMS
         HOISTS & CRANES
 170  02      1430.900 BRIDGE CRANE                EACH      2   18,211   92,610 M 36,422  185,220 M 123,560   247,121
                  ** TOTAL HOISTS & CRANES                                         36,422  185,220 M             247,121

                 *** TOTAL CONVEYING SYSTEMS                                       36,422  185,220 M             247,121

HVAC SYSTEMS & EQUIPMENT
         HVAC SYSTEMS & EQUIP
 172  02      1599.902 MECHANICAL - PLANT          LS        1           101,510 S          101,510 S 101,510   101,510
                  ** TOTAL HVAC SYSTEMS & EQUIP                                             101,510 S             101,510

                 *** TOTAL HVAC SYSTEMS & EQUIPMENT                                         101,510 S             101,510

ELEC SYSTEMS & EQUIPMENT
         ELECTRICAL SYSTEMS
 175  02      1699.902 ELECTRICAL - PLANT          LS        1            42,062 S           42,062 S  42,062    42,062
                  ** TOTAL ELECTRICAL SYSTEMS                                                42,062 S              42,062

                 *** TOTAL ELEC SYSTEMS & EQUIPMENT                                          42,062 S              42,062

                **** TOTAL WAREHOUSE                                              196,168  367,831 M           1,028,742
                                                                                            13,529 E
                                                                                           365,932 S
```

FILE ID - 256A
PROJECT JOB NO -
PROJECT NAME - WHATCO PROJECT
PROJECT SIZE - 14,000 SQFT
SECTION - 03 OFFICE

E S T I M A T E D E T A I L R E C A P
** SECTION, ITEM CODE SEQUENCE **

PAGE - 1
DATE -
TIME -

MANAGEMENT COMPUTER CONTROLS, INC.
2881 DIRECTORS COVE
MEMPHIS, TENNESSEE 38131

DESCRIPTION	LABOR	MATERIAL	EQUIPMENT	SUBCONTRACT	TAX & INS INCLUDED COST/SQFT	TAX & INS INCLUDED TOTAL
GENERAL REQUIREMENTS				186,766	13.340	186,766
CONTINGENCY				52,772	3.769	52,772
EXCAV, GRADING & BACKFILL	7,561	1,895	4,318		1.146	16,050
CONCRETE FINISHING	7,196	238			.676	9,464
FORM WORK	3,049	675			.331	4,633
REINFORCING STEEL	6,538	4,487			.945	13,226
CAST-IN-PLACE CONCRETE	3,715	14,514	761		1.516	21,227
CEMENTITIOUS DECKS	3,248	3,857			.595	8,330
MASONRY	83,287	34,222	396		10.288	144,038
STONE	891	794			.143	2,000
STRUCTURAL METALS	88,903	98,853	11,912		16.624	232,736
MISC & ORNAMENTAL METAL	959	2,646			.292	4,092
WATERPROOF & DAMPPROOF	1,100	215			.117	1,639
INSULATION	4,645	4,511			.773	10,826
ROOFING,SHEETMETAL&ACCESS	13,079	8,632			1.863	26,080
METAL DOORS & FRAMES	2,586	4,234			.564	7,893
WOOD & PLASTIC DOORS	1,431	4,564			.484	6,773
METAL WINDOWS				14,884	1.063	14,884
HARDWARE	3,062	12,128			1.218	17,047
GLASS & GLAZING				2,867	.205	2,867
GYPSUM DRYWALL	31,806	11,356			3.786	53,005
TILE & TERRAZZO	4,842	1,489			.558	7,808
ACOUSTICAL TREATMENT	8,881	8,819			1.494	20,910
FLOORING	7,551	16,705			1.982	27,748
PAINTING & WALL COVERING	36,596	8,795			4.026	56,363
SPECIALTIES	2,159	4,851			.573	8,016
CONVEYING SYSTEMS				38,588	2.756	38,588
HVAC SYSTEMS & EQUIPMENT				71,542	5.110	71,542
ELEC SYSTEMS & EQUIPMENT				40,892	2.921	40,892
***** SECTION TOTAL	323,085	248,480	17,387	408,311	79.158	1,108,215

FILE ID - 256A
PROJECT JOB NO -
PROJECT NAME - WHATCO PROJECT
PROJECT SIZE - 14,000 SQFT
SECTION - 03 OFFICE

ESTIMATE DETAIL REPORT

** SECTION, ITEM CODE SEQUENCE **

PAGE - 1
DATE -
TIME -

MANAGEMENT COMPUTER CONTROLS, INC.
2881 DIRECTORS COVE
MEMPHIS, TENNESSEE 38131

REF NO.	S D SC	ELEM	ITEM CODE	DESCRIPTION	UNIT MEAS	QUANTITY	LAB UNIT PRICE	MAT/EQP/ SUB UNIT	TOTAL LABOR	TOTAL MAT/ EQUIP/SUB	TAX & INS INCLUDED TOT UNIT PRICE	TOTAL PRICE
GENERAL REQUIREMENTS												
				OTHER GENERAL REQUIREMENTS								
3	03		185.903	OH/PROFIT/GC/BOND @ 21.5%	LS	1		186,766 S		186,766 S	186,766	186,766
				** TOTAL OTHER GENERAL REQUIREMENTS						186,766 S		186,766
				*** TOTAL GENERAL REQUIREMENTS						186,766 S		186,766
CONTINGENCY												
6	03		199.903	CONTINGENCY @ 5%	LS	1		52,772 S		52,772 S	52,772	52,772
				*** TOTAL CONTINGENCY						52,772 S		52,772
EXCAV, GRADING & BACKFILL												
				SITE GRADING								
10	03		221.011	EXCAV-LOAD-HAUL SURP 3-4MILES	CUYD	764	3.027	3.638 E	2,313	2,779 E	7.513	5,740
				** TOTAL SITE GRADING					2,313	2,779 E		5,740
				EXCAVATION & BACKFILL								
12	03	0112	222.301	EXCAVATE COLUMN FOOTING	CUYD	8	8.875	1.268 E	71	10 E	12.625	101
16	03	0111	222.308	EXCAVATE GRADE BEAM	CUYD	50	8.875	1.103 E	444	55 E	12.460	623
17	03	0112	222.321	BACKFILL COLUMN FOOTING	CUYD	2	18.529		37		23.500	47
20	03	0111	222.328	BACKFILL GRADE BEAM	CUYD	30	18.529		556		23.733	712
22	03	0112	222.331	FINE GRADE COLUMN FOOTING	SQFT	144	.529		76		.674	97
26	03	0211	222.334	FINE GRADE SLAB ON GRADE	SQFT	7,000	.370		2,590		.474	3,318
32	03		222.900	STRUCTURAL FILL	CUYD	382	3.859	4.961 M 3.859 E	1,474	1,895 M 1,474 E	14.168	5,412
				** TOTAL EXCAVATION & BACKFILL					5,248	1,895 M 1,539 E		10,310
				*** TOTAL EXCAV, GRADING & BACKFILL					7,561	1,895 M 4,318 E		16,050

CONCRETE FINISHING										
		CONCRETE FINISH								
36	03 0211	301.019 TROWEL CEMENT FINISH	SQFT	14,000	.386		5,404		.494	6,916
41	03 0211	301.050 PROTECT & CURE	SQFT	14,000	.128	.017 M	1,792	238 M	.182	2,548
		** TOTAL CONCRETE FINISH					7,196	238 M		9,464
		*** TOTAL CONCRETE FINISHING					7,196	238 M		9,464
FORM WORK										
		FORM WORK								
44	03 0111	310.025 GRADE BEAM FORMS	SQFT	720	4.235	.937 M	3,049	675 M	6.435	4,633
		** TOTAL FORM WORK					3,049	675 M		4,633
		*** TOTAL FORM WORK					3,049	675 M		4,633
REINFORCING STEEL										
		RE-BARS								
49	03 0211	321.009 RE-STEEL @ SLAB ON GRADE	CWT	55	35.412	24.255 M	1,948	1,334 M	71.582	3,937
50	03 0312	321.040 RE-STEEL @ SLAB ON MTL DECK	CWT	109	35.412	24.255 M	3,860	2,644 M	71.587	7,803
52	03 0112	321.110 RE-STEEL @ COLUMN FOOTING	CWT	8	34.798	24.255 M	278	194 M	70.750	566
56	03 0111	321.200 RE-STEEL @ GRADE BEAM	CWT	13	34.798	24.255 M	452	315 M	70.769	920
		** TOTAL RE-BARS					6,538	4,487 M		13,226
		*** TOTAL REINFORCING STEEL					6,538	4,487 M		13,226
CAST-IN-PLACE CONCRETE										
		CAST-IN-PLACE CONCRETE								
58	03 0112	333.005 CONCRETE @ COLUMN FTG	CUYD	7	12.156	51.652 M	85	362 M	71.429	500
62	03 0111	333.025 CONCRETE @ GRADE BEAM	CUYD	25	14.761	51.652 M	369	1,291 M	74.800	1,870
65	03 0211	333.035 CONCRETE @ SLAB ON GRADE	CUYD	134	12.381	51.652 M	1,659	6,921 M	71.761	9,616

FILE ID - 256A
PROJECT JOB NO -
PROJECT NAME - WHATCO PROJECT
PROJECT SIZE - 14,000 SQFT
SECTION - 03 OFFICE

E S T I M A T E   D E T A I L   R E P O R T
** SECTION, ITEM CODE SEQUENCE **

PAGE - 2
DATE -
TIME -

MANAGEMENT COMPUTER CONTROLS, INC.
2881 DIRECTORS COVE
MEMPHIS, TENNESSEE 38131

REF NO.	S D	SC ELEM	ITEM CODE	DESCRIPTION	UNIT MEAS	QUANTITY	LAB UNIT PRICE	MAT/EQP/ SUB UNIT	TOTAL LABOR	TOTAL MAT/ EQUIP/SUB	TAX & INS INCLUDED TOT UNIT PRICE	TAX & INS INCLUDED TOTAL PRICE
CAST-IN-PLACE CONCRETE												
				CAST-IN-PLACE CONCRETE								
66		03 0312	333.080	CONCRETE OVER METAL DECK	CUYD	115	13.929	51.652 M	1,602	5,940 M	80.357	9,241
								6.615 E		761 E		
				** TOTAL CAST-IN-PLACE CONCRETE					3,715	14,514 M		21,227
										761 E		
				*** TOTAL CAST-IN-PLACE CONCRETE					3,715	14,514 M		21,227
										761 E		
CEMENTITIOUS DECKS												
71		03 0321	350.100	LTWT CONC ROOF FILL	SQFT	7,000	.464	.551 M	3,248	3,857 M	1.190	8,330
				*** TOTAL CEMENTITIOUS DECKS					3,248	3,857 M		8,330
MASONRY												
72		03 0611	402.102	8" CONC BLOCK	SQFT	15,370	4.030	1.907 M	61,941	29,311 M	7.222	111,002
73		03 0411	402.300	FACE BRICK	SQFT	2,700	7.022	1.819 M	18,959	4,911 M	10.957	29,584
74		03	430.101	EXT TUBULAR SCAFFOLDING	SQFT	3,600	.663	.110 E	2,387	396 E	.959	3,452
				*** TOTAL MASONRY					83,287	34,222 M		144,038
										396 E		
STONE												
75		03 0411	445.200	PRECAST COPING	LNFT	120	7.426	6.615 M	891	794 M	16.667	2,000
				*** TOTAL STONE					891	794 M		2,000

STRUCTURAL METALS

STRUCTURAL METALS

78	03		501.011	STRUCTURAL STEEL FRAME	CWT	1,580	35.756	38.588 M	56,494	60,969 M	93.053	147,024
								5.513 E		8,711 E		
83	03		501.900	BASE PLATE	LBS	336	.551	.496 M	185	167 M	1.241	417
86	03		510.901	SANDBLAST	SQFT	17,500	.276	.132 M	4,830	2,310 M	.540	9,451
								.044 E		770 E		
89	03		510.902	PRIMER	SQFT	17,500	.088	.055 M	1,540	963 M	.184	3,221
								.011 E		193 E		
92	03		510.903	INTERMEDIATE COAT	SQFT	17,500	.121	.132 M	2,118	2,310 M	.309	5,409
								.011 E		193 E		
95	03		510.904	TOP COAT (ELEVATED)	SQFT	17,500	.276	.165 M	4,830	2,888 M	.543	9,504
								.011 E		193 E		
				** TOTAL STRUCTURAL METALS					69,997	69,607 M		175,026
										10,060 E		
		METAL JOIST										
98	03	0312	521.011	STEEL JOIST	CWT	420	29.447	36.383 M	12,368	15,281 M	81.488	34,225
								4.410 E		1,852 E		
				** TOTAL METAL JOIST					12,368	15,281 M		34,225
										1,852 E		
		METAL DECKING										
101	03	0312	532.010	1-1/2" METAL DECK	SQFT	7,000	.450	.937 M	3,150	6,559 M	1.590	11,130
102	03	0312	532.011	2" METAL DECK	SQFT	7,000	.484	1.058 M	3,388	7,406 M	1.765	12,355
				** TOTAL METAL DECKING					6,538	13,965 M		23,485
				*** TOTAL STRUCTURAL METALS					88,903	98,853 M		232,736
										11,912 E		

MISC & ORNAMENTAL METAL

METAL FABRICATIONS

103	03		551.013	ALUMINUM HORIZ HAND RAIL	LNFT	75	12.790	35.280 M	959	2,646 M	54.560	4,092
				** TOTAL METAL FABRICATIONS					959	2,646 M		4,092
				*** TOTAL MISC & ORNAMENTAL METAL					959	2,646 M		4,092

FILE ID - 256A
PROJECT JOB NO -
PROJECT NAME - WHATCO PROJECT
PROJECT SIZE - 14,000 SQFT
SECTION - 03 OFFICE

E S T I M A T E D E T A I L R E P O R T

** SECTION, ITEM CODE SEQUENCE **

PAGE - 3
DATE -
TIME -

MANAGEMENT COMPUTER CONTROLS, INC.
2881 DIRECTORS COVE
MEMPHIS, TENNESSEE 38131

REF NO.	S D	SC	ELEM	ITEM CODE	DESCRIPTION	UNIT MEAS	QUANTITY	LAB UNIT PRICE	MAT/EQP/ SUB UNIT	TOTAL LABOR	TOTAL MAT/ EQUIP/SUB	TAX & INS INCLUDED TOT UNIT PRICE	TAX & INS INCLUDED TOTAL PRICE
WATERPROOF & DAMPPROOF													
106		03		700.110	DAMPPROOFING	SQFT	367	.460	.110 M	169	40 M	.708	260
109		03	0211	710.000	VAPOR BARRIER @ SLAB	SQFT	7,000	.133	.025 M	931	175 M	.197	1,379
					*** TOTAL WATERPROOF & DAMPPROOF					1,100	215 M		1,639
INSULATION													
112		03	0111	720.030	2" FOUNDATION INSULATION	SQFT	367	.745	.331 M	273	121 M	1.311	481
113		03	0411	720.042	2" RIGID INSULATION	SQFT	8,850	.494	.496 M	4,372	4,390 M	1.169	10,345
					*** TOTAL INSULATION					4,645	4,511 M		10,826
ROOFING,SHEETMETAL&ACCESS													
					ROOFING & ROOF INSULATION								
118		03	0501	751.002	3 PLY TAR& GRAVEL ROOFING	SQS	70	59.515	50.081 M	4,166	3,506 M	130.400	9,128
121		03	0503	755.003	2" FIBER BD ROOF INSULATION	SQFT	7,000	1.037	.659 M	7,259	4,613 M	2.040	14,280
					** TOTAL ROOFING & ROOF INSULATION					11,425	8,119 M		23,408
					FLASHING & SHEETMETAL								
124		03	0504	762.040	24 GA GALV IRON SHEETMETAL	SQFT	140	7.513	1.243 M	1,052	174 M	10.957	1,534
127		03	0504	762.064	6" GALV DOWNSPOUT	LNFT	70	3.807	2.143 M	266	150 M	7.186	503
130		03	0504	762.069	6" GALV GUTTER	LNFT	100	3.356	1.889 M	336	189 M	6.350	635
					** TOTAL FLASHING & SHEETMETAL					1,654	513 M		2,672
					*** TOTAL ROOFING,SHEETMETAL&ACCESS					13,079	8,632 M		26,080
METAL DOORS & FRAMES													
133		03	0616	805.100	HM FRAME	EACH	44	51.543	66.150 M	2,268	2,911 M	137.591	6,054
136		03	0616	805.200	HM DOOR	EACH	8	39.762	165.375 M	318	1,323 M	229.875	1,839
					*** TOTAL METAL DOORS & FRAMES					2,586	4,234 M		7,893

No.	Div.	Item	Description	Unit	Qty	Labor Unit	Material Unit	Labor	Material	Total Unit	Total
WOOD & PLASTIC DOORS											
137	03 0616	823.000	SC WOOD DOOR	EACH	36	39.762	126.788 M	1,431	4,564 M	188.139	6,773
			*** TOTAL WOOD & PLASTIC DOORS					1,431	4,564 M		6,773
METAL WINDOWS											
140	03 0421	851.110	FIXED WINDOWS	SQFT	900		16.538 S		14,884 S	16.538	14,884
			*** TOTAL METAL WINDOWS						14,884 S		14,884
HARDWARE											
143	03 0616	871.027	FINISH HARDWARE ALLOWANCE	OPNG	44	69.583	275.625 M	3,062	12,128 M	387.432	17,047
			*** TOTAL HARDWARE					3,062	12,128 M		17,047
GLASS & GLAZING											
			ENTRANCES & STOREFRONTS								
144	03 0423	885.205	PAIR ALUM GLASS DOOR	EACH	2		1433.250 S		2,867 S	1433.500	2,867
			** TOTAL ENTRANCES & STOREFRONTS						2,867 S		2,867
			*** TOTAL GLASS & GLAZING						2,867 S		2,867
GYPSUM DRYWALL											
145	03 0611	926.620	STANDARD DRYWALL PARTITION	SQFT	8,900	2.374	.882 M	21,129	7,850 M	3.994	35,547
146	03 0611	926.625	MTL FURRING W/GYPSUM BOARD	SQFT	7,950	1.343	.441 M	10,677	3,506 M	2.196	17,458
			*** TOTAL GYPSUM DRYWALL					31,806	11,356 M		53,005
TILE & TERRAZZO											
			CERAMIC TILE								
147	03 0622	931.010	CERAMIC TILE FLOOR	SQFT	900	5.380	1.654 M	4,842	1,489 M	8.676	7,808
			** TOTAL CERAMIC TILE					4,842	1,489 M		7,808
			*** TOTAL TILE & TERRAZZO					4,842	1,489 M		7,808

FILE ID - 256A
PROJECT JOB NO -
PROJECT NAME - WHATCO PROJECT
PROJECT SIZE - 14,000 SQFT
SECTION - 03 OFFICE

E S T I M A T E   D E T A I L   R E P O R T

** SECTION, ITEM CODE SEQUENCE **

PAGE - 4
DATE -
TIME -

MANAGEMENT COMPUTER CONTROLS, INC.
2881 DIRECTORS COVE
MEMPHIS, TENNESSEE 38131

REF NO.	S D	SC	ELEM	ITEM CODE	DESCRIPTION	UNIT MEAS	QUANTITY	LAB UNIT PRICE	MAT/EQP/ SUB UNIT	TOTAL LABOR	TOTAL MAT/ EQUIP/SUB	TAX & INS INCLUDED TOT UNIT PRICE	TAX & INS INCLUDED TOTAL PRICE
ACOUSTICAL TREATMENT													
148		03	0623	950.400	ACOUST CEIL SYS-EXPOSED GRID	SQFT	12,300	.722	.717 M	8,881	8,819 M	1.700	20,910
					*** TOTAL ACOUSTICAL TREATMENT					8,881	8,819 M		20,910
FLOORING													
			RESILIENT FLOORING										
149		03	0622	965.010	VINYL COMPOSITION TILE	SQFT	3,825	.981	.584 M	3,752	2,234 M	1.888	7,221
					** TOTAL RESILIENT FLOORING					3,752	2,234 M		7,221
			CARPETING										
150		03	0622	968.200	CARPET	SQYD	875	4.342	16.538 M	3,799	14,471 M	23.459	20,527
					** TOTAL CARPETING					3,799	14,471 M		20,527
					*** TOTAL FLOORING					7,551	16,705 M		27,748
PAINTING & WALL COVERING													
			PAINTING										
153		03	0616	990.031	PAINT INTERIOR DOOR	SIDE	88	26.865	3.308 M	2,364	291 M	37.966	3,341
156		03	0616	990.032	PAINT DOOR FRAME	EACH	44	23.398	2.756 M	1,030	121 M	32.932	1,449
157		03	0621	990.061	PAINT PLASTER-GYP BD 3 CTS	SQS	515	50.025	12.128 M	25,763	6,246 M	77.159	39,737
160		03	0621	990.081	PAINT MAS-CONC 3 CTS	SQS	107	69.082	19.845 M	7,392	2,123 M	109.907	11,760
163		03	0411	992.000	EXTERIOR PAINTING	SQS	1	46.797	14.333 M	47	14 M	76.000	76
					** TOTAL PAINTING					36,596	8,795 M		56,363
					*** TOTAL PAINTING & WALL COVERING					36,596	8,795 M		56,363

No.	Code	Description	Unit	Qty						
SPECIALTIES										
		COMPARTMENTS & CUBICLES								
164	03 0631	1018.010 TOILET COMPARTMENT	EACH	12	125.435	220.500 M	1,505	2,646 M	399.250	4,791
		** TOTAL COMPARTMENTS & CUBICLES					1,505	2,646 M		4,791
		LOCKERS								
165	03 0631	1050.001 DOUBLE TIER LOCKER	OPNG	40	16.361	55.125 M	654	2,205 M	80.625	3,225
		** TOTAL LOCKERS					654	2,205 M		3,225
		*** TOTAL SPECIALTIES					2,159	4,851 M		8,016
CONVEYING SYSTEMS										
		ELEVATORS								
168	03	1420.900 ELEVATOR	EACH	1		38,587 S		38,588 S	38,588	38,588
		** TOTAL ELEVATORS						38,588 S		38,588
		*** TOTAL CONVEYING SYSTEMS						38,588 S		38,588
HVAC SYSTEMS & EQUIPMENT										
		HVAC SYSTEMS & EQUIP								
173	03	1599.903 MECHANICAL - PLANT	LS	1		71,542 S		71,542 S	71,542	71,542
		** TOTAL HVAC SYSTEMS & EQUIP						71,542 S		71,542
		*** TOTAL HVAC SYSTEMS & EQUIPMENT						71,542 S		71,542
ELEC SYSTEMS & EQUIPMENT										
		ELECTRICAL SYSTEMS								
176	03	1699.903 ELECTRICAL - PLANT	LS	1		40,892 S		40,892 S	40,892	40,892
		** TOTAL ELECTRICAL SYSTEMS						40,892 S		40,892
		*** TOTAL ELEC SYSTEMS & EQUIPMENT						40,892 S		40,892
		**** TOTAL OFFICE					323,085	248,480 M 17,387 E 408,311 S		1,108,215

FILE ID - 256A
PROJECT JOB NO -
PROJECT NAME - WHATCO PROJECT
PROJECT SIZE - 14,000 SQFT
SECTION - 03 OFFICE

E S T I M A T E   D E T A I L   R E P O R T

** SECTION, ITEM CODE SEQUENCE **

PAGE - 5
DATE -
TIME -

MANAGEMENT COMPUTER CONTROLS, INC.
2881 DIRECTORS COVE
MEMPHIS, TENNESSEE 38131

REF NO.	S D	SC	ELEM	ITEM CODE	DESCRIPTION	UNIT MEAS	QUANTITY	LAB UNIT PRICE	MAT/EQP/ SUB UNIT	TOTAL LABOR	TOTAL MAT/ EQUIP/SUB	TAX & INS INCLUDED TOT UNIT PRICE	TAX & INS INCLUDED TOTAL PRICE
ELEC SYSTEMS & EQUIPMENT													
					ELECTRICAL SYSTEMS								
					***** GRAND TOTAL					1239,902	1822,377 M 105,074 E 2389,690 S		6,054,579

File ID - 256A
Client Job No -
Project Name - WHATCO PROJECT
Project Size - 50,000 SQFT
Estimator - DOUG

M I S C E L L A N E O U S   I T E M S   R E P O R T

Page - 1
Date -
Time -

MANAGEMENT COMPUTER CONTROLS, INC.
2881 DIRECTORS COVE
MEMPHIS, TENNESSEE 38131

S D	Sec	Elem	Item Code	Description	Unit Meas	Quantity	Labor	Material	Equipment	Subcontract
	01		185.901	OH/PROFIT/GC/BOND @ 21.5%	LS	1				
	02		185.902	OH/PROFIT/GC/BOND @ 21.5%	LS	1				
	03		185.903	OH/PROFIT/GC/BOND @ 21.5%	LS	1				
	01		199.901	CONTINGENCY @ 5%	LS	1				
	02		199.902	CONTINGENCY @ 5%	LS	1				
	03		199.903	CONTINGENCY @ 5%	LS	1				
	01		211.900	SITE PREP (CLEAR)	ACRE	1	1100.000		1200.000	
	01		221.011	EXCAV-LOAD-HAUL SURP 3-4MILES	CUYD	3,345	2.746		3.300	
	02		221.011	EXCAV-LOAD-HAUL SURP 3-4MILES	CUYD	669	2.746		3.300	
	03		221.011	EXCAV-LOAD-HAUL SURP 3-4MILES	CUYD	764	2.746		3.300	
	01		222.900	STRUCTURAL FILL	CUYD	1,672			3.500	
	01		222.900	STRUCTURAL FILL	CUYD	1,672	3.500	4.500		
	02		222.900	STRUCTURAL FILL	CUYD	334	3.500	4.500		
	02		222.900	STRUCTURAL FILL	CUYD	334			3.500	
	03		222.900	STRUCTURAL FILL	CUYD	382			3.500	
	03		222.900	STRUCTURAL FILL	CUYD	382	3.500	4.500		
	01		333.910	LOADING DOCK SLAB	SQFT	4,000	1.350	1.150		
	02		333.910	LOADING DOCK SLAB	SQFT	400	1.350	1.150		
	03		430.101	EXT TUBULAR SCAFFOLDING	SQFT	3,600	.601		.100	
	01		501.011	STRUCTURAL STEEL FRAME	CWT	6,460	32.432	35.000		
	02		501.011	STRUCTURAL STEEL FRAME	CWT	920	32.432	35.000		
	03		501.011	STRUCTURAL STEEL FRAME	CWT	1,580	32.432	35.000		
	01		501.051	MISC. SUPPORT FRAMING	CWT	122	64.864	45.000		
	02		501.051	MISC. SUPPORT FRAMING	CWT	270	64.864	45.000		

01	501.900	BASE PLATE	LBS	3,708			
02	501.900	BASE PLATE	LBS	504			
03	501.900	BASE PLATE	LBS	336			
01	510.901	SANDBLAST	SQFT	47,632	.250	.120	
02	510.901	SANDBLAST	SQFT	15,160	.250	.120	
03	510.901	SANDBLAST	SQFT	17,500	.250	.120	
01	510.902	PRIMER	SQFT	47,632	.080	.050	
02	510.902	PRIMER	SQFT	15,160	.080	.050	
03	510.902	PRIMER	SQFT	17,500	.080	.050	
01	510.903	INTERMEDIATE COAT	SQFT	47,632	.110	.120	
02	510.903	INTERMEDIATE COAT	SQFT	15,160	.110	.120	
03	510.903	INTERMEDIATE COAT	SQFT	17,500	.110	.120	
01	510.904	TOP COAT (ELEVATED)	SQFT	47,632	.250	.150	
02	510.904	TOP COAT (ELEVATED)	SQFT	15,160	.250	.150	
03	510.904	TOP COAT (ELEVATED)	SQFT	17,500	.250	.150	
03	551.013	ALUMINUM HORIZ HAND RAIL	LNFT	75	11.601	32.000	
01	700.110	DAMPPROOFING	SQFT	1,805	.417	.100	
02	700.110	DAMPPROOFING	SQFT	649	.417	.100	
03	700.110	DAMPPROOFING	SQFT	367	.417	.100	
01	830.900	VERTICAL LIFT DOOR	EACH	4	1200.000	12000.000	
02	830.900	VERTICAL LIFT DOOR	EACH	2	1200.000	12000.000	
03	1420.900	ELEVATOR	EACH	1			35000.000
01	1430.900	BRIDGE CRANE	EACH	1	58264.000	360000.000	
02	1430.900	BRIDGE CRANE	EACH	2	16518.000	84000.000	
01	1599.901	MECHANICAL - PLANT	LS	1			419910.000
02	1599.902	MECHANICAL - PLANT	LS	1			101510.000
03	1599.903	MECHANICAL - PLANT	LS	1			71542.000
01	1699.901	ELECTRICAL - PLANT	LS	1			348754.000
02	1699.902	ELECTRICAL - PLANT	LS	1			42062.000
03	1699.903	ELECTRICAL - PLANT	LS	1			40892.000

File ID - 256A
Client Job No -
Project Name - WHATCO PROJECT
Project Size - 50,000 SQFT
Estimator - DOUG

QUANTITY SURVEY REPORT

Page - 1
Date -
Time -

MANAGEMENT COMPUTER CONTROLS, INC.
2881 DIRECTORS COVE
MEMPHIS, TENNESSEE 38131

1.000 FOUNDATIONS

-----DIMENSIONS----- -----SPECIFICATION SELECTION-----

B-SQFT,LENGTH (FEET & DECIMAL OF G-PILASTERS (WHOLE NUMBER) 9-METHOD OF CALCULATING AREA 4-FOUNDATION INSULATION
C-WIDTH,HEIGHT,SHAFT DIA (FEET & H-PILASTER WIDTH (FEET & DECIMAL 8-TYPE OF FOUNDATION 3-CAISSON
D-DEPTH,THICKNESS (FEET & DECIMA J-PILASTER PROJECTION (FEET & DE 7-TYPE OF FOUNDATION WALL 2-PILING
E-EXCAVATION DEPTH (FEET & DECIM K-NOT USED 6-FORMS
F-REINFORCING STEEL (LBS & DECIM A-QUANTITY 5-REINFORCING

WS	LINE SEQ	B	C	D	E	F	G	H	J	K	A	SPEC. SEL. 9876543210	S/D SEC	ELEM	LOC	MARK FIELD
DW	1	800.00	3.00	1.25	3.00	7.68					1	20342	01		01	
DW	2	1805.00									1	100003	01		01	
DW	3	7.50	7.50	2.50	2.50	378.50					16	23005	01		01	
DW	32	340.00	3.00	1.00	3.00	6.50					1	20342	02		01	
DW	33	649.00									1	100003	02		01	
DW	34	4.00	4.00	1.25	1.25	85.00					12	22005	02		01	
DW	63	120.00	3.00	1.75	3.00	10.40					1	20342	03		01	
DW	64	367.00									1	100003	03		01	
DW	65	4.00	4.00	1.25	1.50	79.00					9	22005	03		01	

S/D	SEC	ELEM	ITEM CODE	DESCRIPTION	U/M	QUANTITY
	02	0112	222.301	EXCAVATE COLUMN FOOTING	CUYD	9
	03	0112	222.301	EXCAVATE COLUMN FOOTING	CUYD	8
	01	0112	222.302	EXCAVATE PILE CAP	CUYD	84
	01	0111	222.308	EXCAVATE GRADE BEAM	CUYD	289
	02	0111	222.308	EXCAVATE GRADE BEAM	CUYD	114
	03	0111	222.308	EXCAVATE GRADE BEAM	CUYD	50
	03	0112	222.321	BACKFILL COLUMN FOOTING	CUYD	2
	01	0111	222.328	BACKFILL GRADE BEAM	CUYD	196
	02	0111	222.328	BACKFILL GRADE BEAM	CUYD	84
	03	0111	222.328	BACKFILL GRADE BEAM	CUYD	30
	02	0112	222.331	FINE GRADE COLUMN FOOTING	SQFT	192
	03	0112	222.331	FINE GRADE COLUMN FOOTING	SQFT	144
	01	0112	222.332	FINE GRADE PILE CAP	SQFT	900
	01	0111	310.025	GRADE BEAM FORMS	SQFT	4800
	02	0111	310.025	GRADE BEAM FORMS	SQFT	2040
	03	0111	310.025	GRADE BEAM FORMS	SQFT	720
	02	0112	321.110	RE-STEEL @ COLUMN FOOTING	CWT	11
	03	0112	321.110	RE-STEEL @ COLUMN FOOTING	CWT	8
	01	0112	321.120	RE-STEEL @ PILE CAP	CWT	63
	01	0111	321.200	RE-STEEL @ GRADE BEAM	CWT	64
	02	0111	321.200	RE-STEEL @ GRADE BEAM	CWT	23
	03	0111	321.200	RE-STEEL @ GRADE BEAM	CWT	13
	02	0112	333.005	CONCRETE @ COLUMN FTG	CUYD	10
	03	0112	333.005	CONCRETE @ COLUMN FTG	CUYD	7
	01	0112	333.010	CONCRETE @ PILE CAP	CUYD	88
	01	0111	333.025	CONCRETE @ GRADE BEAM	CUYD	117
	02	0111	333.025	CONCRETE @ GRADE BEAM	CUYD	40
	03	0111	333.025	CONCRETE @ GRADE BEAM	CUYD	25
	01	0111	720.030	2" FOUNDATION INSULATION	SQFT	1805
	02	0111	720.030	2" FOUNDATION INSULATION	SQFT	649
	03	0111	720.030	2" FOUNDATION INSULATION	SQFT	367

```
    File ID - 256A                QUANTITY SURVEY REPORT                                   Page -      2
Client Job No -                                                                            Date -
Project Name  - WHATCO PROJECT                                                             Time -
Project Size  -    50,000 SQFT                                       MANAGEMENT COMPUTER CONTROLS, INC.
Estimator     - DOUG                                                 2881 DIRECTORS COVE
                                                                     MEMPHIS, TENNESSEE    38131
```

2.000 SUBSTRUCTURE

```
-----------------------DIMENSIONS----------------------------------------------------SPECIFICATION SELECTION----------------

B-SQFT,LENGTH (FEET & DECIMAL OF G-LNFT SHORING OR LAYBACK (FEET  9-METHOD OF CALCULATING AREA      4-REINFORCING
C-SLAB WIDTH,WALL HEIGHT,EDGE FO H-PIT,TRENCH-SLAB THICKNESS,STAI 8-TYPE OF CONSTRUCTION            3-WATERPROOFING
D-SLAB/WALL THK, BSMT/PIT/TRENCH J-PIT,TRENCH-WALL THICKNESS,STAI 7-TYPE OF CONSTRUCTION   (CONT.) 2-SHORING BRACING LAYBACK
E-UNDERSLAB FILL THICKNESS (FEET K-STAIR # OF RISERS              6-FORMS                           1-FOUNDATION DRAINAGE
F-REINFORCING STEEL (LBS & DECIM A-QUANTITY                       5-CONCRETE FINISH                 0-
```

WS	LINE SEQ	B	C	D	E	F	G	H	J	K	A	SPEC. SEL. 9876543210	S/D SEC	ELEM	LOC	MARK FIELD
DW	4	30000.00	800.00	1.00		1.25					1	1102141	01		01	
DW	35	6000.00	340.00	.75		.83					1	1102141	02		01	
DW	66	7000.00		.50		.75					1	1100141	03		01	

S/D	SEC	ELEM	ITEM CODE	DESCRIPTION	U/M	QUANTITY
	01	0211	222.334	FINE GRADE SLAB ON GRADE	SQFT	30000
	02	0211	222.334	FINE GRADE SLAB ON GRADE	SQFT	6000
	03	0211	222.334	FINE GRADE SLAB ON GRADE	SQFT	7000
	01	0211	301.019	TROWEL CEMENT FINISH	SQFT	30000
	02	0211	301.019	TROWEL CEMENT FINISH	SQFT	6000
	03	0211	301.019	TROWEL CEMENT FINISH	SQFT	7000
	01	0312	301.020	POINT & PATCH	SQFT	800
	02	0312	301.020	POINT & PATCH	SQFT	340
	01	0211	301.050	PROTECT & CURE	SQFT	30000
	02	0211	301.050	PROTECT & CURE	SQFT	6000
	03	0211	301.050	PROTECT & CURE	SQFT	7000
	02	0211	310.100	SLAB ON GRADE EDGE FORMS	LNFT	340
	01	0211	310.105	SLAB ON GRADE EDGE FORMS	SQFT	800
	01	0211	321.009	RE-STEEL @ SLAB ON GRADE	CWT	387
	02	0211	321.009	RE-STEEL @ SLAB ON GRADE	CWT	52
	03	0211	321.009	RE-STEEL @ SLAB ON GRADE	CWT	55
	01	0211	333.035	CONCRETE @ SLAB ON GRADE	CUYD	1145
	02	0211	333.035	CONCRETE @ SLAB ON GRADE	CUYD	172
	03	0211	333.035	CONCRETE @ SLAB ON GRADE	CUYD	134
	01	0211	710.000	VAPOR BARRIER @ SLAB	SQFT	30000
	02	0211	710.000	VAPOR BARRIER @ SLAB	SQFT	6000
	03	0211	710.000	VAPOR BARRIER @ SLAB	SQFT	7000

```
     File ID - 256A                 Q U A N T I T Y   S U R V E Y   R E P O R T              Page -      3
Client Job No -                                                                               Date -
Project Name  - WHATCO PROJECT                                                                Time -
Project Size  -    50,000 SQFT                                        MANAGEMENT COMPUTER CONTROLS, INC.
Estimator     - DOUG                                                  2881 DIRECTORS COVE
                                                                      MEMPHIS, TENNESSEE    38131

          2.031     PILINGS & CAISSONS

-----------------------DIMENSIONS--------------------------------------------SPECIFICATION SELECTION----------------

B-NUMBER OF PILES PER PLACE      G-RE-STEEL-LBS/LNFT     9-DRILLED CAISSON                4-WOOD/CONCRETE PILINGS
C-SIZE OF PILE-DIAMETER OR SIZE                          8-QUALITY OF CONCRETE            3-STEEL PILINGS
D-LENGTH OF PILE                                         7-METHOD OF PLACING CONCRETE     2-PIPE PILINGS
E-DIAMETER OF BELL @ CAISSONS                            6-EXCAVATION                     1-MISCELLANEOUS PILINGS
F-RE-STEEL - NO. & SIZE OF BAR   A-QUANTITY              5-REINFORCING STEEL              0-ELEMENT

WS LINE    B       C       D       E       F       G       H       J      K      A   SPEC. SEL.    S/D        MARK FIELD
   SEQ                                                                               9876543210 SEC  ELEM LOC

DW   5    2.00            55.00                                                  16  0000007    01         01

                        S/D  SEC  ELEM  ITEM CODE        DESCRIPTION                 U/M    QUANTITY

                             01   0121   231.046   14X14-89LB H-SEC STL PILE         LNFT     1760
```

QUANTITY SURVEY REPORT

File ID - 256A
Client Job No -
Project Name - WHATCO PROJECT
Project Size - 50,000 SQFT
Estimator - DOUG

Page - 4
Date -
Time -

MANAGEMENT COMPUTER CONTROLS, INC.
2881 DIRECTORS COVE
MEMPHIS, TENNESSEE 38131

3.000 CONCRETE STRUCTURAL SYSTEM

------------------------DIMENSIONS------------------------SPECIFICATION SELECTION----------------

B-SQFT,LENGTH (FEET & DECIMAL OF	G-# OF VOIDS (WHOLE NUMBER)	9-METHOD OF CALCULATING AREA	4-MISC.
C-WIDTH (SLAB,BEAM,COLUMN),HEIGH	H-LENGTH OF VOID (FEET & DECIMAL	8-CONCRETE SLAB CONSTRUCTION	3-FORMS
D-SLAB & WALL THICKNESS,BEAM DEP	J-WIDTH OF VOID (FEET & DECIMAL	7-PRECAST SLAB CONSTRUCTION	2-CONCRETE SLAB FINISH
E-REINFORCING (LBS & DECIMAL OF	K-DIA OF CAPITAL (FEET & DECIMAL	6-BEAMS	1-UNDER SLAB/BEAM/COLUMN/WALL FI
F-DEPTH OF PAN,DOME (FEET & DECI	A-QUANTITY	5-COLUMNS	0-REINFORCING

WS	LINE SEQ	B	C	D	E	F	G	H	J	K	A	SPEC. SEL. 9876543210	S/D SEC	ELEM	LOC	MARK FIELD
DW	94	7000.00		.42	1.50						1	1400000102	03		01	

S/D SEC	ELEM	ITEM CODE	DESCRIPTION	U/M	QUANTITY
03	0211	301.019	TROWEL CEMENT FINISH	SQFT	7000
03	0211	301.050	PROTECT & CURE	SQFT	7000
03	0312	321.040	RE-STEEL @ SLAB ON MTL DECK	CWT	109
03	0312	333.080	CONCRETE OVER METAL DECK	CUYD	115

```
      File ID - 256A                QUANTITY  SURVEY  REPORT                          Page -      5
Client Job No -                                                                        Date -
Project Name  - WHATCO PROJECT                                                         Time -
Project Size  -     50,000 SQFT                                   MANAGEMENT COMPUTER CONTROLS, INC.
Estimator     - DOUG                                              2881 DIRECTORS COVE
                                                                  MEMPHIS, TENNESSEE     38131

        3.005     STEEL & WOOD FRAME STRUCTUAL SYSTEM

------------------------DIMENSIONS------------------------------------------------------SPECIFICATION SELECTION-----------------

B-SQFT,LENGTH (FEET & DECIMAL OF                             9-METHOD OF CALCULATING AREA    4-FIREPROOFING
C-WIDTH  (FEET & DECIMAL OF FOOT                             8-STEEL FRAME                   3-PRE-FAB METAL BUILDING
D-STRUCT STL-LBS/SQFT, WOOD FRM-                             7-WOOD FRAME
                                                             6-METAL DECK
                      A-QUANTITY                             5-WOOD DECK

WS LINE    B        C       D       E     F     G     H     J     K     A   SPEC. SEL.    S/D        MARK FIELD
   SEQ                                                                      9876543210 SEC  ELEM LOC

DW    7 30000.00           8.00                                         1   1201        01         01
DW   38  6000.00           7.33                                         1   1201        02         01
DW   69  7000.00           6.00                                         1   1201        03         01
DW   95  7000.00                                                        1   1002        03         01
```

S/D	SEC	ELEM	ITEM CODE	DESCRIPTION	U/M	QUANTITY
	01	0312	521.011	STEEL JOIST	CWT	2400
	02	0312	521.011	STEEL JOIST	CWT	440
	03	0312	521.011	STEEL JOIST	CWT	420
	01	0312	532.010	1-1/2" METAL DECK	SQFT	30000
	02	0312	532.010	1-1/2" METAL DECK	SQFT	6000
	03	0312	532.010	1-1/2" METAL DECK	SQFT	7000
	03	0312	532.011	2" METAL DECK	SQFT	7000

File ID - 256A
Client Job No -
Project Name - WHATCO PROJECT
Project Size - 50,000 SQFT
Estimator - DOUG

QUANTITY SURVEY REPORT

Page - 6
Date -
Time -

MANAGEMENT COMPUTER CONTROLS, INC.
2881 DIRECTORS COVE
MEMPHIS, TENNESSEE 38131

4.000 EXTERIOR WALLS

------------------------DIMENSIONS------------------------SPECIFICATION SELECTION------------------

B-SQFT,LENGTH (FEET & DECIMAL OF
C-HEIGHT (FEET & DECIMAL OF FOOT
D-THICKNESS OF BLOCK (FEET & DEC

A-QUANTITY

9-METHOD OF CALCULATING AREA
8-EXTERIOR FACING
7-EXTERIOR FACING (CONT)
6-EXTERIOR FACING (CONT)
5-CORE CONSTRUCTION

4-CORE CONSTRUCTION (CONT)
3-INSULATION
2-EXTERIOR SHEATHING
1-SOFFIT
0-BALCONY RAILING

WS	LINE SEQ	B	C	D	E	F	G	H	J	K	A	SPEC. SEL. 9876543210	S/D SEC	ELEM	LOC	MARK FIELD
DW	9	18000.00									1	1005	01		01	
DW	10	4516.00									1	106	01		01	
DW	40	6000.00									1	1005	02		01	
DW	41	2216.00									1	106	02		01	
DW	72	42.00									1	106	03		01	
DW	96	120.00									1	007	03		01	
DW	97	2700.00		.67							1	101001	03		01	
DW	98	8850.00									1	1000004	03		01	

S/D SEC	ELEM	ITEM CODE	DESCRIPTION	U/M	QUANTITY
03	0611	402.102	8" CONC BLOCK	SQFT	2700
03	0411	402.300	FACE BRICK	SQFT	2700
03	0411	445.200	PRECAST COPING	LNFT	120
03	0411	720.042	2" RIGID INSULATION	SQFT	8850
01	0411	740.010	INSULATED METAL SIDING	SQFT	18000
02	0411	740.010	INSULATED METAL SIDING	SQFT	6000
01	0411	992.000	EXTERIOR PAINTING	SQS	46
02	0411	992.000	EXTERIOR PAINTING	SQS	23
03	0411	992.000	EXTERIOR PAINTING	SQS	1

```
     File ID - 256A                 Q U A N T I T Y   S U R V E Y   R E P O R T                 Page -      7
Client Job No -                                                                                 Date -
Project Name  - WHATCO PROJECT                                                                  Time -
Project Size  -     50,000 SQFT                                          MANAGEMENT COMPUTER CONTROLS, INC.
Estimator     - DOUG                                                     2881 DIRECTORS COVE
                                                                         MEMPHIS, TENNESSEE    38131

        4.005     EXTERIOR DOORS & WINDOWS

------------------------DIMENSIONS--------------------------------------------------SPECIFICATION SELECTION-----------------

B-SQFT,LENGTH (FEET & DECIMAL OF                                 9-METHOD OF CALCULATING AREA    4-OVERHEAD DOORS
C-HEIGHT (FEET & DECIMAL OF FOOT                                 8-EXTERIOR GLASS & GLAZING      3-EXTERIOR SILLS
                                                                 7-WINDOWS                       2-INTERIOR SILLS
                                                                 6-GLASS DOORS
                              A-QUANTITY                         5-EXTERIOR DOORS

WS LINE     B       C      D      E      F      G      H      J      K    A   SPEC. SEL.   S/D        MARK FIELD
    SEQ                                                                       9876543210 SEC  ELEM LOC


DW   11                                                                   5   00001       01         01
DW   42                                                                   2   00001       02         01
DW   73                                                                   2   0002        03         01
DW  104   900.00                                                          1   102         03         01


                     S/D  SEC  ELEM  ITEM CODE        DESCRIPTION            U/M    QUANTITY

                           01  0616   805.100   HM FRAME                     EACH        5
                           02  0616   805.100   HM FRAME                     EACH        2
                           01  0616   805.200   HM DOOR                      EACH        5
                           02  0616   805.200   HM DOOR                      EACH        2
                           03  0421   851.110   FIXED WINDOWS                SQFT      900
                           01  0616   871.027   FINISH HARDWARE ALLOWANCE    OPNG        5
                           02  0616   871.027   FINISH HARDWARE ALLOWANCE    OPNG        2
                           03  0423   885.205   PAIR ALUM GLASS DOOR         EACH        2
                           01  0423   990.008   PAINT EXTERIOR DOOR          SIDE       10
                           02  0423   990.008   PAINT EXTERIOR DOOR          SIDE        4
                           01  0616   990.032   PAINT DOOR FRAME             EACH        5
                           02  0616   990.032   PAINT DOOR FRAME             EACH        2
```

```
     File ID - 256A                 Q U A N T I T Y   S U R V E Y   R E P O R T                        Page -       8
Client Job No -                                                                                        Date -
Project Name  - WHATCO PROJECT                                                                         Time -
Project Size  -     50,000 SQFT                                              MANAGEMENT COMPUTER CONTROLS, INC.
Estimator     - DOUG                                                         2881 DIRECTORS COVE
                                                                             MEMPHIS, TENNESSEE    38131

          5.000     ROOFING,SHEETMETAL, & ACCESSORIES

------------------------DIMENSIONS---------------------------------------------------SPECIFICATION SELECTION-----------------

B-SQFT,LENGTH (FEET & DECIMAL OF                                     9-METHOD OF CALCULATING AREA    4-CEMENTITIOUS DECK
C-WIDTH (FEET & DECIMAL OF FOOT)                                     8-TYPE OF ROOF SYSTEM           3-FLASHING
                                                                     7-TYPE OF SHINGLE               2-GUTTER & DOWNSPOUT
                                                                     6-OTHER TYPE OF ROOF            1-MISC
                              A-QUANTITY                             5-SKYLIGHT

WS LINE    B       C       D       E       F       G       H       J       K     A   SPEC. SEL.   S/D        MARK FIELD
   SEQ                                                                               9876543210 SEC  ELEM LOC

DW    6 30000.00                                                                 1   100001     01        01
DW   37  6000.00                                                                 1   100001     02        01
DW   68  7000.00                                                                 1   100001     03        01


                         S/D  SEC  ELEM  ITEM CODE       DESCRIPTION              U/M    QUANTITY

                              01   0321   350 100   LTWT CONC ROOF FILL           SQFT     30000
```

File ID - 256A
Client Job No -
Project Name - WHATCO PROJECT
Project Size - 50,000 SQFT
Estimator - DOUG

QUANTITY SURVEY REPORT

Page - 9
Date -
Time -

MANAGEMENT COMPUTER CONTROLS, INC.
2881 DIRECTORS COVE
MEMPHIS, TENNESSEE 38131

6.200 INTERIOR PARTITIONS

------------------------DIMENSIONS------------------------SPECIFICATION SELECTION----------------

B-SQFT, LENGTH (FEET & DECIMAL O
C-HEIGHT (FEET & DECIMAL OF FOOT
D-THICKNESS OF BLOCK (FEET & DEC
A-QUANTITY

9-METHOD OF CALCULATING
8-MASONARY PARTITION
7-METAL STUD & GYPSUM PARTITION
6-WOOD STUD & GYPSUM PARTITION
5-ADDITIONAL LAYER OF GYPSUM
4-PLASTER PARTITION
3-MISC PARTITION
2-GLASS PARTITION & WINDOWS
1-RAILING
0-MISC. PARTITION

WS	LINE SEQ	B	C	D	E	F	G	H	J	K	A	SPEC. SEL. 9876543210	S/D SEC	ELEM	LOC	MARK FIELD
DW	99	12670.00		.67							1	11	03		01	
DW	100	8900.00									1	101	03		01	
DW	101	7950.00									1	106	03		01	

S/D	SEC	ELEM	ITEM CODE	DESCRIPTION	U/M	QUANTITY
	03	0611	402.102	8" CONC BLOCK	SQFT	12670
	03	0611	926.620	STANDARD DRYWALL PARTITION	SQFT	8900
	03	0611	926.625	MTL FURRING W/GYPSUM BOARD	SQFT	7950

File ID - 256A
Client Job No -
Project Name - WHATCO PROJECT
Project Size - 50,000 SQFT
Estimator - DOUG

QUANTITY SURVEY REPORT

Date -
Time -

MANAGEMENT COMPUTER CONTROLS, INC.
2881 DIRECTORS COVE
MEMPHIS, TENNESSEE 38131

6.205 INTERIOR DOORS & FRAMES

------------------------DIMENSIONS------------------------SPECIFICATION SELECTION----------------

B-# OF DOORS PER OPENING (WHOLE

A-QUANTITY

9-WOOD DOORS
8-METAL DOORS
7-FRAMES
6-HARDWARE ALLOWANCE
5-THRESHOLD
4-MISC DOORS

WS	LINE SEQ	B	C	D	E	F	G	H	J	K	A	SPEC. SEL. 9876543210	S/D SEC	ELEM	LOC	MARK FIELD
DW	102	1.00									8	0115	03		01	
DW	103	1.00									36	1015	03		01	

S/D	SEC	ELEM	ITEM CODE	DESCRIPTION	U/M	QUANTITY
	03	0616	805.100	HM FRAME	EACH	44
	03	0616	805.200	HM DOOR	EACH	8
	03	0616	823.000	SC WOOD DOOR	EACH	36
	03	0616	871.027	FINISH HARDWARE ALLOWANCE	OPNG	44
	03	0616	990.031	PAINT INTERIOR DOOR	SIDE	88
	03	0616	990.032	PAINT DOOR FRAME	EACH	44

File ID - 256A
Client Job No -
Project Name - WHATCO PROJECT
Project Size - 50,000 SQFT
Estimator - DOUG

QUANTITY SURVEY REPORT

Page - 11
Date -
Time -

MANAGEMENT COMPUTER CONTROLS, INC.
2881 DIRECTORS COVE
MEMPHIS, TENNESSEE 38131

6.210 FINISHES

------------------------DIMENSIONS------------------------SPECIFICATION SELECTION----------------

B-SQFT OF FINISH AREA,LENGTH OF
C-LENGTH OF FINISHED PARTITION,W
D-FINISH WALL HEIGHT (FEET & DEC
E-WAINSCOT HEIGHT (FEET & DECIMA
F-ACT TYPE
A-QUANTITY

9-METHOD OF CALCULATING AREAS
8-CORRELATE SIDES (USE SS8 WHEN
7-FLOOR FINISH
6-FLOOR FINISH (CONT)
5-FLOOR FINISH (CONT)
4-BASE
3-WAINSCOT
2-WALL FINISH
1-CEILING FINISHES
0-CEILING SYSTEM

WS	LINE SEQ	B	C	D	E	F	G	H	J	K	A	SPEC. SEL. 9876543210	S/D SEC	ELEM	LOC	MARK FIELD
DW	105	12300.00									1	1000000003	03		01	
DW	106	900.00									1	101	03		01	
DW	107	3825.00									1	1001	03		01	
DW	108	7875.00									1	10004	03		01	

S/D	SEC	ELEM	ITEM CODE	DESCRIPTION	U/M	QUANTITY
	03	0622	931.010	CERAMIC TILE FLOOR	SQFT	900
	03	0623	950.400	ACOUST CEIL SYS-EXPOSED GRID	SQFT	12300
	03	0622	965.010	VINYL COMPOSITION TILE	SQFT	3825
	03	0622	968.200	CARPET	SQYD	875

File ID - 256A
Client Job No -
Project Name - WHATCO PROJECT
Project Size - 50,000 SQFT
Estimator - DOUG

QUANTITY SURVEY REPORT

Page - 12
Date -
Time -

MANAGEMENT COMPUTER CONTROLS, INC.
2881 DIRECTORS COVE
MEMPHIS, TENNESSEE 38131

6.215 GENERAL SPECIALTIES

------------------------DIMENSIONS------------------------SPECIFICATION SELECTION----------------

B-LENGTH (FEET & DECIMAL),#FLOOR
C-WIDTH (FEET & DECIMAL OF FOOT)
D-$ ALLOWANCE,W/SS1

A-QUANTITY

9-CHALK/TACK BOARDS
8-CHUTES
7-WALL & CORNER GUARDS
6-LOCKERS
5-FIRE EXTINGUISHERS & CABINETS
4-PROJECTION SCREEN
3-TRACKS
2-MISC.
1-ALLOWANCES

WS	LINE SEQ	B	C	D	E	F	G	H	J	K	A	SPEC. SEL. 9876543210	S/D SEC	ELEM	LOC	MARK FIELD
DW	110										40	0002	03		01	

S/D	SEC	ELEM	ITEM CODE	DESCRIPTION	U/M	QUANTITY
	03	0631	1050.001	DOUBLE TIER LOCKER	OPNG	40

File ID - 256A
Client Job No -
Project Name - WHATCO PROJECT
Project Size - 50,000 SQFT
Estimator - DOUG

QUANTITY SURVEY REPORT

Page - 13
Date -
Time -

MANAGEMENT COMPUTER CONTROLS, INC.
2881 DIRECTORS COVE
MEMPHIS, TENNESSEE 38131

6.220 TOILET & BATH ACCESSORIES - COMM./INDUSTRIAL/HOSPT

------------------------DIMENSIONS--SPECIFICATION SELECTION-----------------

B-PCS REQUIRED PER PLACE (WHOLE
C-LENGTH OF SHELF (FEET)
D-WIDTH OF MIRROR (INCHES)
E-HEIGHT OF MIRROR (INCHES)
A-QUANTITY

9-COMPARTMENTS
8-GRAB BAR/TOWEL BAR/SHELF
7-TOILET PAPER HOLDER
6-DISPENSER/DISPOSALS
5-SOAP DISPENSERS

4-MIRRORS
3-HOSPITAL ACCESSORIES
2-MISC.

WS	LINE SEQ	B	C	D	E	F	G	H	J	K	A	SPEC. SEL. 9876543210	S/D SEC	ELEM	LOC	MARK FIELD
DW	111	1.00									12	1	03		01	

S/D	SEC	ELEM	ITEM CODE	DESCRIPTION	U/M	QUANTITY
	03	0631	1018.010	TOILET COMPARTMENT	EACH	12

QUANTITY SURVEY REPORT

File ID - 256A
Client Job No -
Project Name - WHATCO PROJECT
Project Size - 50,000 SQFT
Estimator - DOUG

Page - 14
Date -
Time -

MANAGEMENT COMPUTER CONTROLS, INC.
2881 DIRECTORS COVE
MEMPHIS, TENNESSEE 38131

7.010 ROOFING

------------------------DIMENSIONS------------------------SPECIFICATION SELECTION------------------

B-LENGTH OF ROOF AREA	G-LENGTH OF DOWNSPOUT	9-ROOF INSULATION	4-GUTTER
C-WIDTH OF ROOF AREA	H-NOMINAL SIZE OF BLOCKING	8-BUILT-UP ROOF	3-DOWNSPOUT
D-THICKNESS OF INSULATION	J-LENGTH/DIA OF ROOF ACCESSORIES	7-SHINGLE ROOFS	2-ROOF ACCESSORIES
E-WIDTH OF SHEET METAL W/SS-5&6	K-WIDTH OF ROOF ACCESSORIES	6-SIDE REQ. FOR SHT MTL/BLOCKING	1-WALKWAYS
F-LENGTH OF GUTTER	A-QUANTITY	5-SHEET METAL TYPE	0-ELEMENT

WS	LINE SEQ	B	C	D	E	F	G	H	J	K	A	SPEC. SEL. 9876543210	S/D SEC	ELEM	LOC	MARK FIELD
DW	8	300.00	100.00	2.00	1.00	400.00	245.00				1	1201323	01		01	
DW	39	60.00	50.00	2.00	1.00	50.00	35.00				2	1201323	02		01	
DW	70	100.00	70.00	2.00	1.00	100.00	70.00				1	1206323	03		01	

S/D SEC	ELEM	ITEM CODE	DESCRIPTION	U/M	QUANTITY
01	0501	751.002	3 PLY TAR& GRAVEL ROOFING	SQS	300
02	0501	751.002	3 PLY TAR& GRAVEL ROOFING	SQS	60
03	0501	751.002	3 PLY TAR& GRAVEL ROOFING	SQS	70
01	0503	755.003	2" FIBER BD ROOF INSULATION	SQFT	30000
02	0503	755.003	2" FIBER BD ROOF INSULATION	SQFT	6000
03	0503	755.003	2" FIBER BD ROOF INSULATION	SQFT	7000
01	0504	762.040	24 GA GALV IRON SHEETMETAL	SQFT	800
02	0504	762.040	24 GA GALV IRON SHEETMETAL	SQFT	440
03	0504	762.040	24 GA GALV IRON SHEETMETAL	SQFT	140
01	0504	762.064	6" GALV DOWNSPOUT	LNFT	245
02	0504	762.064	6" GALV DOWNSPOUT	LNFT	70
03	0504	762.064	6" GALV DOWNSPOUT	LNFT	70
01	0504	762.069	6" GALV GUTTER	LNFT	400
02	0504	762.069	6" GALV GUTTER	LNFT	100
03	0504	762.069	6" GALV GUTTER	LNFT	100

File ID - 256A
Client Job No -
Project Name - WHATCO PROJECT
Project Size - 50,000 SQFT
Estimator - DOUG

QUANTITY SURVEY REPORT

Page - 15
Date -
Time -

MANAGEMENT COMPUTER CONTROLS, INC.
2881 DIRECTORS COVE
MEMPHIS, TENNESSEE 38131

9.090 FINISHES

------------------------DIMENSIONS------------------------SPECIFICATION SELECTION----------------

B-LENGTH OF AREA
C-WIDTH OF AREA
D-CEILING HEIGHT
E-WAINSCOT HEIGHT
F-ACOUSTICAL TILE SZ LNGTH & WDT
G-TILE/GYPSUM BOARD THICKNESS
A-QUANTITY

9-CORRELATED SIDES
8-TYPE OF FLOOR FINISH
7-TYPE OF FLOOR FINISHES (CONT.)
6-TYPE OF BASE
5-WAINSCOT
4-WALL FINISHES
3-CEILING FINISHES
2-TYPE OF CEILING
1-CEILING SUSPENSION SYSTEMS
0-ELEMENT

WS	LINE SEQ	B	C	D	E	F	G	H	J	K	A	SPEC. SEL. 9876543210	S/D SEC	ELEM	MARK FIELD LOC
DW	12	6000.00		1.00							1	700005	01		01
DW	43	3600.00		1.00							1	700005	02		01
DW	74	10640.00		1.00							1	700005	03		01
DW	109	5150.00		10.00							1	700002	03		01

S/D SEC	ELEM	ITEM CODE	DESCRIPTION	U/M	QUANTITY
03	0621	990.061	PAINT PLASTER-GYP BD 3 CTS	SQS	515
01	0621	990.081	PAINT MAS-CONC 3 CTS	SQS	60
02	0621	990.081	PAINT MAS-CONC 3 CTS	SQS	36
03	0621	990.081	PAINT MAS-CONC 3 CTS	SQS	107

File ID - 256A
Client Job No -
Project Name - WHATCO PROJECT
Project Size - 50,000 SQFT
Estimator - DOUG

QUANTITY SURVEY REPORT

Page - 16
Date -
Time -
MANAGEMENT COMPUTER CONTROLS, INC.
2881 DIRECTORS COVE
MEMPHIS, TENNESSEE 38131

11.000 EQUIPMENT/FURNISHINGS

----------DIMENSIONS---------- ----------SPECIFICATION SELECTION----------

B-LENGTH (FEET & DECIMAL OF FOOT
C-WIDTH (FEET & DECIMAL OF FOOT)
D-$ ALLOWANCE, W/SS1
A-QUANTITY

9-DOCK EQUIPMENT
8-DETENTION EQUIPMENT
7-BANK EQUIPMENT
6-X-RAY
5-FURNISHINGS
4-RECEPTACLES & PLANTERS
3-SEATING
2-MISC
1-ALLOWANCES (LS)
0-RESIDENTIAL KITCHEN EQUIP.

WS	LINE SEQ	B	C	D	E	F	G	H	J	K	A	SPEC. SEL. 9876543210	S/D SEC	ELEM	LOC	MARK FIELD
DW	13										4	2	01		01	
DW	44										2	2	02		01	

S/D	SEC	ELEM	ITEM CODE	DESCRIPTION	U/M	QUANTITY
	01	1116	1116.000	DOCK LEVELER	EACH	4
	02	1116	1116.000	DOCK LEVELER	EACH	2

Chapter

9

Estimating and Value Engineering

Value analysis developed as a specific technique after World War II. The development work was done at the direction of the General Electric Company's vice president of purchasing, Harry Erlicher, who observed that some of the substitute materials and designs utilized as a necessity because of wartime shortages offered superior performance at lower cost. He directed that an in-house effort be undertaken to improve product efficiency by intentionally developing substitute materials and methods to replace the function of more costly components. This task was assigned in 1947 to Lawrence Miles, staff engineer with General Electric Company. Miles researched the techniques and methodology available and utilized a number of proven approaches in combination with his own procedural approach to analysis for value. The value-analysis technique was accepted as a G.E. standard, and gradually other companies and governmental organizations adopted the new approach as a means of reducing costs.

In 1954, the Navy Bureau of Ships became the first Department of Defense organization to set up a formal value-analysis program, and Lawrence Miles was instrumental in the development of that program. The program was retitled "value engineering" (VE) to reflect the engineering emphasis of the Bureau of Ships.

In 1956, the Army Ordnance Corps initiated a value-engineering program, again with the assistance of G.E. personnel. This program has continued over the years, and the Army Management Engineering Training Agency offers value-engineering training as part of its curriculum.

In 1961, the Air Force became interested in the potential of value engineering as a result of the effectiveness of applications by Air Force weapons systems contractors such as G.E. This interest was one of the building blocks which, in 1962, led Secretary of Defense

McNamara to place his prestige behind the use of value engineering, calling it a key element in the drive to reduce defense costs.

To implement the directive of Secretary McNamara, value engineering was included as a mandatory requirement by the Armed Service Procurement Regulations (ASPR). The Armed Forces' definition of value engineering was "a systematic effort directed at analyzing the functional requirement of the Department of Defense systems, equipment, facilities and supplies for the purpose of achieving essential functions at the lowest total cost, consistent with the needed performance, reliability, quality, and maintainability."

Value Analysis in Construction

Prior to the introduction of value engineering into ASPR, applications in the construction field were random and infrequent. Although the adoption of value engineering by the Department of Defense was oriented principally toward the purchase of materials, equipment, and systems, the change in ASPR automatically introduced value engineering to two of the largest construction agencies in the country: the U.S. Army Corps of Engineers and the U.S. Navy Bureau of Yards and Docks. In the first 10 years of use, the Corps of Engineers estimated savings of almost $200 million, with most of the savings the result of its own reevaluations of major projects. In that time period, 2200 contractors had submitted cost-cutting value-engineering suggestions, of which 1400 were accepted with the cash-shared savings of $7 million.

As the success of value engineering, principally in the construction phase, was documented, other federal agencies began tentative steps toward the adoption of value engineering. In 1965, the Bureau of Reclamation undertook value-engineering training for its engineering staff and in 1966 placed a value-engineering incentive clause in its construction contracts.

In 1967, the Post Office Department set up a value-engineering staff in its Bureau of Research and Engineering. In that same year, the Senate Committee on Public Works held hearings on the use of value engineering in the government at which many of the major agencies exchanged information on their utilization of value engineering. In 1969, the Office of Facilities of the National Aeronautics and Space Administration (NASA) began formal value-engineering studies and training. The U.S. Department of Transportation published a value-engineering incentive clause to be used by its agencies. In that same year, the Public Building Service of the General Services Administration (PBS/GSA) set up its value-engineering staff.

Until 1972, the construction industry, in general, had only limited interest in value analysis or value engineering. In 1972, the twelfth annual conference of the Society of American Value Engineers (SAVE) emphasized the application of value analysis in the construction industry. Chartered in 1959, SAVE has had much to do with the evolution of value analysis, particularly in the federal establishment. Key SAVE members in the Washington chapter are predominantly from federal departments. The SAVE conference provided a forum where some 400 engineers, architects, and other industry members heard the specifics of progress which had been made. This conference is now an annual event.

Role of Estimators

Prior to the VE review, team estimators do one of several things:

- Review and verify (spot check) an estimate by the architect, the construction manager, or independent estimator
- Prepare an independent estimate
- Collect cost information relevant to the project to be reviewed

which will be done as a function of time and funding available.

During the 40-hour workshop, the estimators participate in all phases—both to contribute ideas and to understand the ideas to be evaluated.

In the evaluation of potential recommendations, estimators are key to the decision process. Their calculation of "as planned" and "as proposed" costs are usually the deciding factor in whether (or not) to recommend a VE change.

Function Analysis

The first phase of the 40-hour workshop (performed by a multidiscipline team of architects, engineers, specialists, and estimators) is *function analysis:*

Lawrence Miles described value analysis as "a disciplined action system, attuned to one specific need: accomplishing the functions that the customer needs and wants...." He further indicated that "the basic function is straightforward to determine, and any work done before that has been accomplished is wasted effort...."

The Public Building Service describes function analysis in this way: "A user purchases an item or service because it will provide certain functions at a cost he is willing to pay. If something does not do what it is intended to do, it is of no use to the user and no amount of cost reduction will improve its value. Actions that sacrifice needed utility of an item actually reduce the value to the user. On the other hand, expenditures to increase the functional capacity of an item beyond that which is needed are also of little value to the user...." The value-management office of PBS sees anything less than the necessary functional capability as unacceptable—and anything more as unnecessary and wasteful. A project or part of a project which is to receive function analysis is addressed with the six basic questions of value analysis:

What is it?

What does it do?

What is its worth?

What does it cost?

What else would work?

What does that cost?

Within the area to be analyzed, selection of the functions which should be considered can be guided by the Pareto distribution shown in Fig. 9.1. Pareto's law of distribution, often referred to by value analysts, suggests that 20 percent of the items in any complex thing account for 80 percent of the cost. This distribution does not exactly hold for construction projects. However, it is axiomatic that a minority of the components of a project make up the majority of the cost. Within this minority of high-cost areas, the best opportunities can be found for value analysis. The identification of the high-value items narrows the number of viable opportunities for value analysis which remain in a project. Analysis of the high-value areas produces the "best bang for the buck." Standard cost-breakdown models for types of projects can be utilized in identifying the potential areas for best results.

It is the consensus of value specialists that the most productive areas of opportunity should be selected for analysis—and these areas are inherently the most expensive ones.

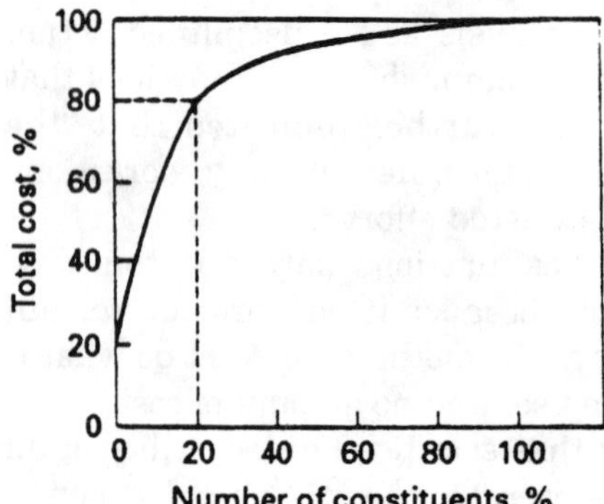

Figure 9.1 Pareto's law of distribution.

What Does It Do?

This is the key value-analysis question. It requires definition of the function of the item under study. The method of this definition is prescribed by established value-analysis procedure as two words, a *noun* and a *verb*.

The simplicity of the approach is deceptive. Selection of the proper two-word description is often quite different, requiring comprehensive understanding of the item under study.

The functional description is not necessarily correct when it is the most obvious—and therein lies the potential for successful analysis. The analyst is not limited to a single two-word description. In fact, a series of descriptions can apply to the same item. However, *only one* of these descriptions is the basic function. The other descriptions, necessarily, become secondary functions. Secondary functions may be important, but they are not controlling.

It is quite usual to find that many have assumed that the basic function is really one which had always been considered a secondary one.

Some typical verbs which can be used to describe construction functions are the following:

absorb	enclose	protect
amplify	filter	reduce
apply	generate	reflect
change	hold	reject
collect	improve	separate
conduct	increase	shield
control	insulate	support
create	interrupt	transmit
decrease	prevent	

Some typical nouns that are used in construction that are measurable are the following:

circuit	force	power
contamination	friction	pressure
current	heat	protection
damage	insulation	radiation
density	light	repair
energy	liquid	voltage
flow	noise	water
fluid	oxidation	weight

Some typical nouns that are used in construction that are aesthetic are the following:

appearance	effect	prestige
beauty	features	style
convenience	form	symmetry

The difficulty of identifying in a two-word phrase is often substantial. Accordingly, nouns such as thing, part, article, device, or component generally indicate a failure to conclusively define. The noun should be measurable or capable of quantification.

For instance, a water service line to a building could have the function "provide service," but a term "service" is not readily measurable. A better definition would be "transport water." The noun in this definition is measurable, and alternatives can be determined in terms of the quantity of water to be transported or, in special situations such as laboratories, in terms of the quality of water being transported.

The functional definition must concern itself with the type of use as well as the indemnification of the item. A piece of wire might "conduct current," "fasten part," or "transfer force," depending upon the specific utilization.

The spartan restraint involved in holding the description to two words can be difficult to apply. There is often a temptation to utilize a slightly broader definition by insertion of additional adjectives to condition the noun. Some value analysts permit this variation, but the tried and tested approach insists upon the two-word abridgment. The rigorous application of the two-word functional definition often discloses a factual view of the project category which was subconsciously available but not consciously realized.

For instance, one perception would be that it is not the door itself but the doorway or the accessway which is the true basic purpose for the specification of a door at a given location. The overall context of the application has to be considered. In an office building, the basic purpose of a doorway is to "provide access," while in a prison this might be to "control access." An interior door at a fire wall does "provide access," but its basic purpose is to "control fire." A doorway at a classroom entrance might be there to "control noise."

At a detail level, doors throughout a building would be separate categories. To evaluate the generic category, the subbreakdown would be an important factor. Figure 9.2 describes some of the function definitions for doorways in a complex courthouse structure. The cost of certain doors, such as the elevator shaft doors, would be included in other systems. However, a subset of the system of doors is the door

VALUE ENGINEERING FUNCTION ANALYSIS WORKSHEET

PROJECT: Courthouse NAME: ____________

ITEM: Doors TEL. NO.: ____________

BASIC FUNCTION: Control Access DATE: ____________

ELEMENT DESCRIPTION				
Exterior doors	Provide	Security	B	Control of access manditory
	Exclude	Weather	S	Alternates possible
	Express	Prestige	S	Image to the public
	Provide	Visibility	S	Entry doors
Ofice doors	Exclude	Noise	B	Judge's chambers
	Contain	Noise	B	Typing pool
	Enclose	Space	B	Public areas
	Provide	Visibility	S	Public areas
Area doors	Exclude	Fire	B	Exit passageways
	Contain	Fire	B	Mechanical rooms
	Control	Air	B	HVAC zone perimeter
	Control	Traffic	B	Key zone points
Security doors	Control	Prisoners	B	Secure passage to courtrooms
	Contain	Prisoners	B	Cell doors; cell block
	Exclude	Public	B	Doors from public areas
	Exclude	People	B	Elevator shaft doors
Other	Provide	Privacy	B	Bathroom doors

Figure 9.2 Function definitions for doorways in a complex courthouse structure.

bucks in which they mount. That system would be included in the hollow metal or miscellaneous metal subcontract.

The item under study can, and should, have different levels of indenture as identified in the work-breakdown structure. Figure 9.3 illustrates a breakdown structure for a fire alarm system prepared by PBS. The second level of indenture breaks down the system into two sectors: person and equipment. This identifies the system as semiautomatic. The table in Fig. 9.3 shows the function definition developed with these various levels of indenture. At the third, or detailed, level the basic function of the bell was selected as "make noise," which would permit a greater latitude in the development of alternative ways of effectively making noise not limited to the bell.

What Else Would Work?

After the establishment of the function definition, the next key stage is the application of creativity to determine alternatives which would

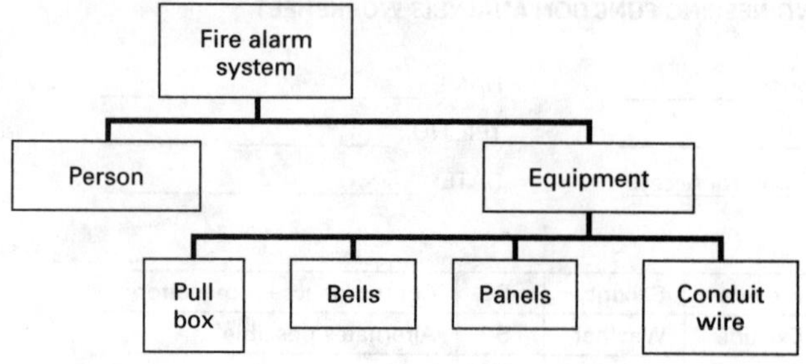

Level of indentation	Component	Functions	Classification B = Basic S = Secondary
1	Fire alarm system	Make noise Detect fire Protect building	B B S
2	Person	Detect fire Pull lever	B S
	Equipment	Make noise Transfer signals	B S
3	Pull boxes	Break circuit	S
	Bells	Make noise	B
	Panels	Provide power Control circuits	S S
	Conduit & wire	Transmit signal Transmit power	S S

Figure 9.3 Breakdown structure for a fire alarm system. (Public Building Service GSA.)

also perform the same function. Creativity is one of the essential techniques of value analysis. It is one of those existing attitudes important to value analysis which value analysts have incorporated into the total value-management approach.

It is human nature for people to resist change—to resist either stopping things which they are doing or initiating new ideas. The development of alternative approaches to meet functional analysis requires a breakdown of this inertial resistance to change.

There are many roadblocks to the creative approach, including but not limited to the following:

1. Fear of making a mistake or appearing to make a mistake
2. Unwillingness to change the accepted norm
3. A desire to conform or adapt to standard patterns
4. Overinvolvement with the standard conceptions of function
5. Unwillingness to consider new approaches

6. Unwillingness to be considered rash or unconservative
7. Unwillingness to appear to criticize, even constructively
8. Lack of confidence resulting from lack of knowledge
9. Overconfidence because of experience, however limited
10. Unwillingness to reject a solution which has previously been shown to be workable
11. Fear of authority and/or distrust of associates
12. Unwillingness to be different
13. Desire for security
14. Difficulty in isolating the true function requirements
15. Inability to distinguish between cause and effect
16. Inability to collect complete information

The roadblocks often manifest themselves in the form of subtle objections or standard platitudes, such as:

1. It may be okay, but management will never buy it.
2. We tried that approach before, and it didn't work.
3. Somebody tried that approach before, and it didn't work.
4. It might work, but it's not our responsibility.
5. It might work for someone else, but our requirements are unique.
6. It sounds okay, but we're not ready to progress that rapidly.
7. It sounds good, but we don't have enough time to implement the new approach.
8. Let's assign it to a committee.
9. Management might agree, but the union would be against it.
10. It's against company policy (we think).
11. We don't have the authority to make this change.
12. It's a good idea, but it would never work.
13. It's never been tried before.

A key technique in the development of creative alternatives is the use of a think-tank or brainstorming approach. The group is used to develop ideas without evaluating them. The creative session is not limited to ideas which will probably work but to a free-form type of idea listing. The idea is to break through typical inhibitions and restraints and list even outlandish ideas. Rube Goldberg would feel right at home in this kind of creative session.

In brainstorming, the starting point is the two-word abridged function definition. Ideas which are abstract, even humorous, are encouraged, because through stream-of-consciousness thinking and free associating, these ideas may lead to more practical considerations.

The creative session has its own set of ground rules to encourage free and open thinking. Not only is evaluation and judgment deferred, but also critical response is discouraged until the entire listing has been developed.

Judicial Analysis Phase

The ideas developed in the creative phase are screened to eliminate those which do not meet the function requirements. The viable alternatives are identified and ranked according to feasibility and cost. Various techniques can be used, most based on a matrix comparison, to develop and select the best solution.

Development Phase

The alternatives selected for potential implementation are now reviewed in depth. The ideas are checked out with purchasing departments, vendors, and other specialists in the field. Cost factors are verified. Ideas are reviewed for flaws or problem areas, and a development plan is established. The plan is based upon the alternative or alternatives which are economically viable.

Presentation

The recommendations are presented to the owner and design professional, usually in the afternoon of the fifth day. The presentation is accompanied by a preliminary VE report including:

- Function analysis
- Speculation phase
- Idea evaluation
- Idea development, including cost
- Recommendations

Figure 9.4 is a schematic of the 40-hour job plan.

40-hour VE workshop

O'Brien-Kreitzberg uses the following job plan for our 40-hour workshop. This job plan conforms with the Society of American Value Engineers (SAVE) guidelines for a 40-hour VE workshop:

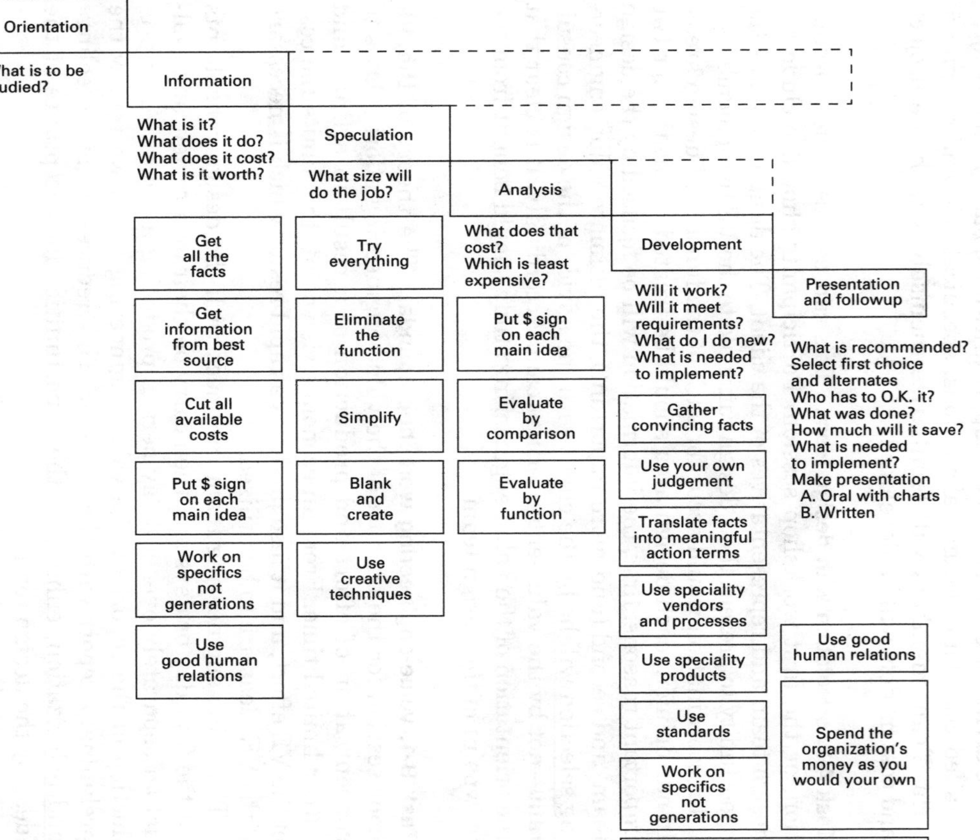
VALUE ENGINEERING JOB PLAN CHART
Orientation
What is to be studied?
Information
What is it?
What does it do?
What does it cost?
What is it worth?
Get all the facts
Get information from best source
Cut all available costs
Put $ sign on each main idea
Work on specifics not generations
Use good human relations
Speculation
What size will do the job?
Try everything
Eliminate the function
Simplify
Blank and create
Use creative techniques
Analysis
What does that cost?
Which is least expensive?
Put $ sign on each main idea
Evaluate by comparison
Evaluate by function
Development
Will it work?
Will it meet requirements?
What do I do new?
What is needed to implement?
Gather convincing facts
Use your own judgement
Translate facts into meaningful action terms
Use speciality vendors and processes
Use speciality products
Use standards
Work on specifics not generations
Presentation and followup
What is recommended?
Select first choice and alternates
Who has to O.K. it?
What was done?
How much will it save?
What is needed to implement?
Make presentation
A. Oral with charts
B. Written
Use good human relations
Spend the organization's money as you would your own
Setting the proposal

Figure 9.4 Value-engineering job-plan chart. (U.S. Department of Defense.)

Task A-1, precoordination (preworkshop). During this stage, the value-engineering team coordinator (VETC) collects drawings and specifications, reviews information, and subdivides into projects. The coordinator (VETC) makes arrangements with key members, arranges logistics for the actual study, and schedules various participants.

The architect-engineer would supply drawings, background reports, detailed cost data, design calculations, specifications, and design criteria. Team members will familiarize themselves with the project, and cost models will be developed.

Task A-2, orientation with design team. This can be part of precoordination, or the first workshop session. Participants should include the design team and representatives of the client. The design team has the opportunity to present their design rationale to the VE team members.

Precoordination would occur for each of the functional design areas. In our opinion, each design area should be treated separately for two important reasons: first, the information will be focused by the design team, and second (and more important), the ultimate value-engineering selection will be by the client with the advice of the design consultant—not by the value-engineering team. Accordingly, it is clear that the orientation of the value-engineering study should benefit from the viewpoint of the design team.

Task B-1, value-engineering workshop agenda. The theory of the 40-hour session for the conduct of the value-engineering effort relates to concentration of effort and production of a result in a finite and rather limited time. From an economic viewpoint, this limits the cost of the VE effort, and it also produces a rapid result which is necessary if the VE ideas are to be utilized.

The value-engineering team can meet with the design consultants for the first day or days and then carry on their own review and evaluation separately over a 3-day period, producing a report. After production of the report by the value-engineering team, we review the preliminary report with the designer. This reduces negative feelings and polarization, enhancing the opportunities for acceptance of the ideas by the design team.

The workshop is divided into five phases:

1. *The information phase.* The value-engineering team becomes familiar with all detailed data important to the design of the project. As noted, the design engineer may present the project at this point, earlier (precoordination), or both.
2. *The creative phase.* The purpose of this phase is to create an extensive list of alternative ways to perform the essential func-

tions of the proposed facility. There is no evaluation of alternatives in this phase in order not to inhibit the flow of ideas.

3. *The analytical phase.* During this phase, the value-engineering team considers the feasibility of the various alternatives developed during the creative phase. Each alternative is evaluated positively rather than critically. The best ideas are selected for further investigation.
4. *The investigation phase.* In this phase, the most feasible alternatives selected in the analytical phase are evaluated in detail. Factors such as cost, performance, reliability, aesthetics, and flexibility are considered.
5. *Recommendation phase.* In this phase draft recommendations are prepared; these are finalized in the postworkshop stage.

Typical problems

Some problems which have been identified in typical value-engineering programs where either the designer or the VE team are inexperienced include:

1. *Inadequate cost and design data prior to the value-engineering study.* Without adequate cost information, it is difficult to compare and analyze value-engineering proposals with the original design. The VE team must identify exactly what additional data are needed to perform the study.
2. *Lack of constructive atmosphere.* The VE team must make realistic proposals that will improve the cost effectiveness of the design without adversely affecting the project schedule. It is important that the value-engineering team set the proper atmosphere and climate for a constructive value-engineering workshop. Value engineering is a fresh look at the project. It should not be a criticism of the design.
3. *Ill-defined constraints.* Major problems can occur in a value-engineering study if the project constraints are not well established in the prestudy stage. Many worker-hours of time can be wasted analyzing systems that were not intended for evaluation. As part of the precoordination and/or the designer's briefing of the team, unchangeable project criteria must be identified.
4. *Inadequate detail in the preliminary report.* Probably the most important aspect of the value-engineering study is the preliminary value-engineering report, which is the basis for the design engineer's and the owner's acceptance or rejection of the value-engineering proposals. Each recommendation must be supported

by before and after design criteria, sketches, and costs. The source of this information must be fully documented so the design engineer will have a thorough basis for acceptance or rejection of the proposal.

5. *Lack of communication with the designer.* It is important, as demonstrated in the creative phase, that the VE team not limit itself by the design approach. However, it is necessary to convince the designer that the change proposed is worthwhile. Informal oral presentation of proposals at the conclusion of the workshop can enhance the better proposals (and may delete some which won't work). This courtesy not only is within the framework of the no fault concept but is obvious common sense.

WHATCO project

Based on the design development estimate, the WHATCO projected construction costs are about 30% over budget. The owner gives O'Brien-Kreitzberg (OK) the assignment to perform a value-engineering review. The review is to take 2 weeks and will be concurrent with two other review teams, one for design review and the other for constructibility review.

Value-engineering team coordinator (VETC) Jim O'Brien, CVS (Certified Value Specialist), assembles a multidisciplined team. Since there are no electrical, mechanical, or site work details to review, the team does not include specialists in those areas. The team includes:

VETC, James J. O'Brien, CVS, P.E.
Architect, Michael Simmons, AIA
Civil Engineer, Ellis Serdikoff, P.E.
Structural Engineer Consultant, P.E.
Chief Estimator, Vincent Pagliaro
Senior Estimator, Lillian Watson
Material Handling Consultant, P.E.
Cranes/Hoists Consultant, P.E.

In the first week, O'Brien arranges to reserve the meeting area which includes:

- Working conference room 20 ft × 20 ft with whiteboards, conference table, and chairs
- Two smaller working areas for separate task team assignments

He also arranges for four sets of the design development plans and outline specifications to be provided. Also several copies of the design development estimate are provided. Xerox and telephone services are also made available. An overhead viewgraph projector is scheduled for days 4 and 5 of the second week.

During week 1, estimators Pagliaro and Watson prepare the following cost models from the estimate:

Figure 9.5, cost summary ($) by building by category

Figure 9.6, cost summary (%) by building by category

Figure 9.7, cost model (%), plant

Figure 9.8, cost model (%), warehouse

Figure 9.9, cost model (%), office

Figure 9.10, cost model ($), plant, warehouse, office

Figure 9.10 includes only cost areas which are potentially controllable at this stage. That is, general conditions, sitework, and mechanical and electrical costs are not shown because they are still undefined.

Figure 9.10 clearly points to cost areas which must be reviewed. First, the majority of cost is clearly in the plant area. Second, three cost categories (structural steel, cranes, and wall panels) make up 70% of the potentially controllable costs. Clearly, the value engineer-

WHATCO Project
Cost Summary by Category ($)

	Plant	Warehouse	Office	Total
General Conditions	$573,058	$151,680	$182,093	906,831
Excavation	59,818	11,619	12,665	84,102
Piles	72,160	-	-	72,160
Concrete	256,943	44,406	44,367	345,716
Brick/Block	-	-	128,009	128,009
Structural Steel Joists/Deck	1,381,465	255,225	307,037	1,943,727
Wall Panels	259,740	95,898	00	355,638
Insulation/Roofing	93,437	20,888	33,036	147,361
Doors & Glaze	88,694	44,076	46,535	179,305
Partitions/Finishes Ceilings	5,679	3,141	224,094	232,914
Conveyances	418,264	201,036	51,200	670,500
Equipment	29,184	29,200	-	58,384
Div. 15 - 16	768,664	143,572	112,434	1,024,670
	4,007,106	1,000,741	1,141,470	6,149,317

Figure 9.5 WHATCO project cost summary by category ($).

WHATCO Project
Cost Summary by Category (%)

	Plant	Warehouse	Office	Total
General Conditions	14.3%	15.2%	16.0%	14.6%
Excavation	1.5	1.2	1.1	1.4
Piles	1.8	--	--	1.2
Concrete	6.4	4.4	3.9	5.7
Brick/Block	--	--	11.2	2.1
Structural Steel Joists/Deck	34.5	25.5	26.9	31.6
Wall Panels	6.5	9.6	--	5.8
Roofing/Insulation	2.3	2.1	2.9	2.4
Paritions/Finishes Ceilings	0.1	0.1	19.6	3.8
Doors/Glaze	2.2	4.4	4.1	2.9
Conveyances	10.4	20.1	4.5	10.9
Equipment	0.7	2.9	--	0.9
Division 15 - 16	19.2	14.4	9.8	16.7
	99.9%	99.9%	100.0%	100.0%

Figure 9.6 WHATCO project cost summary by category (%).

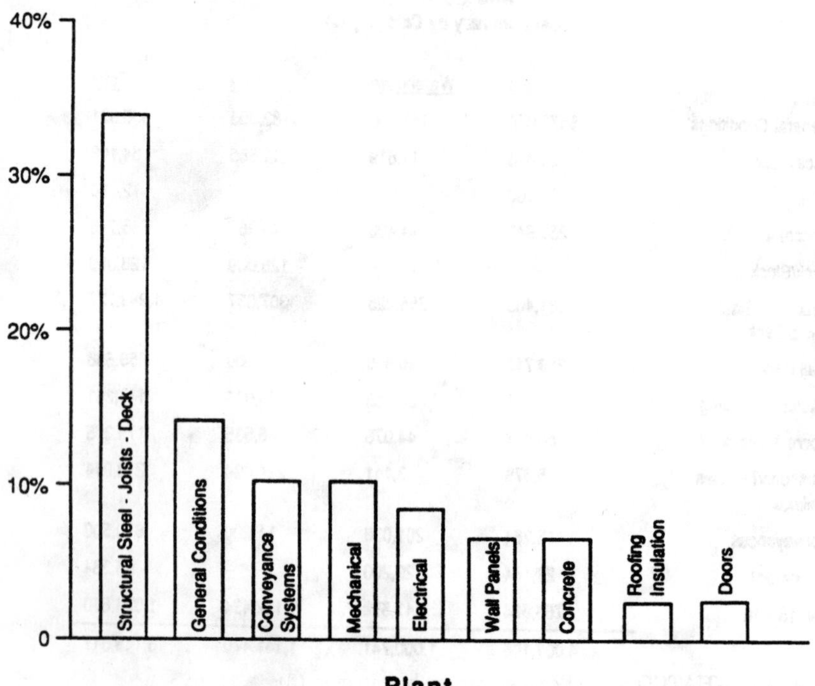

Figure 9.7 Cost model (%), plant.

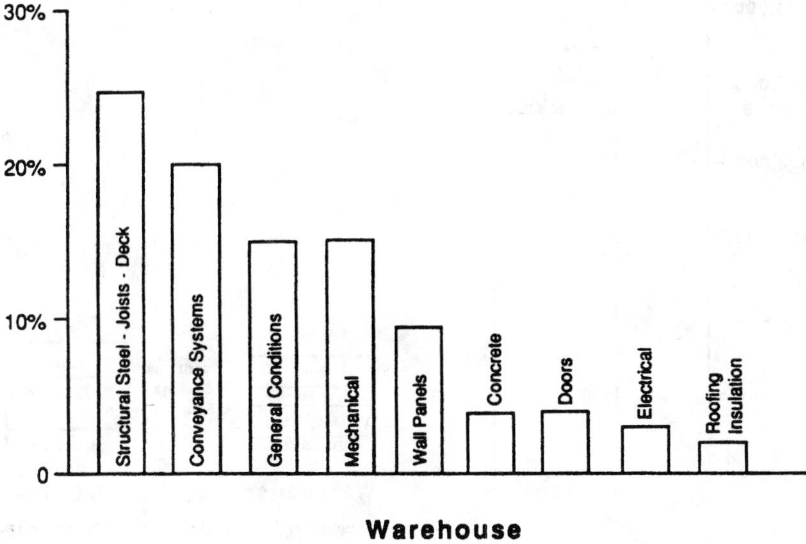

Figure 9.8 Cost model (%), warehouse.

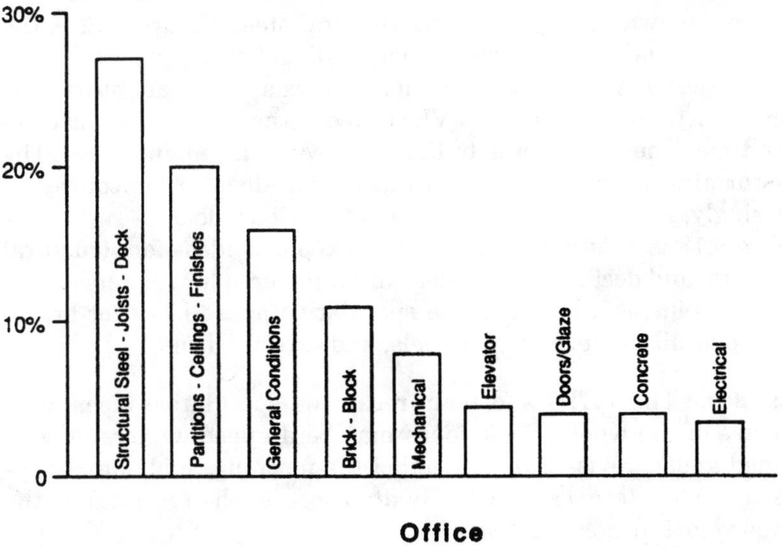

Figure 9.9 Cost model (%), office.

ing must focus on these three categories. This does not preclude the team from recognizing "targets of opportunity."

First day. The architect-engineer team briefed the VE team and answered questions. After they left, O'Brien (the VETC) led the VE

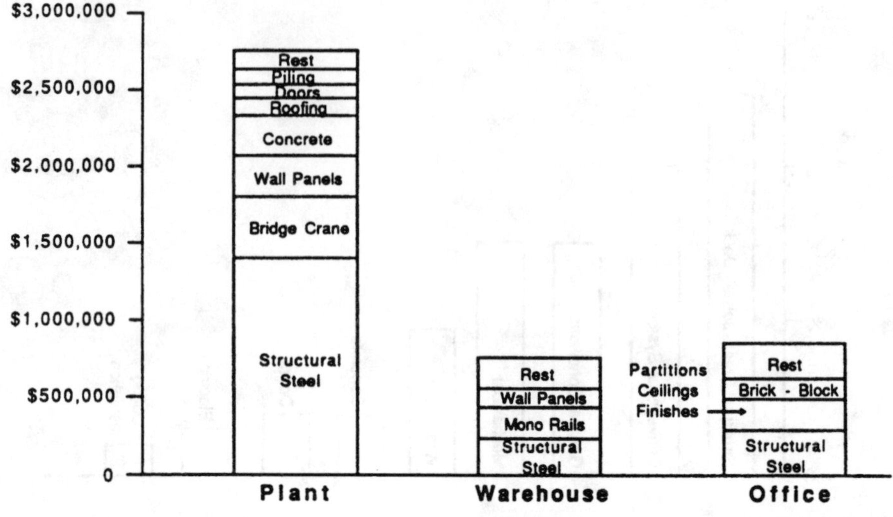

Figure 9.10 Cost model ($) (excluding: general conditions, Divisions 15 and 16).

team in function analysis of major categories of the project. The results are shown in Fig. 9.11 for structural steel. Figure 9.12 is the function analysis for bridge crane, monorail, and wall panels.

After discussion about the function analysis, the team starts the creative speculation phase. The VETC keeps the focus on one category at a time. The approach is to list ideas without judging them. The brainstorming is free form; even humorous ideas are encouraged. Surprisingly, some humorous ideas lead to serious ideas.

Figure 9.13 is a listing of the speculative phase ideas for structural steel, joists, and deck. Note the ideas are numbered for reference.

Figure 9.14(*a*) is a listing of the speculative phase ideas for bridge crane, monorails, exterior wall panels, and exterior doors.

Second day. The VETC and team recap the speculative ideas, as a committee of the whole. Some ideas may be dropped. Others may be combined and/or revised. By the end of the morning, subteam assignments are made (see Table 9.1). By afternoon of the second day, the subteams start idea evaluation.

Third day. Subteams continue idea evaluation. In the afternoon, the team meets as a group to review the idea evaluation forms. The results are shown in Table 9.2.

Fourth day. The subteams continue development of the remaining 13 ideas. In addition to developing their own four ideas, the estimators

ITEM:	
BASIC FUNCTION:	

FUNCTION ANALYSIS

FUNCTION			COMMENTS
VERB	NOUN	KIND *	
			Beams/Girders/Joists
Carry	Loads	B	Live and Dead Load
Span	Space	B	Horizontal
			Columns
Transfer	Loads	B	Live, Dead, and Crane Loads Vertical
			Deck
Carry	Loads	B	Live and Dead Loads
Provide	Surface	B	For Roofing/Insulation
Exclude	Elements	B	Rain/Snow/Wind

* B = BASIC
S = SECONDARY

Figure 9.11 Function analysis for structural steel.

must support the other subteams by developing their nine ideas. During this development, some of the ideas may be demonstrated to be impractical.

S1. Review structural loads. Review of the structural calculations reveals that live load of 100 psf was assumed for the roofs. From

ITEM:	
BASIC FUNCTION:	

FUNCTION ANALYSIS

FUNCTION			COMMENTS
VERB	NOUN	KIND *	
			Bridge Crane
Deliver	Subassemblies	B	Horizontal and Vertical
Move	Product	B	Horizontal and Vertical
			Monorails
Receive	Parts	B	Vertical - Horizontal (Rail)
Store	Parts	B	Vertical - Horizontal (Rail)
Move	Parts	B	Vertical - Horizontal (Rail)
			Wall Panels
Exclude	Elements	B	Rain/Snow/Wind
Support	Insulation	B	Included or Attached
Provide	Aesthetics	S	Paint/Metal Surface/Coating
Carry	Load	B	Wind
			Concrete
Carry	Load	B	Grade Beam/Slab-On-Grade
Distribute	Load	B	Spread Footings/Slab-On-Grade
Transfer	Load	B	Pile Caps

* B = BASIC
S = SECONDARY

Figure 9.12 Function analysis for bridge crane, monorails, wall panels, and concrete.

the BOCA National Building Code (Fig. 9.19) the live load factors are:

$$\text{Minimum} = 12 \text{ psf}$$

$$\text{Wind} = 21 \text{ psf}$$

$$\text{Snow} = 15 \text{ psf}$$

LIST ALL IDEAS-
EVALUATE LATER

SPECULATIVE PHASE

ITEM:

BASIC FUNCTION:

S1. Review loads to see if member sizes can be downsized.

S2. Add central column line to cut Plant span to 50 feet. (Requires two bridge cranes.

S3. Use precast roof plank instead of deck.

S4. Galvanize joists to avoid painting in air.

S5. Finish coat paint on girders on ground/touch-up in air.

S6. Add two column lines to office building to reduce girder spans (change to beams)

S7. Use precast tees for loading docks.

Foundations/Concrete

F1. Delete non-load bearing grade beams.

F2. Use curb (3'-0" deep) in place of non-load grade beams.

F3. Use welded wire fabric in place of rebar in Slab-On-Grade.

F4. Use spread footings in place of piling.

OKA FORM 124

Figure 9.13 Speculative phase: structural steel and joists deck.

LIST ALL IDEAS-
EVALUATE LATER

SPECULATIVE PHASE	
ITEM:	
BASIC FUNCTION:	

Cranes and Monorails

C1. Use 50 ton mobile straddle carrier instead of bridge crane (see Fig. 9.14A)

C2. Use hi-reach forklifts plus motorized hoists in place of monorails.

Exterior Wall Panels

W1. Examine alternate wall panels.

W2. Examine tilt-up walls.

Exterior Doors

D1. Use roll-up steel doors instead of vertical multi-leaf.

OKA FORM 124

Figure 9.14a Speculative phase: cranes, monorails, and exterior panels and doors.

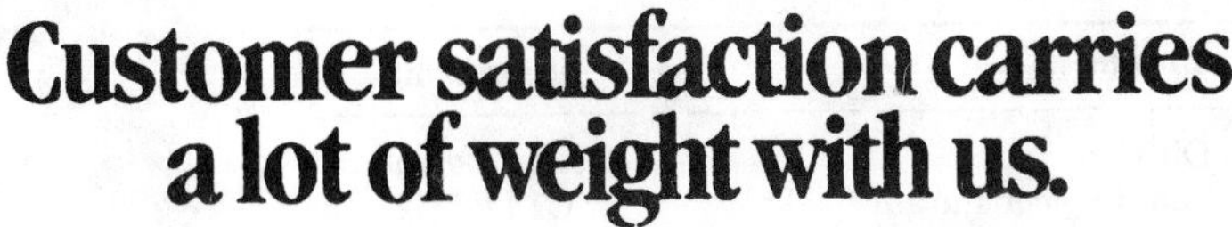

Figure 9.14b Mobile hoist.

The live load to use is the greater of:

$$\text{Live load} = (\text{minimum} + \tfrac{1}{2}\,\text{wind} + \text{snow})\ \text{(a)}$$

or

$$= (\text{minimum} + \text{wind} + \tfrac{1}{2}\,\text{snow})\ \text{(b)}$$

$$= (12 + \tfrac{1}{2} \times 21 + 15) = 37.5\ \text{psf (a)}$$

or

$$= (12 + 21 + \tfrac{1}{2} \times 15) = 40.5\ \text{psf (b)}$$

Use 40.5 psf as live load.

TABLE 9.1 Subteam Assignments

Subteam	Assignments
A. VETC O'Brien Material handling consultant Crane and hoist consultant	*Cranes and monorails* C1 and C2 *Structural* Part of S2
B. Serdikoff Structural consultant	*Structural* S1, S2, S6 *Concrete* F3, F4
C. Simmons	*Concrete* F1, F2 *Exterior doors* D1 *Exterior wall panels* W1, W2
D. Pagliaro Watson	*Structural* S3, S4, S5, S7

TABLE 9.2 Results of Idea Evaluation

Subteam	Figure	Continue	Drop
A	9.15	C2. Delete monorails	C1. Mobile carrier
B	9.15	F3. Welded wire fabric	
	9.15	F4. Spread footings	
	9.16	S1. Downsize loads	
	9.16	S2. Add column line, plant	S6. Add column lines, office
C	9.17	F1. Delete grade beams	F2. Curb instead of grade beam
	9.17	D1. Exterior doors	
	9.17	W1. Menu of wall panels	
	9.17	W2. Tiltup walls	
D	9.18	S3. Precast roof plank	
	9.18	S4. Galvanized joists	
	9.18	S5. Finish coat on ground	
	9.18	S7. Precast at loading dock	

Structural calculations determine the changes in Table 9.3 permitted in the beam and girder plan (Fig. 6.11). In turn, the lighter structural load permits a reduction in joist size (and weight). From Fig. 6.17, the joist weight is 60% of previous, or savings of 40% (163) tons. This is distributed as follows:

$$\text{Plant, } 40\% \text{ (120 tons)} = 48 \text{ tons}$$

$$\text{Office, } 40\% \text{ (21 tons)} = 8.4 \text{ tons}$$

$$\text{Warehouse, } 40\% \text{ (22 tons)} = 8.8 \text{ tons}$$

IDEA EVALUATION	
ITEM:	
BASIC FUNCTION:	

IDEA	ADVANTAGES	DISADVANTAGES	IDEA RATING*
C1. 50 ton moble straddle carrier instead of bridge crane	* Delete bridge crane - save money. * Reduce Plant vertical height by 5 feet (+).	* Increased horizontal space for circulation. * Less operational flexibility. * More potential for accidents.	3
C2. Use hi-reach forklifts plus motorized hoists in place of monorails.	* Delete monorails - save money. * Increase horizontal flexibility	* None	9
F3. Use welded wire fabric instead of rebar in slab-on-grade.	* Save money.	* None	10
F4. Spread footings instead of piles.	* Save time. * Save money.	* None	10
* 10 = MOST DESIRABLE, 1 = LEAST DESIRABLE			

Figure 9.15 Evaluations, subteams A and B.

IDEA EVALUATION

ITEM:	
BASIC FUNCTION:	

IDEA	ADVANTAGES	DISADVANTAGES	IDEA RATING*
S1. Review loads, downsize.	* Reduce dead weight. * Lower cost.	* None	10
S2. Add column line between C and D, Plant.	* Cut 100' span to 50' span. * Can use beams instead of girders. * Reduce dead weight. * Reduce cost.	* Requires one additional bridge crane. * Reduces operational flexibility. * Adds cost of new column line.	6
S.6 Add two column lines to Office at column lines.	* Cut 100' span. * Use beams instead of girders. * Reduce cost. * Reduction of vertical space.	* Makes space less flexibile. * Columns reduce net square footage. * Columns must be fireproofed.	3

* 10 = MOST DESIRABLE, 1 = LEAST DESIRABLE

Figure 9.16 Evaluations, subteam B.

IDEA EVALUATION	
ITEM:	
BASIC FUNCTION:	

IDEA	ADVANTAGES	DISADVANTAGES	IDEA RATING*
F1. Delete non-load bearing grade beams.	* Save money. * Save time.	* Moisture inflitration. * Temperature infiltration.	8
F2. Use curb instead of grade beams in non-load bearing situations.	* Save money. * Save time.	* Not necessary.	3
D1. Exterior roll-up doors instead of vertical multi-leaf	* Save money.	* Possible slight deterioration in operating time.	8
W1. Examine various types of exterior wall panels.	* Identify menu of opportunities.	* None	10
W2. Tilt-up walls.	* Save money. * Reduce girts.	* Aesthetics * Qualities	6

* 10 = MOST DESIRABLE, 1 = LEAST DESIRABLE

Figure 9.17 Evaluations, subteam C.

IDEA EVALUATION	
ITEM:	
BASIC FUNCTION:	

IDEA	ADVANTAGES	DISADVANTAGES	IDEA RATING*
S3. Use precast roof plank in place of deck.	* Lower cost.	* Increases dead weight.	6
S4. Galvanize joists to avoid painting in air.	* Lower cost.	* None	9
S5. Finish coat on girders on ground.	* Lower cost.	* Need to touch-up in air.	6
S7. Use precast vs. deck at loading docks.	* Lower cost.	* None	9

* 10 = MOST DESIRABLE, 1 = LEAST DESIRABLE

Figure 9.18 Evaluations, subteam D.

Table 1110
MINIMUM ROOF LINE LOADS
IN POUND-FORCE PER SQUARE FOOT OF HORIZONTAL PROJECTION

Roof slope	Tributary loaded area in square feet[a] for any structural member		
	0 to 200	201 to 600	Over 600
Flat or rise less than 4 inches per foot (1:3) Arch or dome with rise less than ⅛ of span	20	16	12

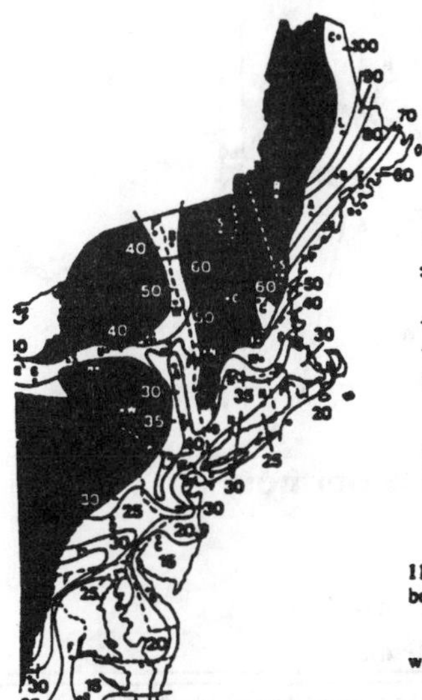

Table 1111.4a
SNOW EXPOSURE FACTOR (C_e)

Roofs located in generally open terrain extending one-half mile or more from the structure	0.6
Structure located in densely forested or sheltered areas	0.9
All other structures	0.7

Table 1111.4b
IMPORTANCE FACTOR (I)

All buildings and structures not listed below	1.0
Buildings and structures where the primary occupancy is one in which more than 300 people congregate in one area	1.1

1111.4 Flat-roof snow loads: The snow load on an unobstructed flat roof shall be calculated in pounds-force per square foot using the following formula:

$$P_f = C_e\, I\, P_g$$

where C_e is determined by Table 1111.4a and I is determined by Table 1111.4b.

Figure 1111.2a
GROUND SNOW LOADS, P_g, FOR THE EASTERN UNITED STATES
(pounds-force per square foot)

Figure 9.19 Extracts from BOCA code (2 pages).

The columns can all be changed to W12, 53 lb. From Figs. 6.13 and 6.14, the weight savings will be:

Plant:

14W, 145 lb = 92 lb/lf = 22 × 30 ft × 92 lb = 60,720 lb = 30.4 tons

12W, 170 lb = 117 lb/lf = 8 × 30 ft × 117 lb = 28,080 lb = 14 tons

Warehouse:

12W, 170 lb = 117 lb/lf = 6 × 30 ft × 117 lb = 21,060 lb = 10.5 tons

Office:

12W, 170 lb = 117 lb/lf = 4 × 30 ft × 117 lb = 14,040 lb = 7 tons

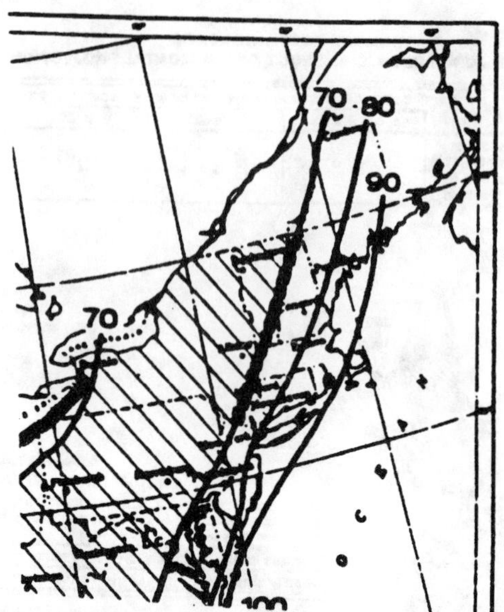

BASIC WIND SPEED (miles per hour)²

Table 1112.3.3a
EFFECTIVE VELOCITY PRESSURES P_e (lb/ft²) FOR BUILDINGS AND STRUCTURES (EXPOSURE B)[a]

Height above grade (ft)	Basic wind speed (mph)				
	70	80	90	100	110
0-20	9	12	15	18	22
20-40	10	13	17	21	25
40-60	13	16	21	26	31
60-100	14	18	23	28	34
100-150	16	21	27	33	40
150-200	18	23	29	36	43
200-300	20	26	33	41	50
300-400	23	30	37	46	56
> 400	Per ANSI A58.1 listed in Appendix A.				

Note a. 1 pound per square foot = 47 88 P 1 mile per hour = 0.447 m/s; 1 foot= 304.8 mm.

Figure 9.19 *(Continued)* Extracts from BOCA code (2 pages).

Also from Figs. 6.18 and 6.19 the plant cross bracing weight will be reduced to the following:

Figure 6.18 from W12, 210 lb to W12, 50 lb or 50/210 (35.6 tons)
= 8.5 tons (savings 27.1 tons)

Figure 6.19 from W12, 170 lb to W12, 50 lb or 50/170 (24.5 tons)
= 7.2 tons (savings 17.3 tons)

TABLE 9.3 Downsizing Weight Reduction

Old	New	Difference, lb per lf
Plant		
G57, 267 lb	G53, 173 lb	94
G45, 183 lb	G45, 115 lb	68
or G45, 183 lb	WF24, 84 lb	99 (for 30-ft span)
Warehouse		
G45, 183 lb	G45, 115 lb	68
Office		
G45, 183 lb	G45, 115 lb	68

Using Fig. 6.11, the weight savings are:
Plant:

G57, 267 = (11 × 100 × 94 lb) = 103,400 lb

G45, 183 = (600 × 99 lb) = 59,400 lb

162,800 lb

or 81.4 tons

Warehouse:

G45, 183 = (100 + 120) 2 (68 lb) = 29,920 lb

or 15 tons

Office (roof):

G45, 183 = (70 + 70 + 100) (68 lb) = 16,320 lb

or 8.2 tons

Figure 9.20 summarizes the cost savings in downsizing for the plant, Fig. 9.21 summarizes for the warehouse, and Fig. 9.22 for the office:

Plant = $593,505

Warehouse = $94,911

Office = $61,356

All figures include all markups. Note that the VE team included a total allowance of $60,560 for redesign (i.e., this figure already deducted from savings).

S2. Plant central column line. By adding a central column line in the plant, the girders can be downsized to beams. Assuming that S1 will be accepted, the value of this change is measured from the recommended S1 values (Table 9.4). However, 11 columns must be added:

COST WORKSHEET

Item	Units	No. Units	ORIGINAL ESTIMATE Cost/Unit	Total	Savings NEW ESTIMATE Cost/Unit	Total
Plant						
Structural Steel	Ton	323	2,240	723,520		
Savings						
Beams & Girders	Ton	81.4				
Columns	Ton	44.4				
Cross Bracing	Ton	44.4				
Subtotal	Ton	170.2			2,240	381,248
Joists	Ton	120	1,504	180,480		
Savings	Ton	48			1,504	72,192
Sandblast/Prime/						
Intermediate/Finish						
Paint	Ton	323	498	161,000		
Savings	Ton	170.2			498	76,194
Subtotal Savings						529,634
OH/Profit/GC/Bond/HOE @	21.5%					113,871
Subtotal Savings						643,505
Less Redesign Cost	LS	1			(50,000)	(50,000)
Total Savings						$593,505

Figure 9.20 Cost savings due to downsizing plant (S1).

$$11 \times 30 \text{ ft} \times (\text{W12, 53 lb}) = 17{,}490 \text{ lb or } 8.75 \text{ tons}$$

The cost of two 50 ft × 50 ton bridge cranes is assumed to be 75% of the 100 ft × 50 ton (each) or 1.5 × the installed cost, or 1.5 ($418,264), or $627,396. Figure 9.23 summarizes these figures. The new scheme is not desirable from an operations viewpoint, and costs $154,077 more. Accordingly, *drop* idea S2.

COST WORKSHEET

Item	Units	No. Units	ORIGINAL ESTIMATE Cost/Unit	Total	Savings NEW ESTIMATE Cost/Unit	Total
Warehouse						
Structural Steel	Ton	46	2,240	103,040		
Savings						
Girders	Ton	15				
Columns	Ton	10.5				
Subtotal	Ton	25.5			2,240	57,120
Joists	Ton	22	1,504	33,088		
Savings	Ton	8.8			1,504	13,235
Sandblast/Paint						
Structural Steel	Ton	46	498	22,908		
Savings	Ton	25.5			498	12,699
Savings Subtotal						83,054
OH/Profit/GC/Bond/HOE @	21.5%					17,857
Subtotal Savings						100,911
Less Redesign Cost	Ton	20			300	(6,000)
Total Savings						94,911

Figure 9.21 Cost savings (warehouse) due to downsizing (S1).

S3. Precast roof plank

Design:

$$\text{Base cost of metal deck} = \$5.12/\text{sf}$$

$$\text{Joists (downsized) } 48\% \times \$1504 = \underline{\$2.41/\text{sf}}$$

$$30{,}000 \text{ sf} = \$7.53$$

COST WORKSHEET

Item	Units	No. Units	ORIGINAL ESTIMATE Cost/Unit	ORIGINAL ESTIMATE Total	Savings NEW ESTIMATE Cost/Unit	Savings NEW ESTIMATE Total
Office						
Structural Steel	Tons	79	2,240	176,960		
Savings						
Girders	Tons	8.2				
Columns	Tons	7.0				
Subtotal	Tons	15.2			2,240	34,048
Joists	Tons	21	1,504	31,584		
Savings	Tons	8.4			1,504	12,634
Sandblast/Paint	Ton	79	498	39,342		
Savings	Ton	15.2			498	7,570
Subtotal Savings						54,252
OH/Profit/GC/HOE/Bond @	21.5%					11,664
Subtotal						65,916
Less Redesign	Ton	15.2			300	(4,560)
Net Savings						61,356

Figure 9.22 Cost savings (office) due to downsizing (S1).

VE:

$$\text{Precast panel } 16 \text{ in} \times 8 \text{ ft} \times 30 \text{ ft} = \$8.33$$

Drop: Precast idea S3.

S4. Galvanized joists vs. painting. The cost standoff is an option, not a VE recommendation.

TABLE 9.4 Weight Savings, 50-ft Beams versus Girders

Plant girders (after downsizing)	Beams, 50-ft span
G-53, 173 lb	W30, 90 lb
Weight savings 11 × 100 ft × (173 − 90 lb) = 1100 (83 lb) =	91,300 lb or 45.7 tons

COST WORKSHEET

Item	Units	No. Units	ORIGINAL ESTIMATE Cost/Unit	ORIGINAL ESTIMATE Total	NEW ESTIMATE Cost/Unit	NEW ESTIMATE Total
Girders	Tons	95	2,240	212,800		
Beams	Tons	49.5			2,240	110,880
New Columns	Tons	8.75			2,240	19,600
Single Bridge Crane	EA	1	418,264	418,264		
Two 50' Bridge Cranes	EA	2			313,698	627,396
Subtotal				631,064		757,876
Increase in Cost						126,812
OH/Bond/HOE/Profit @	21.5%					27,265
Total Increase						154,077

Figure 9.23 Two bridge cranes at 50-ft span, cost summary (S2).

S5. Structural steel, finish paint on ground. Almost a standoff. The premium for painting a finish coat in the air is about $25,000. This is an option only, no claim for net gain. Therefore, it is not a VE recommendation.

S7. Precast for loading docks. A wash—not a VE recommendation.

F1. Non-load-bearing grade beams. The grade beam is non-load-bearing at the plant and warehouse perimeters.

$$\text{Plant, } [(50 \text{ ft} + 100 \text{ ft})\, 2 + 300 \text{ ft}] = 600 \text{ ft}$$

$$\text{Warehouse, } (60 \text{ ft} + 50 \text{ ft})\, 2 = 220 \text{ ft}$$

Figure 9.24 tabulates the savings (from Table 8.1):

$$\text{Plant} = \$74{,}389$$

$$\text{Warehouse} = \$22{,}042$$

$$\text{Office} = 0$$

F3. Use welded wire fabric instead of rebar in slabs-on-grade. Figure 9.25 summarizes the costs for the design vs. welded wire fabric (WWF). The potential cost savings are small. Accordingly, this is *dropped* as a VE recommendation.

F4. Use spread footings rather than piling (plant). Using the downsized loads for F1, the load at each column is as follows:

$$\text{Girder} = \frac{173 \text{ lb} \times 100 \text{ ft}}{2 \times 1000} = 8.7 \text{ kips}$$

$$\text{Joists} = \frac{144{,}000}{10 \times 2 \times 1000} = 7.2 \text{ kips}$$

$$\text{Deck and roof} = \frac{3000 \times 5}{2 \times 1000} = 7.5 \text{ kips}$$

$$\text{Live load} = \frac{3000 \times 40}{2 \times 1000} = 60.0 \text{ kips}$$

$$\text{Crane load} = \frac{50 \times 2000}{2 \times 1000} = \underline{50.0 \text{ kips}}$$

$$\text{Total column load} = 133.4 \text{ kips}$$

COST WORKSHEET

			ORIGINAL ESTIMATE		NEW ESTIMATE	
Item	Units	No. Units	Cost/Unit	Total	Cost/Unit	Total
Plant						
Peinforcing	LB	6144	0.67	4,116		0
Concrete	CY	111	486	53,946		0
Dampproof	SF	1805	0.45	812		0
Insulation	SF	1805	1.28	2,310		0
Subtotal - Savings				61,184		0
Add OH/Profit/GC/HOE/Bond @ 21.5%				13,155		0
Savings				74,389		0
Warehouse						
Peinforcing	LB	2208	0.67	1,479		0
Concrete	CY	40	486	19,440		0
Dampproofing	SF	649	0.45	292		0
Insulation	SF	649	1.28	831		
Subtotal Savings				22,042		
Add OH/Profit/GC/HOE/Bond @ 21.5%				4,739		
Savings				26,781		

Figure 9.24 Non-load-bearing grade beam savings (F1).

$$\text{Design load-bearing capacity} = 3 \text{ tons/sf}$$

$$\text{Footing area} = \frac{133{,}000}{6000} = 22 \text{ sf}$$

$$\text{Footing sizes: } 4 \text{ ft} \times 6 \text{ ft} = 24 \text{ sf}$$

$$5 \text{ ft} \times 5 \text{ ft} = 25 \text{ sf}$$

WWF vs. Rebar for Slabs-On-Grade (F3)

			Original Estimate		New Estimate	
Item	Units	No. Units	Cost/Unit	Total	Cost/Unit	Total
Plant						
Rebar - SOG	LB	37,328	0.60	22,397		
- SOG	CSF	320			$57	18,240
Warehouse						
Rebar - SOG	LB	5,002	0.60	3,001		
- SOG	CSF	64			$48	3,072
Office						
Rebar - SOG	LB	15,698	0.60	9,419		
- SOG	CSF	140			$48	6,720

Figure 9.25 WWF vs. rebar for slabs-on-grade (F3).

Therefore, delete piles.

$$\text{Savings (from Table 8.1)} = \$72{,}160$$
$$\times\ 1.215$$
$$\text{Savings} = \$87{,}674 \text{ (plant)}$$

C2. Delete monorails. The team materials handling consultant conceptualized a combination of electric (or LNG) high-reach fork lifts to mesh with a company-furnished shelf storage system. The fork-lift system is costed at $25,000 each, total $50,000.

Monorail system price (Table 8.1) =	$201,036
HOE, OH, profit, and GC at 21.5% =	43,223
Subtotal =	$244,259
Less high-lift fork lifts =	50,000
HOE, GC, profit, and OH at 21.5% =	10,750
Subtotal =	$60,750
Net delete =	$183,509

W1/W2. (Combine) wall panels. Architect Simmons assembled a list of five alternate (to the design) exterior wall panels. These (plus the

ALTERNATIVE EVALUATION

ITEM:

		FACTORS *										
		CAPITAL COST	O&M COST	REDESIGN	IMPLEMENTATION TIME	PERFORMANCE	RELIABILITY	SAFETY	AESTHETICS		TOTAL	RANK
	FACTOR WEIGHT 10 = MAXIMUM ▶	10	10	0	10	8	8	10	10			
1.	Metal Facing Panel - 16 ga. Al/18 ga.	3	4		4	4	4	4	4			
	galvanized - insul. 0742020300 $14.19	30	40		40	32	32	40	40		254	1
2.	Metal Facing Panel - Stainless Steel	2	4		4	4	4	4	4			
	insul. $15.69 0742020700	20	40		40	32	32	40	40		244	2
3.	Metal Facing Panel - Baked Enamel 20g	3	3		4	3	3	4	3			
	$12.19 0742020900	30	30		40	24	24	40	30		218	3
4.	Exposed Aggregate with Insulation	3	2		4	2	2	4	2			
	$12.41 0742010500	30	20		40	16	16	40	16		182	4
5.	Factor Fab. 10 mil Al on Plywood-Foil	4	1		4	1	1	2	1			
	Back $8.29 0742021200	40	10		40	8	8	16	8		136	6
6.	Tilt-up Concrete	4	2		2	2	2	4	1			
	$7.38 0347040100	40	20		20	16	16	32	8		162	5

* EXCELLENT = 4, GOOD = 3, FAIR = 2, POOR = 1

Figure 9.26 Exterior wall panel evaluation.

design panel) are listed in Fig. 9.26. Six factors are listed at the top of the chart. Each of these is assigned a weight. Each panel is then scored on each of the six factors on the following basis: excellent = 4; good = 3; fair = 2; poor = 1. Each score is then multiplied by the weight for each factor. The weighted scores are totaled.

The evaluation shows the design panel to be the best, with only one close runner-up. Accordingly, *no VE recommendation* is made.

D1. Exterior doors. Figure 9.27 is a cost comparison between insulated, steel, roll-up exterior doors vs. the design multileaf vertical doors. The roll-up door price is based upon a vendor quote on insulated, motor-operated, roll-up steel doors.

COST WORKSHEET

Item	Units	No. Units	ORIGINAL ESTIMATE Cost/Unit	ORIGINAL ESTIMATE Total	NEW ESTIMATE Cost/Unit	NEW ESTIMATE Total
Plant						
Multi-leaf	EA	4	21,400	85,600		
Roll-ups	EA	4			12,500	50,000
Savings Subtotal						35,600
OH/Profit/HOE/GC/Bond @	21.5%					7,654
Total Savings						43,254
Warehouse						
Multi-leaf	EA	2	21,400	42,800		
Roll-ups	EA	2			12,500	25,000
Savings Subtotal						17,800
OH/Profit/HOE/GC/Bond @	21.5%					4,450
Total Savings						22,250

Figure 9.27 Exterior door cost comparison.

V.E. RECOMMENDATION

ITEM:

PROPOSED CHANGE:

VE Recommendation	Plant	Warehouse	Office	Total
S1. Structural downsize	$593,505	$ 94,911	$ 61,356	$749,772
F1. Delete grade beams	74,389	22,042	—	96,431
F4. Delete piles	87,674	—	—	87,674
C2. Delete monorails	—	183,509	—	183,509
D1. Exterior doors	43,254	22,250	—	65,504
	$798,822	$322,712	$ 61,356	$1,182,890

COST SUMMARY	
INITIAL - ORIGINAL	
INITIAL - PROPOSED	
INITIAL SAVINGS	
TOTAL ANNUAL COSTS - ORIGINAL	
TOTAL ANNUAL COSTS - PROPOSED	
ANNUAL SAVINGS	
PRESENT WORTH - ANNUAL SAVINGS	

Figure 9.28 Summary of VE suggestions.

The roll-up doors are clearly a VE recommendation, with the following savings:

Plant = $43,254

Warehouse = $22,250

Fifth day. During the development phase, only five of thirteen ideas become recommendations. Two were dropped as more costly. The remaining six are essentially a "wash," so no purpose is served by keeping them in the VE report. In fact, it should be emphasized by the VETC, in the briefing to the owner and design professionals, that this VE review has been a validation of the design to date. This briefing will be on the afternoon of the fifth day.

Figure 9.28 is a summary of the VE suggestions. Implementation will decrease the total construction cost from $6,360,207 to $5,177,317, or 23%. This returns the project to within the budget contingency. Table 9.5 shows the adjusted design development estimate.

TABLE 9.5 Design Development Estimate Adjusted for VE Changes

	Initial design development estimate	VE change	Revised design development estimate
Plant	$4,140,375	$798,822	$3,341,551
Warehouse	1,036,015	322,712	713,303
Office	1,183,817	61,356	1,122,461
Total	$6,360,207	$1,182,890	$5,177,317

Chapter

10

Contract Document Estimate

The contract documents describe the contract between the contractor and the owner; they generally include the contract itself, plus working drawings and specifications. The working drawings describe in graphic form, with dimensions, the location, size, arrangement, layout, and spatial relationships of the work to be installed. Their organization is determined to a large degree by the organization of the disciplines within the architect-engineering organization that prepares them. Typically, the *drawing set* is divided as previously described in the design development phase.

In many public jobs, local or state law requires the awarding of separate contracts, so that this breakdown is also useful in subdividing the work into the separate prime contracts.

The *architectural drawings* are generally shown in plan view at a scale of either ⅛ in to 1 ft or ¼ in to 1 ft, depending on the size of the project. For a large-area project, the plans may be subdivided into several sheets, using major geometric shapes as dividing points, or match lines. Typically, the column lines are marked with letter designations along one axis and numbers along the opposite, to form a location grid pattern for reference purposes. Another approach is the use of an alphanumeric grid on the drawing margin itself. Many design teams establish the format in the design development phase at the same scale so that the design development effort is evolved and not wasted.

The plans are further developed by appropriate elevations showing the outside of buildings or structures, and sections showing a cutaway view. At strategic locations, special sections may be cut to show details. These are often shown at a substantially larger scale, such as ½ in = 1 ft.

In the *mechanical-electrical* sections, it is usual to use the same outline plan with the same scale as in the architectural to present the various systems. The systems may be shown in heavy line, superimposed upon a light-line architectural format. As in the architectural section, the mechanical-electrical sections show elevations and cross sections. An important addition in the mechanical-electrical areas is the use of single-line diagrams to indicate functional flow of fluids, gases, or current.

Notes are utilized on the drawings to explain special conditions, and owners having large building programs may reference standard details previously established within their design organization.

Specifications

The role of specifications is to describe the quality of the materials to be placed in the project. Whereas the drawings present scope in terms of quantities, dimensions, form, and building details, the specifications provide the qualities of materials for construction. Specifications, even for a small project, tend to be voluminous, and in large projects can extend to several thousand pages. Because of their legal connotations, the specifications tend to be loaded with all possible information.

Designers tend to evolve their own specification clauses or to use computer-generated draft specifications. In either case there is a tendency to include all previous applicable clauses, just in case they are needed. Accordingly, something as straightforward as 3000 psi concrete may require several pages of description, including factors such as size and gradation of sand and aggregate, origin of the cement, and detailed information describing the reinforcing bars. Inevitably, some human omissions become apparent in the course of construction. To counter this, specifications often have substantial escape or cover-all clauses to protect the designer. Further, the specifications typically include massive amounts of boilerplate—that is, standard insert material used by the particular architect or engineering organization, often carried forward from prior projects, and quite often including ambiguous or even incorrect material.

The CSI 16 Division format set up in the design development phase is continued in completing the detailed specifications.

General Conditions

Traditionally, the "General Conditions" section preceded the technical portion of the specifications; now it is Division 1 of the CSI format. This section included definitions of responsibilities of the owner, the architect, the contractor, and other parties, such as the construction manager. Organizations with ongoing building programs, such as the state of New Jersey or the New York State Dormitory Authority, have a standard set of General Conditions. The American Institute of Architects (AIA) publishes a series of recommended General Conditions (the A201 series) with versions tailored to general contractors or construction management contracts. A typical set of General Conditions (paraphrased from the Standard General Conditions published by the Division of Building and Construction of the State of New Jersey) follows:

1. Definitions—Notices
   A. The Contract Documents consist of the Agreement, Instructions to Bidders and General Conditions of the Contract, the Drawings and Specifications, Bulletins, and Addenda, including all modifications thereof incorporated in the documents before their execution. Whenever the word "Contract" is used herein, it means all of the above documents or such parts of them as are clearly indicated.
   B. Whenever the word "Owner" is used herein, it means the John Doe Company.
   C. Whenever the word "Director" is used herein, it means the Director of Construction of the John Doe Company, or his duly authorized representative.
   D. Whenever the word "Contractor," "Prime Contractor," or "Single Contractor" is used herein, it means the individual or firm undertaking to do all work contracted for under the Contract.
   E. Whenever the word "Architect" or "Engineer" is used herein, it means the Architect or Engineer engaged by the Owner and acting as the duly authorized representative of the Director.
   F. The term "Subcontractor," as employed herein, includes an individual or firm having a direct contract with the Contractor, Prime Contractor, or Single Contractor, and it includes one who furnishes labor or material worked to a special degree according to the drawings or specifications of this work, but does not include one who merely furnishes material not so worked.
   G. When the term "acceptable" or "approved" is used herein, it means that the material or work shall be acceptable to or approved by the Director.
   H. The term "Work" of the Contractor, Prime Contractor, Single Contractor, or Subcontractor, as used herein, includes labor, materials, plant, and equipment required to complete the contract.

2. Architect's Status
   The Architect shall interpret the drawings and specifications, and as agent for the Director shall judge the quantity, quality, fitness, and acceptability of all parts of the work. The Director shall certify contractor's invoices for work performed and materials delivered to the site, and shall be given access to part of the work for inspection at all times. The Architect shall not have authority to give approval or

order changes in work that alters the terms of conditions of the Contract, or that involves additional costs. The Architect may, however, make recommendations to the Director for such changes, whether or not costs are revised, and the Director may act on the basis of the Architect's recommendations.

3. Intention
   A. The drawings and specifications, with the Contract of which they form part, are intended to provide for and comprise everything necessary to the proper and complete finishing of the work in every part, notwithstanding that each and every item necessary may not be shown on the drawings or mentioned in the specifications.
   B. Each Contractor shall abide by and comply with the true intent and meaning of all the drawings and the specifications taken as a whole, and shall not utilize any unintentional error or omission should any exist.
   C. Should any error or discrepancy appear, or should any doubt exist or any dispute arise as to the true intent and meaning of the drawings or of the specifications, or should any portion of same be obscure or capable of more than one construction, the Contractor shall immediately apply to the Architect for the corrections or explanation thereof, and, in case of dispute, the Director's decision shall be final.
   D. Determinations will be in the form of Bulletins to the Specifications which will be forwarded to all affected contractors.
   E. The sequence of precedence pertaining to contract documents is as follows:
      1. Agreement
      2. Addenda, Bulletins
      3. Specifications (including General Conditions)
      4. Details
      5. Figured Dimensions
      6. Scaled Dimensions
      7. Drawings
   F. Any provision in any of the Contract Documents that may be in conflict or inconsistent with any of the paragraphs in these General Conditions shall be void to the extent of such conflict or inconsistency unless the provision is specifically referenced as a supplement or change thereto.
   G. Each and every provision of law and clause required by law to be inserted in this contract shall be deemed to be inserted herein, and the contract shall be read and enforced as though it were included herein, and if, through mistake or otherwise, any such provision is not inserted, or is not correctly inserted, then upon the application of either party the contract shall forthwith be physically amended to make such insertion or correction.

4. Contract Limit Lines
   A. Whenever the words "contract limit lines" are used herein, they refer to the lines shown on the drawings, surrounding the con-

tract work, beyond which no construction work shall be performed unless otherwise noted on the drawings or specified. Each Contractor shall check and verify conditions outside of the contract limit lines to determine whether or not any conflict exists between elevations or other data shown on the drawings and existing elevations or other data outside of the contract limit lines.

B. Whenever the words "construction site" or "project site" are used herein, they refer to the geographical grounds of the entire Institution, college campus, or other Using Agency property at which contract work is performed.

5. Responsibility for Work

A. Each Contractor shall be responsible for all damages due to its operations; to all parts of the work, both temporary and permanent; and to all adjoining property.

B. Each Contractor shall protect the Joe Doe Company from all suits, actions, damages, and costs of every name and description resulting from the work under this contract.

C. Each Contractor shall provide, in connection with its own work, all safeguards, rails, night lights, and other means of protection against accidents.

D. Each Contractor shall make, use, and provide all proper, necessary, and sufficient precautions, safeguards, and protections against the occurrence of any accident, injuries, damage, or hurt to any person or property during the progress of the work.

E. Each Contractor shall, at its own expense, protect all finished work liable to damage, and keep the same protected until the project is completed and accepted by the Director.

6. Superintendence—Supervision—Laying Out

A. At the site of the work the Contractor shall employ a construction superintendent or foreman who shall have full authority to act for the Contractor. It is understood that such representative shall be acceptable to the Architect/Engineer and the Director and shall be one who is to be continued in that capacity for the particular job involved unless he ceases to be on the Contractor's payroll.

B. The various subcontractors shall have competent foremen in charge of their respective part of the work at all times. They are not to employ on the work any unfit person or anyone not skilled in the work assigned.

C. Each Contractor shall give the work its special supervision, lay out its own work, do all the necessary leveling and measuring, or employ a competent licensed engineer or land surveyor satisfactory to the Director to do so.

D. If, due to trade agreement, additional standby personnel are required to supervise equipment or temporary services used by other trades, the Contractor providing such standby services shall evaluate requirements and include the cost thereof in his bid.

E. All Contractors and Subcontractors shall rely on their own measurements for the performance of their work.

7. Subcontractor—Material Approval
   A. Each Prime Contractor shall submit through the Architect or Engineer to the Director for approval a list of all subcontractors, manufacturers, materials, and equipment, whether specified or not, within thirty (30) days after award of contract. No contract shall be entered into with any Subcontractor before his name has been approved in writing by the Director.
8. Subcontractors—Equipment and Materials
   A. Each Contractor shall, within thirty (30) days after award of the contract, notify the Director through the Architect or Engineer in writing of the names of subcontractors proposed for the principal parts of the work and for such others as the Director may direct, and shall not employ any without prior written approval of the Director or any that the Director may within a reasonable time reject.
   B. Each Contractor agrees that it is as fully responsible to the John Doe Company for the acts and omissions of its subcontractors and of persons either directly or indirectly employed by them, as it is for the acts and omissions of persons directly employed by it.
9. Material—Workmanship-Labor
   A. All material and work shall conform to the best practice. Only the best of the several kinds of materials shall be used, and the work carefully carried out in strict accordance with the general and detail drawings, under the supervision of the Director, or accredited representative, who shall have full power at any time to reject such work or material that does not, in The Director's opinion, conform to the true intent and meaning of the drawings and specifications.
   B. All work, when completed in a substantial and workmanlike manner, to the satisfaction of the Director, shall be accepted in writing. Unless otherwise specified, all materials used shall be new.
   C. Each Contractor shall furnish and pay all necessary transportation, scaffolding, forms, water, labor, tools, light and power, mechanical appliances, and all other means, materials, and supplies for properly prosecuting its work under contract, unless expressly specified otherwise.
10. Defective Work and Materials
   A. Any materials or work found to be defective, or not in strict conformity with the requirements of the drawings and specifications, or defaced or damaged through the negligence of any Contractor or its Subcontractors or employees, or through action of fire or the weather or any causes, shall be removed immediately and new materials or work substituted therefor without delays by the Contractor involved.
11. Inspection of Work
   A. The Director, or authorized representative, shall at all times have access to the work whether it is in preparation or in progress, and

the Contractor shall provide proper facilities for such access for inspection.

Business Considerations

Business items are also included, providing the basis for payment and progress payment—perhaps one of the best means of controlling contractor cooperation. Some typical information would include:

1. Unit Schedule Breakdown
   The Contractor shall file with the Director a unit schedule breakdown in sufficient detail to include the following, which will be used as the basis for determining the amount of payment to be made on a periodic basis for work completed and installed in accordance with Contract Documents:
   (1) Description of material or equipment and number of units involved.
   (2) Lump sum price for labor and lump sum price for equipment and/or material listed.
   (3) Lump sum allowances included in the specifications.
   The total of items shall equal the lump sum contract price.

2. Payment
   A. The basis for computing monthly progress payments shall be the Unit Schedule breakdown. The Unit Schedule breakdown shall be submitted for approval to the Director through the Architect within ten (10) days from date of written notice to proceed by the John Doe Company.
   B. In making such partial payments for work, there will be retained 10% of estimated amount until final acceptance and completion of all work covered by the contract: provided, however, that after 50% of the work has been completed, if the Director determines that Contractor's performance and progress are satisfactory, the John Doe Company may make remaining partial payments in full for work subsequently completed.

3. Additions—Deductions—Deviations
   The Director, at his discretion, may at anytime during the progress of the work authorize additions, deductions, or deviations from the work described in the specifications as herein set forth; and the contract shall not be vitiated or the surety released thereby.
   A. Additions, deductions, deviations may be authorized as follows at Director's option:
   (1) On the basis of unit prices specified.
   (2) On a lump sum basis.
   (3) On a time and material basis.
   (4) On standby time or overtime basis.

4. Payments Withheld
   A. The Director may withhold or, on account of subsequently discovered evidence, nullify the whole or a part of any certificate for pay-

ment to such extent as may be necessary to protect the owner from loss on account of:

(1) Defective work not remedied.

(2) Claims filed, or reasonable evidence indicating probable filing of claims.

(3) Failure of any Contractor to make payments promptly to subcontractors or for material or labor.

(4) A reasonable doubt that the contract can be completed for the balance then unpaid.

(5) Damage to another contractor.

B. When all the above grounds are removed, certificates will at once be issued for amounts withheld because of them.

5. Construction Progress Schedule

A. The Contractor for the General Construction Contract shall be responsible for preparing and furnishing, before the first contract requisition date, a coordinated single progress schedule that incorporates progress schedules for all Prime Contractors engaged on the project. The schedule shall be an arrow diagram network, bar chart, or other recognized graphic progress schedule in a form and in sufficient detail satisfactory to the Director.

Other Topics

Special topics that are project-specific would not be found in a "packaged" set of general conditions. These may include items such as:

Construction sign

Temporary drives and walks

Temporary building and sanitary facilities

Temporary water

Temporary light and power

Temporary heat

Temporary enclosures, glass breakage, and cleaning

Hoisting facilities

Fire protection

Shop drawings

As-built drawings

Samples

Testing of equipment

Concrete and other structural testing

Photographs

Job meetings

There may be special sections on business matters such as:

Federal taxes
Guarantees
Right of owner to terminate contract
Lands and rights-of-way
Disputes, claims, appeals
Withholding of payments
Payrolls and basic records
Minimum wages
Anti-kickback act
Statute prohibiting discrimination—civil rights
Equal employment opportunity
Certification of nonsegregated facilities

Supplemental Conditions

Particularly where the general conditions are a standard package evolved over time and through experience, it may be necessary to add a section entitled "Supplemental Requirements." This section supplements or modifies the general conditions to reflect the specific and special circumstances of the project.

The Contract Document Estimate

This estimate, sometimes known as the cost confidence estimate, is usually performed at about 90% completion of design. It is usually performed by the construction manager or a cost and estimating consultant.

The principal purpose is to determine that the design is within budget. If the estimate identifies probable cost overruns, there are several alternatives:

- High-cost items can be deleted.
- Lower-cost materials can be substituted.
- Optional items or systems can be listed as add alternatives, to be accepted if the bids permit. (Deduct alternatives are not recommended.)
- The design team can redesign areas for lower cost.

At this 90% of design stage, any major redesign will cost major time delays.

Material takeoff

Material takeoff can be 100% "from scratch." A shortcut can use the design development drawings and estimate for the following items:

- Foundations
- Structure
- Concrete work
- Exterior shell

If there are only minor changes in those items, this estimate can be the design development figure adjusted by costing the changes.

Suggestions for estimators

1. Mark areas or systems in different highlighter colors as items are taken off. Never use a black felt-tip pen.
2. Keep similar items together, different items separate.
3. Identify location and drawing numbers to aid in future checking for completion.
4. Measure or list everything on the drawings or mentioned in the specifications.
5. List items not called for needed to make the job complete.
6. Be alert for notes on plans such as N.T.S. (not to scale); changes in scale throughout the drawings; reduced size drawings; discrepancies between the specifications and drawings.
7. Develop a consistent pattern of performing an estimate. For example:
   a. Start the quantity takeoff at the lower floor and move to the next higher floor.
   b. Proceed from the main section of the building to the wings.
   c. Proceed from the south to north or vice versa, clockwise or counterclockwise.
   d. Take off floor plan quantities first, elevations next, then detail drawings.
   e. Write measurements horizontally and then vertically. You will be able to determine where measurements came from at a later date. This also expedites figuring perimeter-related quantities.
8. List all gross dimensions that can be either used again for different quantities or used as a rough check for other quantities for verification (exterior perimeter, gross floor area, individual floor areas, etc.).
9. Utilize design symmetry or repetition (repetitive floors, repetitive wings, symmetrical design around a centerline, similar room lay-

outs, etc.). *Note:* Extreme caution is needed here so as not to omit or duplicate an area.

10. Do not convert units until the final total is obtained. For instance, when estimating concrete work, keep all units to the nearest cubic foot, then summarize and convert to cubic yards.
11. When figuring alternatives, it is best to total all items involved in the basic system, then total all items involved in the alternates. Thus you work with the positive numbers in all cases. When adds and deducts are used, it often is confusing whether to add or subtract a portion of an item, especially on a complicated or involved alternate.
12. Develop a checklist for your specific discipline.
13. Review job specifications, taking notes of all clauses that affect bid costs.
14. Contact suppliers as soon as possible in order to get the best price for special material or to get price protection for materials.
15. Use a red arrow to call attention to a potential problem area of construction or a conflict with another trade.
16. Check off everything you have completed. If there is more than one person working on a project, a single initial can be used to signify who did the work.
17. Never use a long word to describe the work item when a short word will do. Abbreviate words whenever possible using standard abbreviations for words pertaining to the construction industry.
18. Question any detail or condition which contradicts standard practices of the trade.
19. Tab subcontractor and supplier quotation recap.
20. Tab unit price and alternate summary sheets.

Digitizers

Digitizers can support the takeoff process, combining computers with electronic takeoff devices. The takeoff stylus can record line lengths, measure cross sections, act as a counter, and measure areas (such as earthwork contours).

The following are available digitizer systems:

Company	Trademark	Location
Advanced Grade Technology (AGTEK)	Earthworks	Livermore, CA
GTCo	Rollups	Columbia, MD
Science Accessories Corp. (SAC)	GP-9	Stamford, CT
Software Shop Systems	Site work	Farmingdale, NY

WHATCO project. Table 10.1 is a summary of quantities and costs for Divisions 1 to 14 (not including site work). These are the same as the design development quantities (Table 8.1) *except* that the table has been adjusted for the value-engineering changes.

Table 10.2 is a summary of quantities and costs for Division 15 (mechanical), and Table 10.3 is the same for Division 16 (electrical).

TABLE 10.1 Summary of Quantities and Costs, Divisions 1 to 14

Item	CSI No.	R.S. Means I.D. No.	Unit	No.	Unit price 1995	Cost
		Plant				
Site preparation (clear)	02110	021 140 2000	acre	0.7	4,640	3,248
Excavation, footprint	02220	022 246 4220	cy	3,345	5.36	17,929
Excavation, spread footings	02220	022 250 0100	cy	86	70	6,020
Borrow	02220	022 216 0020	cy	1,672	3.10	5,183
Deliver borrow	02220	022 216 3800	cy	1,672	9.34	15,616
Spread	02220	022 216 2500	cy	1,672	5.22	8,728
Compaction	02220	022 226 5040	cy	1,672	0.88	1,471
Fill	02220	022 262 0170	cy	112	2.64	296
Concrete reinf., footing	03200	032 107 0500	lb	6,056	0.78	4,724
Concrete reinf., SOG, load deck	03200	032 107 0750	lb	37,328	0.60	22,397
Slab-on-grade 12 in	03310	033 130 4900	sf	30,000	3.85	115,500
Spread footers	03310	033 130 3850	cy	86	160	13,760
Loading dock concrete	03310	033 130 3150	sf	4,000	2.07	8,280
Insulating lightweight fill	03500	035 212 0250	sf	30,000	1.14	34,200
Structural steel, heavy	05120	051 255 4900	ton	1.53	2,240	342,720
Structural steel, light	05120	051 255 4600	ton	06.1	2,560	153,856
Base plates	05120	051 255 4300	lb	3,708	1.05	3,893
Joists	05210	052 110 0030	tons	72	1,504	108,288
Metal deck	05310	053 104 0440	sf	31,000	5.12	158,720
Sandblast	05120	051 255 6130	sf	22,387	0.81	18,133
Primer	05120	051 255 6520	sf	22,387	0.54	12,089
Intermediate coat	05120	051 255 6630	sf	22,387	0.78	17,462
Top coat (elevated)	05120	051 255 7030	sf	22,387	1.19	26,641
Insulation roof	07220	072 203 0100	sf	30,000	1.04	31,200
MFG wall panels	04710	074 202 0400	sf	18,000	14.53	259,740
Roofing	07510	075 102 0200	square	300	170	48,000
Flashing	07620	076 201 5800	lf	800	4.44	3,552
Ext. downspouts	07630	076 201 2500	lf	245	9.14	2,239
Ext. gutters	07630	076 210 3400	lf	400	13.31	5,324
Steel doors and frames	08110	081 103 0100	each	5	209	1,045
Steel rollup doors	08365	083 604 0010	each	4	15,188	60,750
Door hardware	08710	087 125 3110	each	5	429	2,145
Exterior paint	09910	099 106 2400	sf	4,516	0.54	2,439
Interior paint, block	09920	099 106 2400	sf	6,000	0.54	3,240
Dock levelers	11161	111 601 4800	each	4	7,296	29,184
Bridge crane	14630	Quote	each	1	NA	418,264
Total						$1,966,276

TABLE 10.1 Summary of Quantities and Costs, Divisions 1 to 14 (Continued)

Item	CSI No.	R.S. Means I.D. No.	Unit	No.	Unit price 1995	Cost
		Warehouse(s)				
Site preparation (clear)	02110	021 104 2000	acre	0.14	4,640	650
Excavation, footprint	02220	022 246 4220	cy	669	5.36	3,586
Excavation, spread footing	02220	022 250 0100	cy	9	70	630
Borrow	02220	022 216 0020	cy	334	3.10	1,035
Deliver borrow	02220	022 216 3800	cy	334	9.34	3,120
Spread	02220	022 216 2500	cy	334	5.22	1,743
Compaction	02220	022 226 5040	cy	334	0.88	294
Fill	02220	022 262 0170	cy	39	2.64	103
Concrete reinf., footings	03200	032 107 0500	lb	1,018	0.78	794
Concrete reinf., SOG, load deck	03200	032 107 0750	lb	5,002	0.60	3,001
Slab-on-grade, 9 in	03310	033 130 4840	sf	6,000	2.79	16,740
Spread footers	03310	033 130 3850	cy	9	160	1,440
Loading dock concrete	03310	033 130 3150	sf	400	2.07	828
Insulating lightweight fill	03500	035 212 0250	sf	6,000	1.14	6,840
Structural steel, heavy	05120	051 255 4900	ton	20	2,240	44,800
Structural steel, light	05120	051 255 4600	ton	13.5	2,560	34,560
Base plates	05120	051 255 4300	lb	504	1.05	529
Joists	05210	052 110 0030	tons	13	1,504	19,552
Metal deck	05310	053 104 0440	sf	6,400	5.12	32,768
Sandblast	05120	051 255 6130	sf	6,519	0.87	5,672
Primer	05120	051 255 6520	sf	6,519	0.54	3,520
Intermediate coat	05120	051 255 6630	sf	6,519	0.78	5,085
Top coat (elevated)	05120	051 255 7030	sf	6,519	1.19	7,758
Insulation roof	07220	072 203 0100	sf	6,000	1.04	6,240
MFG wall panels	07410	074 202 0400	sf	6,000	14.53	95,898
Roofing	07510	075 102 0200	square	60	160	9,600
Flashing	07620	076 201 5800	lf	440	4.44	1,954
Ext. downspouts	07630	076 201 2500	lf	70	9.14	640
Ext. gutters	07630	076 210 3400	lf	100	13.31	1,331
Steel doors and frames	08110	081 103 0100	each	2	209	418
Steel rollup doors	08365	083 604 0010	each	2	15,188	30,376
Door hardware	08710	087 125 3110	each	2	429	858
Exterior paint	09910	099 106 2400	sf	2,216	0.54	1,197
Interior paint, block	09920	099 106 2400	sf	3,600	0.54	1,944
Dock levelers	11161	111 601 4800	each	4	7,300	29,200
Fork lifts	NA	Quote	each	2	20,000	40,000
Total						$414,704
		Office				
Site preparation (clear)	02110	021 104 2000	acre	0.16	4,640	742
Excavation, footprint	02220	022 246 4220	cy	764	5.36	4,095
Excavation, spread footing	02220	022 250 0100	cy	6	70	420
Excavation, grade beam	02220	022 254 0050	cy	45	5.95	268
Borrow	02220	022 216 0020	cy	382	3.10	1,184
Deliver borrow	02220	022 216 3800	cy	382	9.34	3,568
Spread	02220	022 216 2500	cy	382	5.22	1,994
Compaction	02220	022 226 5040	cy	382	0.88	336
Fill	02220	022 262 0170	cy	22	2.64	58

(Continued)

TABLE 10.1 Summary of Quantities and Costs, Divisions 1 to 14 (Continued)

Item	CSI No.	R.S. Means I.D. No.	Unit	No.	Unit price 1995	Cost
		Office (*Continued*)				
Concrete reinf., footings	03200	032 107 0500	lb	710	0.78	554
Concrete reinf., SOG, load deck	03200	032 107 0750	lb	5,170	0.60	3,102
Concrete reinf., grade beam	03200	032 107 0100	lb	1,248	0.67	836
Concrete reinf., second floor	03200	032 107 0750	lb	10,528	0.60	6,317
Slab-on-grade, 6 in	03310	033 130 4820	sf	7,000	1.92	13,440
Spread footings	03310	033 130 3850	cy	6	160	960
Grade beams	03310	033 130 4260	cy	23	486	11,178
Insulating lightweight fill	03500	035 212 0250	sf	7,000	1.14	7,980
Scaffold, 2-story	01525	015 254 4100	csf	36	82	2,952
Coping, precast, 14 in wide	04210	042 116 0150	lf	120	35	4,200
Face brick, common, 8 in long, 4 in thick	04210	042 184 0800	sf	2,700	9.98	26,946
Backup wall, masonry block, 8 in thick, reinforced	04220	042 216 1150	sf	2,700	6.11	16,497
Block partitions, 8 in, block, reinforced	04220	042 216 1150	sf	12,670	6.11	77,414
Structural steel, heavy	05120	051 255 4900	ton	64	2,240	143,360
Base plates	05120	051 255 4300	lb	336	1.05	353
Joists	05210	052 110 0030	tons	13	1,504	19,552
Metal deck	05310	053 104 0440	sf	7,000	5.12	35,840
Handrails, aluminum 3-rail anodized	05520	055 203 0220	lf	75	42	3,150
Sandblast	05120	051 255 6130	sf	14,175	0.87	12,332
Primer	05120	051 255 6520	sf	14,175	0.54	7,655
Intermediate coat	05120	051 255 6630	sf	14,175	0.78	11,057
Top coat (elevated)	05120	051 255 7030	sf	14,175	1.19	16,868
Dampproof grade beam	07160	071 602 0300	sf	367	0.45	165
Insulate grade beam	07210	072 203 1740	sf	367	1.28	470
Insulation building	07210	072 203 0100	sf	8,850	1.28	11,328
Insulation roof	07220	072 203 0100	sf	7,000	1.04	7,280
Roofing	07510	075 102 0200	square	70	160	11,200
Flashing	07620	076 201 5800	lf	140	4.44	622
Ext. downspouts	07630	076 201 2500	lf	70	9.14	640
Ext. gutters	07630	076 201 3400	lf	100	13.31	1,331
Steel doors (3 ft × 7 ft)	08110	081 103 0100	each	8	209	1,672
Wood doors (3 ft × 6 ft 8 in)	08205	08 082 1640	each	36	314	11,304
Entrance doors	08410	084 105 1000	sf	42	21.82	916
Hardware	08710	087 125 0030	each	45	262	11,790
Glass windows	08920	089 204 5100	sf	900	23.17	20,853
Acoustical suspension system	09130	091 304 0050	sf	12,300	0.92	11,316
Drywall partitions, metal stud, 2 sides	09260	092 608 0350	sf	17,800	1.05	18,690
Metal studs	09260	092 612 2200	sf	8,900	1.19	10,591
Drywall on block (drywall only)	09260	092 608 0150	sf	7,950	0.59	46,905
Ceramic tile	09310	093 102 0010	sf	900	14.85	13,365

TABLE 10.1 Summary of Quantities and Costs, Divisions 1 to 14 (Continued)

Item	CSI No.	R.S. Means I.D. No.	Unit	No.	Unit price 1995	Cost
		Office (*Continued*)				
Acoustic tile	09510	095 106 0810	sf	12,300	2.46	30,258
Resilient tile	09660	096 601 0100	sf	3,825	1.98	7,574
Carpet	09685	096 852 1100	sy	875	38	33,250
Exterior paint	09910	099 106 2400	sf	42	0.54	23
Interior paint, masonry	09920	099 106 2400	sf	10,640	0.54	5,746
Interior paint, drywall	09920	099 106 2400	sf	51,500	0.54	27,810
Interior paint, doors	09920	099 204 2500	sf	3,444	0.91	3,134
Metal toilet partitions	10160	101 602 0400	each	12	1,146	13,752
Lockers	10505	105 054 0300	each	20	84	1,680
Elevator	14240	142 011 1000	each	1	51,200	51,200
Total						$789,473

TABLE 10.2 Division 15, Mechanical, Summary of Quantities and Costs

Item	CSI No.	R.S. Means I.D. No.	Quantity	Unit	Unit price 1995	Total
		Plant, Mechanical, HVAC				
Water pipe, 3-in steel, Sch. 40 on hangers	16060	151 701 2090	700	lf	26.25	18,375
Gate valves, 3-in steel, flanged	15100	151 980 0850	4	each	1,725.00	6,900
Check valves, 3-in steel, flanged	15100	151 980 1450	2	each	950.00	1,900
Unit heaters, 140 MBH, vertical flow	15620	155 630 5080	20	each	995.00	19,900
Heating pipe (HW), 2-in steel	15060	151 701 2070	1200	lf	16.81	20,172
Gate valves, 2-in FLG	15100	151 980 0830	4	each	1,550.00	6,200
Insulation (2-in pipe)	15260	155 651 9360	1200	lf	7.00	8,400
Air handling units, 18,620 cfm, 15 hp	15850	157 290 4220	10	each	5,530.00	55,300
Subtotal						$137,147
		Warehouse, Mechanical, HVAC				
Water pipe, 2-in steel, Sch. 40	15060	151 701 2070	200	lf	16.81	3,362
2-in gate valves, steel, flanged	15100	151 980 0830	4	each	1,550.00	6,200
Unit heaters, 140 mbh, vertical flow	15620	155 630 5020	8	each	995.00	7,960
HW pipe, 2-in steel	15060	151 701 6070	400	lf	16.81	6,724
2-in gate valves, steel flanged	15100	151 980 0830	4	each	1,550.00	6,200
Insulation, 2-in pipe	15260	155 651 9360	400	lf	7.00	2,800
Air handling units, 18,620 cfm, 15 hp	15850	157 290 4220	4	each	5,530.00	22,120

(*Continued*)

TABLE 10.2 Division 15, Mechanical, Summary of Quantities and Costs (Continued)

Item	CSI	R.S. Means I.D. No.	Quantity	Unit	Unit price 1995	Total
		Warehouse, Plumbing				
Water cooler, nonrecess, wall mount, 8 gph, wheelchair type	15450	153 105 2600	2	each	1375.00	2,750
Subtotal						$58,116
		Office, Plumbing				
Support and carriers						
Drinking fountain	15400	151 170 0500	2	each	73.00	146
Water closet, concealed arm, flat slab	15400	151 170 3200	12	each	175.00	2,100
Urinal, wall	15400	151 170 4200	2	each	295.00	590
Lavatories, wall	15400	151 170 5400	9	each	156.00	1,404
Lavatories, vanity top, porcelain enamel on C.I., 20 in × 18 in	15440	152 136 0600	9	each	330.00	2,970
Kitchen counter top, 32 in 3 21 in double bowl	15440	152 152 220	1	each	440.00	440
Urinals, wall-hung	15440	152 168 3000	2	each	660.00	1,320
Urinals, rough-in	15440	152 168 330	2	each	330.00	660
Wash fountain, S/S 54 in	15440	152 176 3100	1	each	2,625.00	2,625
Wash fountain, rough-in	15440	152 176 5700	1	each	520.00	520
Hot water heater, gas fired, 100 gal	15450	153 110 2120	1	each	1,250.00	1,250
Water closets, floor mounted	15440	152 180 1000	12	each	706.00	8,472
Water closet, rough-in	15440	152 180 1980	12	each	475.00	5,700
Water closet, seats	15440	152 164 0150	12	each	40.00	480
Water cooler, wall mounted nonrecessed, 8 gph, wheelchair type	15450	153 150 2600	2	each	1,375.00	2,750
Steel piping, 1 in	15060	151 401 2200	250	lf	8.75	2,188
Steel piping, ½ in	15060	151 401 2140	100	lf	7.00	700
Cast iron, 2 in	15060	151 301 2120	40	lf	13.88	555
Cast iron, 8 in	15060	151 301 2220	100	lf	38.75	3,875
WC bends, 16 in × 16 in	15060	151 320 0260	12	each	94.00	1,128
P traps, 2 in, pipe size	15060	151 181 3000	9	each	52.50	472
General service PVC, Sch. 40, 1½ in	15060	151 551 0710	150	lf	11.94	1,910
Vent flashing, 1½ in copper	15060	151 195 1430	4	each	37.50	150
Subtotal						$41,001
		Office, Mechanical, HVAC				
Fire extinguisher cabinet, single, S/S door and frame	10522	154 115 1200	8	each	287.00	2,296
Fire extinguisher, 20 lb	10522	154 125 0180	8	each	206.00	1,648
Boiler, hot water, gas fired, 4488 mbh	15550	155 155 3420	1	each	35,400.00	35,400
Fin tube radiation, 1½-in copper tube, 4¼-in Al fin	15515	155 630 1200	600	lf	60.00	36,000

TABLE 10.2 Division 15, Mechanical, Summary of Quantities and Costs (Continued)

Item	CSI No.	R.S. Means I.D. No.	Quantity	Unit	Unit price 1995	Total
		Office, Mechanical, HVAC (*Continued*)				
Steel pipe, 2 in	15060	151 701 2070	300	lf	16.81	5,043
Insulation, 2 in	15260	155 651 9360	300	lf	7.00	2,100
Chimney vent, prefab, 18 in D, metal, gas, double wall, Galv	15575	155 680 0280	30	lf	90.00	2,700
Chimney vent elbow	15575	155 680 0760	1	each	194.00	194
Chiller, 30-ton, 12,000 cfm, pkg.	12,000	15655157 125	3400	1	8,095.00	8,095
Ductwork, Al	15880	157 250 0155	8,920	lb	10.00	89,200
Air handling units, 18,620 cfm	15850	157 290 4220	4	each	5,531.00	22,124
Registers, 10 in × 10 in	15880	157 470 1080	30	each	46.00	1,380
Air balancing	15880	NA	1	ls	30,000.00	30,000
Wet pipe sprinkler, steel, black, Sch. 40	15330	8.2-111	14,000	sf	3.25	45,500
Subtotal						$281,680

TABLE 10.3 Division 16, Electrical, Summary of Quantities and Costs

Item	CSI No.	R.S. Means I.D. No.	Quantity	Unit	Unit price 1995	Total
		Plant , Electrical				
Rigid galv. steel conduit, 3 in	16110	160 205 1930	1800	lf	24.63	44,334
Rigid galv. steel conduit, 2 in	16110	160 205 1870	1500	lf	12.88	19,320
Motor connections, 15 hp	16050	160 275 0150	10	each	125.00	1,250
Motor connections, 100 hp	16050	160 275 1590	1	each	181.00	181
Wire, 3-conductor, 600 V, No. 4	16120	161 140 3000	2	clf	1,594.00	3,188
Wire, 3-conductor, 600 V, No. 12	16120	161 140 2200	32	clf	763.00	24,416
Wire, 2-conductor, 600 V, No. 12	16120	161 155 1500	30	clf	650.00	19,500
Terminations, No. 4	16120	161 520 4800	40	each	19.30	772
Terminations, No. 12	16120	161 520 4520	20	each	7.88	1,576
Receptacles duplex, 120 V, grounded, 20 A	16050	162 320 2470	60	each	19.38	1,163
Grounding	16050	NA	1	ls	10,000.00	10,000
Outlet boxes, pressed steel, octagon	16050	162 110 0020	60	each	19.88	1,193
Pull boxes and cabinets, 12 in × 12 in × 8 in	16050	162 130 0350	12	each	105.00	1,260
Telephone, transition, wall fitting	16700	161 160 2510	10	each	38.00	380
Thermostat, No. 18, 2-conductor	16120	161 155 0500	6	clf	55.00	330

(*Continued*)

TABLE 10.3 Division 16, Electrical, Summary of Quantities and Costs (Continued)

Item	CSI No.	R.S. Means I.D. No.	Quantity	Unit	Unit price 1995	Total
		Plant, Electrical (*Continued*)				
Fire alarm wire	16120	161 155 1600	6	clf	175.00	1,050
Starters, MCC class 1, type B, combined MCP, FVNR, 25 hp	16300	163 110 0200	10	each	1,125.00	11,250
Starters, MCC class 1, type B, combined MCP, FVNR, 100 hp	16300	163 100 0400	1	each	3,220.00	3,220
Circuit breakers, enclosed, NEMA 1, 3-pole, 15 hp	16300	163 205 2040	10	each	320	3,200
Circuit breakers, enclosed, NEMA 1, 3-pole, 100 hp	16300	163 205 2200	1	each	1,725.00	1,725
Lights, metal halide, 1500 W	16500	166 115 7650	36	each	445.00	16,020
Fixture whips, ⅜-in Greenfield, 2 connectors, 6 ft long	16500	166 125 0300	36	each	21.88	788
Exit lights	16500	166 110 0080	8	each	110.00	880
Total						$166,996
		Warehouse, Electrical				
Rigid galv. steel conduit, 2 in	16110	160 205 1870	1000	lf	24.63	2,463
Motor connections, 15 hp	16050	160 275 0150	4	each	125.00	500
Wire, 2-conductor, 600 V, No. 12	16120	161 155 1500	15	clf	650.00	9,750
Termination, No. 12	16120	161 520 4520	120	each	7.88	9,456
Receptacles, duplex, 120 V, grounded, 20 A	16050	162 320 2470	30	each	19.38	581
Grounding	16050	NA	1	ls	20,000	20,000
Outlet boxes, pressed steel, octagon	16050	162 110 0020	30	each	19.88	596
Pull boxes and cabinets, 12 in × 12 in × 8 in	16050	162 130 0350	4	each	105.00	420
Telephone, transition, wall fitting	16700	161 160 2510	10	each	38.00	380
Thermostat, No. 18, 2-conductor	16120	161 155 0500	2	clf	55.00	110
Fire alarm wire	16120	161 155 1600	2	clf	175.00	350
Starters, MCC class 1, type B, combined MCP, FVNR, 25 hp	16300	163 110 0200	4	each	1,125.00	4,500
Circuit breakers, enclosed, NEMA 1, 3-pole, 15 hp	16300	163 205 2040	4	each	320.00	1,280
Lights, metal halide, 1500 W	16500	166 115 7650	12	each	445.00	5,340
Fixture whips, ⅜-in Greenfield, 2 connectors, 6 ft long	16500	166 125 0300	12	each	21.88	263
Exit lights	16500	166 110 0080	6	each	110.00	660
Total						$56,649

TABLE 10.3 Division 16, Electrical, Summary of Quantities and Costs (Continued)

Item	CSI No.	R.S. Means I.D. No.	Quantity	Unit	Unit price 1995	Total
		Office, Electrical				
Rigid galv. steel conduit, 4 in	16110	160 205 2220	100	lf	175.00	17,500
Rigid galv. steel conduit 1 ½ in	16110	160 205 1850	2000	lf	10.56	21,120
Motor connections, 15 hp	16050	160 275 0150	4	each	125.00	500
Motor connections, 50 hp	16050	160 275 1560	1	each	88.00	88
Wire, 3-conductor, 600 V, No. 4	16120	161 140 3000	2	clf	1,594.00	3,188
Wire, 2-conductor, 600 V, No. 12	16120	161 155 1500	40	clf	650.00	26,000
Terminations, No. 4	16120	161 520 4800	4	each	19.30	77
Terminations, No. 12	16120	161 520 4520	200	each	7.88	1,576
Grounding	16050	NA	1	ls	10,000.00	10,000
Outlet boxes, pressed steel, octagon	16050	162 110 0020	60	each	19.88	1,193
Pull boxes and cabinets, 12 in × 12 in × 8 in	16050	162 130 0350	8	each	105.00	840
Telephone, transition, wall fitting	16700	161 160 2510	40	each	38.00	1,520
Thermostat, No. 18, 2-conductor	16120	161 155 0500	2	clf	55.00	110
Fire alarm wire	16120	161 155 1600	2	clf	175.00	350
Starters, MCC class 1, type B, combined MCP, FVNR, 25 hp	16300	163 110 0200	4	each	1,125.00	4,500
Starters, MCC class 1, type B, combined MCP, FVNR, 50 hp	16300	163 100 0300	1	each	1,800.00	1,800
Circuit breakers, enclosed, NEMA 1, 3-pole, 200 A	16300	163 205 1260	1	each	9,125.00	9,125
Circuit breakers, enclosed, NEMA 1, 3-pole, 15 hp	16300	163 205 2040	4	each	320.00	1,280
Transformer, 1500 kVA, oil-filled, pad-mounted, 15KV/277-480, secondary, 3-phase	16400	164 160 0600	1	each	37,375.00	37,375
Main service switchgear	16320	Quote	1	each	15,000.00	15,000
Switchboard	16320	Quote	1	each	12,500.00	12,500
Lighting load centers	16320	163 230 0700	4	each	788.00	3,152
Fluorescent CW lamps, Troffer grid, 2 ft × 4 ft × 3 lamps	16500	166 130 0500	150	each	155.00	23,250
Fixture whips	16500	166 125 0300	150	each	21.50	3,225
Surface incandescent	16500	166 130 4920	12	each	94.00	1,128
Receptacles duplex, 120 V, grounded, 15 A	16050	162 320 2460	60	each	11.44	686
Total						$198,083

TABLE 10.4 Comparison of Results, Construction Estimate vs. Prior Estimates, Divisions 1 to 14 (Not Including Site)

	Construction estimate (manual)		Design development (manual)
Divisions 1–14	Base cost (from Table 10.1)	Markup (1.215)	Marked up (from Table 8.2 and Fig. 8.4)
Plant	$1,966,276	$2,389,025	$3,371,711
Warehouse	414,707	503,869	862,963
Office	789,473	959,210	1,071,383
Subtotal	$3,170,453	$3,852,104	$5,306,057

Table 10.4 compares the construction estimate with the design development for Divisions 1 to 14 (not including site), reflecting a 27% cost reduction as a result of the value engineering.

Table 10.5 compares the Division 15 and Division Construction estimate with the prior estimates. Cost increases are 6 and 18%, for an average increase of 11%.

Table 10.6 summarizes the three elements into the cost estimates for plant, warehouse, and office, plus a grand total.

Table 10.7 compares the cost estimates between the post-VE design development estimate (manual) and the construction documents estimate (manual). Note that the construction document estimate is in 1995 dollars and has *no contingency*.

TABLE 10.5 Comparison of Results, Construction Estimate vs. Prior Estimates, Divisions 15 and 16 (Not Including Site)

	Construction estimate (manual)		Design development (manual)
	Base cost (from Table 10.2)	Markup (1.215)	Marked up (from Fig. 8.4)
Division 15:			
Plant	$137,147	$166,634	$419,910
Warehouse	58,116	70,611	101,510
Office	322,681	392,057	71,542
Subtotal	$517,944	$629,302	$592,962
Division 16	(from Table 10.3)		
Plant	$166,996	$202,900	$348,754
Warehouse	56,649	68,829	42,062
Office	198,083	240,671	40,892
Subtotal	$421,728	$512,400	$431,708

TABLE 10.6 Construction Estimate (Manual) (Not Including Site Work)

	Base cost	Marked up (1.215)
Plant:		
Divisions 1–14	$1,966,276	$2,389,025
Division 15, Mechanical	137,147	166,634
Division 16, Electrical	166,996	202,900
Subtotal	$2,270,419	$2,758,559
Warehouse:		
Divisions 1–14	$414,707	$503,869
Division 15, Mechanical	58,116	70,611
Division 16, Electrical	56,649	68,829
Subtotal	$529,472	$643,309
Office:		
Divisions 1–14	$789,473	$959,210
Division 15, Mechanical	322,681	392,057
Division 16, Electrical	198,083	240,691
Subtotal	$1,310,237	$1,591,958
Project total	$4,110,128	$4,993,826

TABLE 10.7 Comparison of Construction Cost Estimates, Design Development vs. Construction Cost (Not Including Site)

	Revised design development estimate (from Table 9.5)	Construction documents estimate (from Table 10.6)
Plant	$3,341,551	$2,758,559
Warehouse	713,303	643,309
Office	1,122,461	1,591,958
Total	$5,177,317	$4,993,826

Computerized construction document estimate (Composer Gold). Using the quantities (only) from Tables 10.1, 10.2, and 10.3, Fred M. Seidell III (Certified Cost Consultant) of Building Systems Design (BSD) prepared the following construction document estimate. Seidell described the process as follows:

> We used the R. S. Means Facilities Database Unit price cost records inside of Composer Gold without any of the Means published Overhead and Profit factors. The Means Unit Price Book has its labor based on a 30 City mean. We used Philadelphia Davis Bacon wage rates to reprice the labor. We have refined the proper contractor and subcontractor tiering in this estimate to ensure that the cost estimate is as real looking while still using this corrected estimate.
>
> To further assist in helping you to differentiate between this estimate stage and the Design Development Cost Estimate we have had Gold gen-

erate a comparison report. This report points out immediately to the educated eye that the lack of piling and lower structural steel quantities made a significant cost effect.

The *Gold* output is as follows:

- Comparison report
- Basis of report
- Table of contents
- Summary reports
- Detailed estimate
  10 Plant
  20 Warehouse(s)
  30 Office building
- Labor backup

There is zero contingency, but the *Gold* estimate is escalated at 5% annual rate to a March 1996 completion.

Building Systems Design

PROJECT WHATC2: Whatco Manufacturing-Whse-Office - Demonstration Project

Pre-Bid Design Review Cost Estimate — SUMMARY PAGE 3

** 2ND VIEW SUMMARY (Rounded to 100's) **

		** PROJECT **		** WHATCO **	
	QUANTITY UOM	UNIT	TOTAL	TOTAL	COL%
			95% Design.	35% Design	
10 Plant Building					
10.01 Substructure	30000.00 SF	2.13	63,900	210,600	5.4%
10.02 Structural Frame	30000.00 SF	30.12	903,700	1,607,600	41.4%
10.03 Roofing	30000.00 SF	4.17	125,100	137,600	3.5%
10.04 Exterior Closure	30000.00 SF	10.17	305,000	447,800	11.5%
10.06 Interior	30000.00 SF	0.13	3,800	4,200	0.1%
10.07 Specialties	30000.00 SF	0.75	22,600	24,800	0.6%
10.10 Mechanical Systems	30000.00 SF	6.27	188,100	458,600	11.8%
10.11 Electrical	30000.00 SF	8.13	243,800	381,500	9.8%
10.13 Equipment & Conveying	30000.00 SF	18.65	559,500	609,000	15.7%
10 Plant Building	30000.00 SF	80.52	2,415,500	3,881,600	60.7%
20 Warehouse Buildings "A" & "B"					
20.01 Substructure	6000.00 SF	2.01	12,100	19,100	1.9%
20.02 Structural Frame	6000.00 SF	33.49	201,000	334,200	32.4%
20.03 Roofing	6000.00 SF	4.52	27,100	29,800	2.9%
20.04 Exterior Closure	6000.00 SF	17.25	103,500	170,000	16.5%
20.06 Interior	6000.00 SF	0.38	2,300	2,500	0.2%
20.07 Specialties	6000.00 SF	3.76	22,600	24,800	2.4%
20.09 Plumbing Systems	6000.00 SF	0.59	3,600	0	%
20.10 Mechanical Systems	6000.00 SF	13.32	79,900	110,900	10.8%
20.11 Electrical	6000.00 SF	14.82	88,900	45,900	4.5%
20.13 Equipment & Conveying	6000.00 SF	8.99	54,000	292,700	28.4%
20 Warehouse Buildings "A" & "B"	6000.00 SF	99.13	594,800	1,029,900	16.1%

30 Office Building					
30.01 Substructure	7000.00 SF	2.11	14,800	18,900	1.6%
30.02 Structural Frame	7000.00 SF	42.60	298,200	392,200	33.1%
30.03 Roofing	7000.00 SF	9.41	65,900	72,500	6.1%
30.04 Exterior Closure	7000.00 SF	32.33	226,300	248,900	21.0%
30.06 Interior	7000.00 SF	25.40	177,800	195,600	16.5%
30.07 Specialties	7000.00 SF	5.98	41,900	46,000	3.9%
30.08 Fire Protection	7000.00 SF	4.80	33,600	0	%
30.09 Plumbing Systems	7000.00 SF	7.52	52,700	0	%
30.10 Mechanical Systems	7000.00 SF	46.33	324,300	78,100	6.6%
30.11 Electrical	7000.00 SF	29.83	208,800	44,700	3.8%
30.13 Equipment & Conveying	7000.00 SF	11.52	80,700	88,700	7.5%
30 Office Building	7000.00 SF	217.84	1,524,900	1,185,700	18.5%
40 Site Support					
40.10 Site Preparation	6.00 ACR		0	37,000	12.4%
40.20 Site Electrical	6.00 ACR		0	108,800	36.4%
40.30 Site Water	250000.00 GAL		0	152,900	51.2%
40 Site Support	6.00 ACR		0	298,600	4.7%
Whatco Manufacturing-Whse-Office	1.00 EA	4535162	4,535,200	6,395,900	100%

Building Systems Design

PROJECT WHATC2: Whatco Manufacturing-Whse-Office - Demonstration Project

Pre-Bid Design Review Cost Estimate

TITLE PAGE 1

Whatco Manufacturing-Whse-Office

Demonstration Project

Designed By: James J. O'Brien, P.E.
Estimated By: Building Systems Design

Prepared By: Fred M. Seidell III, C.C.C.
Senior Technical Consultant

Date: 02/03/93
Est Construction Time: 360 Days

Building Systems Design
PROJECT WHATC2: Whatco Manufacturing-Whse-Office - Demonstration Project
PROJECT NOTES Pre-Bid Design Review Cost Estimate TITLE PAGE 2

BASIS OF ESTIMATE

This estimate was produced from Design Development information received from James J O'Brien, P.E. of O'BREIN - KREITZBERG & ASSOCIATES, INC. on January 27th, 1993.

Sketch Plans & Designs	: Dated	"NOT DATED"
Original Specifications	: Dated	"NOT DATED"

The estimate has been complited using the 1993 R.S.Means Facilities Data and Davis Bacon wage rates for Philadelphia, Pennsylvania.

BASIS FOR PRICING

Pricing shown reflects probable construction costs obtainable in the Southeastern Pennsylvania area on the date of this statement of costs. This estimate is a determination of fair market value for the construction of this project. It is not a prediction of low bid. Pricing assumes competitive bidding for every portion of the construction work for all subcontractors, as well as the general contractor; that is to mean 6 to 7 bids. If less bids are received, bid results can be expected to be higher.

Length of construction is assumed to be 12 months. Any costs for excessive overtime to meet stringent milestone dates are not included in this estimate.

Bid date is assumed to be April 1995. A value of 5% per annual escalation is added to the cost for the construction which is assumed to be completed March 1996.

The General Contractor's Overhead is set at 8% and his Profit margins are set at 7% on all of his work, and 5% on all of his Subcontractors. Bond is set at 1.5%. The Subcontractors have been properly teired for this project and this is shown in the Indirect Cost Summary Report.

PROJECT DESCRIPTION

This project consists of : The construction of a Plant Office Warehouse on Property in Philadelphia, Pennsylvania.

STATEMENT OF PROBABLE COST

Building Systems Design has no control over the cost of labor and materials, the general contractor's or any subcontractor's method of determining prices, or competitive bidding and market conditions. This opinion of probable cost of construction is made on the basis of experience, qualifications, and best judgement of Building Systems Design estimators familar with the construction industry. We cannot and do not guarantee that proposals, bids or actual construction costs will not vary from this or subsequent cost estimates.

Building Systems Design

PROJECT WHATC2: Whatco Manufacturing-Whse-Office - Demonstration Project

TABLE OF CONTENTS — Pre-Bid Design Review Cost Estimate — CONTENTS PAGE 1

SUMMARY REPORTS — SUMMARY PAGE

DETAILED ESTIMATE — DETAIL PAGE

* * * END TABLE OF CONTENTS * * *

Building Systems Design

PROJECT WHATC2: Whatco Manufacturing-Whse-Office - Demonstration Project

Pre-Bid Design Review Cost Estimate SUMMARY PAGE 1

** PROJECT DIRECT SUMMARY - LEVEL 1 (Rounded to 100's) **

		QUANTITY	UOM	Labor	Equipmnt	Material	SubContr	TOTAL COST	UNIT COST
10	Plant Building	30000.00	SF	528,700	72,300	1,211,300	11,600	1,824,000	60.80
20	Warehouse Buildings "A" & "B"	6000.00	SF	131,700	15,100	278,400	23,200	448,400	74.74
30	Office Building	7000.00	SF	467,700	23,100	575,700	81,600	1,148,200	164.02
	Whatco Manufacturing-Whse-Office	1.00	EA	1,128,100	110,500	2,065,400	116,500	3,420,600	3420556
	Overhead @ 8.0 %							273,600	
	SUBTOTAL							3,694,200	
	Profit @ 7.0 %							191,100	
	SUBTOTAL							3,885,300	
	Bond @ 1.5 %							58,300	
	TOTAL INCL INDIRECTS							3,943,600	
	Escalation @ 5.0 % / Annum							591,500	
	TOTAL INCL OWNER COSTS							4,535,200	

Building Systems Design

PROJECT WHATC2: Whatco Manufacturing-Whse-Office - Demonstration Project

Pre-Bid Design Review Cost Estimate

SUMMARY PAGE 2

** DIVISION OWNER SUMMARY (Rounded to 100's) **

	CONTRACT COST	Escalatn	Contincy	TOTAL COST
01 General Requirements	3,000	400		3,400
02 Site Work	72,000	10,800		82,800
03 Concrete	310,800	46,600		357,400
04 Masonry	133,900	20,100		153,900
05 Metals	951,400	142,700		1,094,100
07 Moisture-Thermal Control	500,600	75,100		575,700
08 Doors, Windows, & Glass	68,400	10,300		78,600
09 Finishes	163,600	24,500		188,200
10 Specialties	13,600	2,000		15,600
11 Equipment	105,800	15,900		121,600
14 Conveying Systems	556,700	83,500		640,200
15 Mechanical	581,700	87,200		668,900
16 Electrical	482,400	72,400		554,800
	3,943,600	591,500	0	4,535,200

Building Systems Design

PROJECT WHATC2: Whatco Manufacturing-Whse-Office - Demonstration Project

Pre-Bid Design Review Cost Estimate

** CONTRACTOR INDIRECT SUMMARY (Rounded to 100's) **

	TOTAL DIRECT	Overhead	Profit	Bond	TOTAL COST	UNIT COST
GM General Markup						
AT Acoustical Contractor	31,700	1,900	2,700	0	36,300	36323.45
CA Carpet & Resilent Contractor	27,200	1,600	1,700	0	30,600	30595.45
CW Concrete Subcontractor						
WP Water Proofing Subcontractor	200	0	0	0	200	219.37
Subtotal Subcontract Work	200	0	0	0	200	219.37
Indirect on Subcontracts	200	0	0	0	200	244.16
Indirect on Own Work	236,100	18,900	15,300	0	270,200	270248.32
CW Concrete Subcontractor	236,300	18,900	15,300	0	270,500	270492.48
EL Electrical Subcontractor	216,500	16,200	18,600	0	251,400	251365.01
EV Elevator & Coveying Subcontractr	51,300	5,100	4,500	0	60,900	60941.34
GL Glazing Subcontractr	18,200	1,500	1,400	0	21,000	21042.87
GW Gypsum Subcontractr	20,200	1,200	1,300	0	22,700	22720.74
HT Hard Tile Subcontractr	5,200	300	300	0	5,900	5876.11
MA Masonry Subcontractr	107,300	8,600	7,000	0	122,900	122861.10
ME Mechanical Subcontractr						
PL Plumbing Subcontractr	91,900	7,400	6,900	0	106,200	106208.45
SM SheetMetal Subcontractr						
TB Test & Balance Contractor	30,000	3,000	3,300	0	36,300	36300.00

Subtotal Subcontract Work	30,000	3,000	3,300	0	36,300	36300.00
Indirect on Subcontracts	36,300	1,800	1,900	0	40,000	40020.75
Indirect on Own Work	74,900	7,500	5,800	0	88,100	88099.61
SM SheetMetal Subcontractr	111,200	9,300	7,700	0	128,100	128120.36
IN Insulation SubContractor	10,000	600	1,100	0	11,700	11675.38
Subtotal Subcontract Work	213,100	17,300	15,700	0	246,000	246004.19
Indirect on Subcontracts	246,000	12,300	12,900	0	271,200	271219.62
Indirect on Own Work	188,100	14,500	16,200	0	218,700	218749.96
ME Mechanical Subcontractr	434,100	26,800	29,100	0	490,000	489969.58
MR Membrane Roofing Subcontractr	112,700	13,500	10,100	0	136,300	136342.04
MW Metal Wall Panel Subcontractr	249,800	22,500	21,800	0	294,100	294069.27
PS Painting Subcontractor	112,600	6,200	7,700	0	126,600	126568.62
SD Speciality Subcontractor						
CR Crane & Monorail Supplier	392,700	15,700	14,300	0	422,700	422717.62
Subtotal Subcontract Work	392,700	15,700	14,300	0	422,700	422717.62
Indirect on Subcontracts	422,700	0	0	0	422,700	422717.62
Indirect on Own Work	48,800	4,900	5,400	0	59,000	59031.31
SD Speciality Subcontractor	471,500	4,900	5,400	0	481,700	481748.93
SS Structural Steel Subcontractor	652,100	47,000	47,500	0	746,600	746626.13
FP Fire Protection Subcontractor	21,200	1,900	2,300	0	25,400	25382.22

Building Systems Design

PROJECT WHATC2: Whatco Manufacturing-Whse-Office - Demonstration Project

Pre-Bid Design Review Cost Estimate

** CONTRACTOR INDIRECT SUMMARY (Rounded to 100's) **

	TOTAL DIRECT	Overhead	Profit	Bond	TOTAL COST	UNIT COST
Subtotal Subcontract Work	2,768,100	178,100	176,800	0	3,122,900	3122925
Indirect on Subcontracts	3,122,900	249,800	168,600	53,100	3,594,500	3594518
Indirect on Own Work	297,600	23,800	22,500	5,200	349,100	349100.65
GM General Markup	3,420,600	273,600	191,100	58,300	3,943,600	3943619

Building Systems Design

PROJECT WHATC2: Whatco Manufacturing-Whse-Office - Demonstration Project

DETAILED ESTIMATE | Pre-Bid Design Review Cost Estimate | DETAIL PAGE 1

10. Plant Building

10.01. Substructure	QUANTY UOM	Labor	Equipmnt	Material	SubContr	TOTAL COST

10. Plant Building

The Plant Building is 300' X 100' rectangle area of 30,000 Gross Square Feet.

SUBSTRUCTURE AND STRUCTURE:

One Story Steel Frame Structure With Heavy Wide Flange Columns On Concrete Foundations Supported by Driven Piles. This system supports The Roof Structure and the overhead Bridge Crane. The Exterior Walls Are Supported By Grade Beams And Spread Footings. The Slab On Grade is a 12 inch Thicknesses. A Monorail Crane Runway Structure, and A Bridge Crane Runway Structure are also provided for.

ROOFING

The Roofing is a Built-up Roofing Over Metal Decking. Rigid Insulation Covers The Entire Roof Area and Roof sloping is accomplished with the insulation.
Gutters And Downspouts Are Used To Channel Rain Water Off The Roof Area.

EXTERIOR WALL

Consists Of Structural Steel and Girt System. Insulated Metal Panels Are Used As A Veneer. The Soffit And Fascia Consists Of Insulated Metal Panels On Metal Studs. Hollow Metal Doors Are Used To Enter And Exit The Facility.

INTERIOR

There are no Interior Partitions.

SPECIALITIES

Includes Dock Levelers, And a Bridge Crane.

MECHANICAL

The Mechanical estimate was compiled from detail.

ELECTRICAL

The Electrical estimate was compiled from detail.

Building Systems Design

PROJECT WHATC2: Whatco Manufacturing-Whse-Office - Demonstration Project

DETAILED ESTIMATE — Pre-Bid Design Review Cost Estimate — DETAIL PAGE 2

10. Plant Building

10.01. Substructure	QUANTY UOM	Labor	Equipmnt	Material	SubContr	TOTAL COST
10.01. Substructure						
02110 40000 Clear And Grub						
RSM GM Site Preparation (Clear)		1954.19	1497.99	0.00	0.00	3452.17
	0.70 ACR	1,368	1,049	0	0	2,417
02220 80000 Backfill, Structural						
RSM GM Borrow; Loaded At Pit, No Haul		0.00	0.00	2.59	0.00	2.59
	1672.00 TON	0	0	4,330	0	4,330
RSM GM Spread W/200Hp Dozer, Bank Yards		1.33	2.68	0.00	0.00	4.01
	1672.00 CY	2,230	4,480	0	0	6,710
RSM GM Borrow Delivery Charge, Min 12		2.55	4.62	0.00	0.00	7.17
Cy, 1 Hr Round Trip	1672.00 CY	4,266	7,719	0	0	11,985

Item	Quantity					
02222 60000 Compaction Vibrating						
RSM GM Compaction, Riding Vibrating Roller, 6" Lift, 4 Pass	1672.00 CY	0.29 478	0.24 407	0.00 0	0.00 0	0.53 885
02224 60000 Excavation, Bulk, Scrapers						
RSM GM Excavating Bulk - Footprint	3345.00 CY	0.67 2,252	3.04 10,184	0.00 0	0.00 0	3.72 12,435
02225 00000 Excavating, Structural						
RSM GM Excavating, Structural, Hand, Spread Footings	86.00 CY	54.03 4,646	0.00 0	0.00 0	0.00 0	54.03 4,646
02226 20000 Fill						
RSM GM Spread Fill W/ Ldr, Crwlr, 300' Haul 130 Hp	112.00 CY	0.62 69	1.27 143	0.00 0	0.00 0	1.89 212
03210 70000 Reinforcing In Place						
RSM CW Reinforcing - Footings	6056.00 LB	0.32 2,230	0.00 0	0.23 1,595	0.00 0	0.55 3,825
		----------	----------	----------	----------	----------
Substructure		17,540	23,980	5,925	0	47,445

Building Systems Design
PROJECT WHATC2: Whatco Manufacturing-Whse-Office - Demonstration Project
DETAILED ESTIMATE — Pre-Bid Design Review Cost Estimate — DETAIL PAGE 3
10. Plant Building

10.02. Structural Frame	QUANTY UOM	Labor	Equipmnt	Material	SubContr	TOTAL COST
10.02. Structural Frame						
03210 70000 Reinforcing In Place						
RSM CW Reinforcing In Place Slab On		0.32	0.00	0.25	0.00	0.57
Grade, & Load Deck	37328 LB	13,747	0	10,683	0	24,431
03313 00000 Concrete In Place						
RSM CW Conc. In Place, Load Deck		0.62	0.20	0.70	0.00	1.52
	4000.00 SF	2,837	925	3,205	0	6,968
RSM CW Concrete In Place,Spread Footing		36.40	5.44	64.00	0.00	105.83
	86.00 CY	3,583	535	6,301	0	10,419
RSM CW Conc In Place, Ground Slab, 12"		0.58	0.19	1.94	0.00	2.71
Thick, Trowel Finish	30000 SF	19,868	6,481	66,627	0	92,976
03521 20000 Insulating						
RSM CW Insul Lightweight Fill 3" Thick		0.22	0.07	0.75	0.00	1.04
	30000 SF	7,418	2,421	25,758	0	35,598

05125 50000 Structural Steel Projects

Item	Quantity					
RSM SS Structural Steel Base Plates		0.34	0.00	0.49	0.00	0.83
	3708.00 LB	1,434	0	2,080	0	3,514
RSM SS Structural Steel Beams Light		205.65	92.90	1300.00	0.00	1598.55
	6.10 TON	1,436	649	9,079	0	11,164
RSM SS Structural Steel Beams Heavy		188.08	84.96	1125.00	0.00	1398.04
	153.00 TON	32,945	14,883	197,065	0	244,893
RSM PS Sand Blast		0.43	0.06	0.14	0.00	0.63
	22687 SF	10,938	1,557	3,569	0	16,064
RSM PS Paints & Protect Coats, Alkyds,		0.14	0.01	0.06	0.00	0.21
Primer	22387 SF	3,547	191	1,509	0	5,247
RSM PS Paints & Protect Coats, Epoxy,		0.18	0.01	0.14	0.00	0.33
Intermediate Coat	22387 SF	4,560	247	3,521	0	8,328
RSM PS Top Coat (Elevated)		0.28	0.02	0.18	0.00	0.48
	22387 SF	7,093	382	4,528	0	12,003

05211 00000 Open Web Joists

Item	Quantity					
RSM SS Open Web Joists, H Series,		257.55	95.11	750.00	0.00	1102.66
	72.00 TON	21,231	7,840	61,824	0	90,895

05310 40000 Metal Decking

Building Systems Design

PROJECT WHATC2: Whatco Manufacturing-Whse-Office - Demonstration Project

DETAILED ESTIMATE — Pre-Bid Design Review Cost Estimate — DETAIL PAGE 4

10. Plant Building

10.02. Structural Frame	QUANTY UOM	Labor	Equipmnt	Material	SubContr	TOTAL COST
RSM SS Metal Deck Galvanized 2" 20-20Ga		0.94	0.05	2.40	0.00	3.39
	31000 SF	33,234	1,785	85,180	0	120,200
Structural Frame		163,873	37,897	480,931	0	682,701
10.03. Roofing						
07220 30000 Roof Deck Insulation						
RSM MR Roof Deck Insulation, Fiberboard		0.36	0.00	0.60	0.00	0.96
	30000 SF	12,951	0	21,773	0	34,724
07510 20000 Built-Up Roofing						
RSM MR Built Up Roofing		91.56	6.75	33.00	0.00	131.31
	300.00 SQ	33,226	2,448	11,975	0	47,649
07620 10000 Downspouts						
RSM MR Downspout, Galvanized, 3" X 4"		2.16	0.00	1.25	0.00	3.41
	800.00 LF	2,092	0	1,210	0	3,302

Item	Quantity					
07620 40000 Flashing						
RSM MR Flashing		2.11	0.00	3.15	0.00	5.26
	800.00 LF	2,046	0	3,048	0	5,095
07620 50000 Gutters						
RSM MR Gutter, 5" X 6"		5.22	0.00	2.50	0.00	7.72
	400.00 LF	2,528	0	1,210	0	3,737
		----------	----------	----------	----------	----------
Roofing		52,844	2,448	39,215	0	94,507
10.04. Exterior Closure						
07420 20000 Metal Facing Panels						
RSM MW MFG Wall Panels-Galv. Steel Int.		5.43	0.13	4.85	0.00	10.41
	18000 SF	115,096	2,687	102,770	0	220,552
08110 30000 Commercial Steel Doors						
RSM GM Steel Doors & Frame 3' X 7'		32.87	0.97	146.00	0.00	179.84
	5.00 EA	164	5	730	0	899
08360 40000 Overhead, Commercial						

Building Systems Design
PROJECT WHATC2: Whatco Manufacturing-Whse-Office - Demonstration Project
DETAILED ESTIMATE — Pre-Bid Design Review Cost Estimate — DETAIL PAGE 5
10. Plant Building

10.04. Exterior Closure	QUANTY UOM	Labor	Equipmnt	Material	SubContr	TOTAL COST
RSM SD Steel Rollup Doors		372.48	0.00	715.00	0.00	1087.48
	4.00 EA	1,803	0	3,461	0	5,263
08712 50000 Mortise Lockset						
RSM GH Door Hardware		31.04	0.00	263.00	0.00	294.04
	5.00 EA	155	0	1,315	0	1,470
09910 60000 Siding						
RSM PS Painting Exterior Siding		0.29	0.00	0.14	0.00	0.43
	4516.00 SF	1,450	0	710	0	2,160
Exterior Closure		118,668	2,692	108,986	0	230,345
10.06. Interior						
09910 60000 Siding						
RSM PS Interior Painting		0.29	0.00	0.14	0.00	0.43
	6000.00 SF	1,926	0	944	0	2,870
Interior		1,926	0	944	0	2,870

10.07. Specialties						
11160 10000 Loading Dock						
RSM SD Dock Platform Levelers, 7'X 8', 10 Tons, Hinged	4.00 EA	553.94 2,681	68.00 329	2900.00 14,036	0.00 0	3521.94 17,046
Specialties		2,681	329	14,036	0	17,046
10.10. Mechanical Systems						
15170 10000 Pipe, Steel						
RSM ME Heating Pipe (HW) 2" Steel	1200.00 LF	8.75 12,207	0.75 1,049	3.69 5,150	0.00 0	13.19 18,407
RSM PL Water Pipe, Steel Sch 40 on Hangers	700.00 LF	12.41 11,065	1.07 951	5.70 5,083	0.00 0	19.17 17,099
15198 00000 Valves, Steel						
RSM ME Gate Valves, 2" Flanged	4.00 EA	37.05 172	0.00 0	1275.00 5,932	0.00 0	1312.05 6,104

Building Systems Design
PROJECT WHATC2: Whatco Manufacturing-Whse-Office - Demonstration Project
DETAILED ESTIMATE Pre-Bid Design Review Cost Estimate DETAIL PAGE 6
10. Plant Building

10.10. Mechanical Systems	QUANTY UOM	Labor	Equipmnt	Material	SubContr	TOTAL COST
RSM PL Gate Valves, 3" Steel, Flanged		118.55	0.00	1325.00	0.00	1443.55
Flanged	4.00 EA	604	0	6,752	0	7,357
RSM PL Check Valves, 3" Steel, Flanged		118.55	0.00	690.00	0.00	808.55
	2.00 EA	302	0	1,758	0	2,060
15563 00000 Hydronic Heating						
RSM ME Unit Heaters, 140MBH, Vertical		133.37	0.00	860.00	0.00	993.37
Flow	20.00 EA	3,103	0	20,006	0	23,109
15565 10000 Insulation						
RSM IN Pipe Covering 1"Thick (2" Pipe)		2.89	0.00	2.38	0.00	5.27
Urthane Ultravoilet Resistant	1200.00 LF	4,458	0	3,671	0	8,130
15729 00000 Fans						
RSM ME Air Handling Units 18,620 CFM		282.13	0.00	4250.00	0.00	4532.13
15Hp No Motor	10.00 EA	3,282	0	49,434	0	52,716

Item	Quantity					Total
RSM ME Motor 230/460V 1800 Rpm 15Hp	10.00 EA	99.29 1,155	0.00 0	515.00 5,990	0.00 0	614.29 7,145
Mechanical Systems		36,348	2,000	103,779	0	142,127

10.11. Electrical

16020 50000 Conduit

Item	Quantity					Total
RSM EL Rigid Galv. Steel Conduit, 2"GRC	1500.00 LF	7.06 12,296	0.00 0	4.00 6,966	0.00 0	11.06 19,262
RSM EL Rigid Galv. Steel Conduit, 3"GRC	1800.00 LF	12.71 26,559	0.00 0	8.60 17,972	0.00 0	21.31 44,532

16027 50000 Motor Connections

Item	Quantity					Total
RSM EL Motor Conn Flex Cond & Ftg 15 Hp 230 Volts 3 Ph	10.00 EA	96.28 1,118	0.00 0	10.55 122	0.00 0	106.83 1,240
RSM EL Motor Conn. 3 Pole 100Hp Motor, 460V, 3 Ph	1.00 EA	127.09 148	0.00 0	30.00 35	0.00 0	157.09 182
OBK EL Grounding Allowance 120 Volt Grounded 20 Amp	1.00 LS	0.00 0	0.00 0	0.00 0	10000.00 11,610	10000.00 11,610

16114 00000 Mineral Insulated Cable

Building Systems Design
PROJECT WHATC2: Whatco Manufacturing-Whse-Office - Demonstration Project
DETAILED ESTIMATE — Pre-Bid Design Review Cost Estimate — DETAIL PAGE 7
10. Plant Building

10.11. Electrical	QUANTY UOM	Labor	Equipmnt	Material	SubContr	TOTAL COST
RSM EL Wire, 2 Conductor, 600V #12		226.95	0.00	360.00	0.00	586.95
	30.00 CLF	7,905	0	12,539	0	20,443
RSM EL Wire, 3 Conductor, 600V, #12		264.77	0.00	415.00	0.00	679.77
	32.00 CLF	9,837	0	15,418	0	25,255
RSM EL Wire, # Conductor, 600V, #4		353.03	0.00	1075.00	0.00	1428.03
	2.00 CLF	820	0	2,496	0	3,316
16115 50000 Special Wires & Fittings						
RSM EL Wire Thermostat No Jac Twist		39.72	0.00	8.00	0.00	47.72
#18-2 Conductor	6.00 CLF	277	0	56	0	332
RSM EL Wire Fire Alrm Fep Teflon #22		39.72	0.00	93.00	0.00	132.72
2Pr	6.00 CLF	277	0	648	0	924
16116 00000 Undercarpet						
RSM EL Transition Fitting Wall Box		13.24	0.00	19.00	0.00	32.24
Surface	10.00 EA	154	0	221	0	374

Item	Quantity					
16152 00000 Cable Terminations						
RSM EL Terminations, 600V # 12		6.35	0.00	0.41	0.00	6.76
	20.00 EA	148	0	10	0	157
RSM EL Cable Term Crimp 2 Way Conn #4		13.81	0.00	2.40	0.00	16.21
	40.00 EA	642	0	111	0	753
16211 00000 Outlet Boxes						
RSM EL Box, 4" Octagon Pressed Steel		15.89	0.00	1.15	0.00	17.04
	60.00 EA	1,107	0	80	0	1,187
16213 00000 Pull Boxes & Cabinets						
RSM EL Pull Box Sc Nema 1 12"W-12"H-8"D		63.55	0.00	26.00	0.00	89.55
	12.00 EA	885	0	362	0	1,248
16232 00000 Wiring Devices						
RSM EL Wiring Device Receptacle Duplex		11.77	0.00	5.00	0.00	16.77
120 Volt Grounded 20 Amp	60.00 EA	820	0	348	0	1,168
16311 00000 Motor Control Center						
RSM EL MCC Starter FVNR 25 Hp Type B		158.86	0.00	760.00	0.00	918.86
	10.00 EA	1,844	0	8,824	0	10,668

Building Systems Design
PROJECT WHATC2: Whatco Manufacturing-Whse-Office - Demonstration Project
DETAILED ESTIMATE Pre-Bid Design Review Cost Estimate DETAIL PAGE 8
10. Plant Building

10.11. Electrical	QUANTY UOM	Labor	Equipmnt	Material	SubContr	TOTAL COST
RSM EL MCC Starter FVNR 100 Hp Type B		453.90	0.00	2275.00	0.00	2728.90
	1.00 EA	527	0	2,641	0	3,168
16320 50000 Circuit Breakers						
RSM EL Crkt Brkr Disco 240V 3Pl 15Hp		113.47	0.00	150.00	0.00	263.47
Motor	10.00 EA	1,317	0	1,742	0	3,059
RSM EL Crkt Brkr Disco 240V 3Pl 100Hp		397.16	0.00	1050.00	0.00	1447.16
Motor	1.00 EA	461	0	1,219	0	1,680
16611 00000 Exit And Emergency Lighting						
RSM EL Exit Lights		39.72	0.00	50.00	0.00	89.72
	8.00 EA	369	0	464	0	833
16611 50000 Exterior Fixtures						
RSM EL Lights, Metal Halide, 1500W		171.71	0.00	595.00	0.00	766.71
	36.00 EA	7,177	0	24,869	0	32,045

```
16612 50000 Fixture Whips

        RSM EL Fixture Whips 6'Long, Greenfield                         9.93      0.00       8.35       0.00        18.28
              THHN 3 Conductors #14                 36.00 EA             415         0        349          0          764
                                                                 ----------- --------- ---------- ----------  -----------
              Electrical                                              75,100         0     97,492     11,610      184,202


     10.13. Equipment & Conveying

14605 40000 Crane Rail

       B RSM CR Crane, Bridge No Equip                              55488.00   2776.25  334450.00       0.00    392714.25
                                                     1.00 EA          59,727     2,988    360,002          0      422,718

                   *** MATERIAL BACKUP FOR  14605 QUOTE Crane, Bridge No Equip ***
     ID        Name                          City/State                      Date     Telephone      Quote
     ---------------------------------------------------------------------------------------------------------
     BRIDGEMAN1 Bridge Crane Quote Company   Annapolis Maryland              12/28/92 (212)456-1234 360000

                                                                 ----------- --------- ---------- ----------  -----------
              Equipment & Conveying                                   59,727     2,988    360,002          0      422,718

                                                                 ----------- --------- ---------- ----------  -----------
              Plant Building                                         528,707    72,334  1,211,310     11,610    1,823,960
```

Building Systems Design

PROJECT WHATC2: Whatco Manufacturing-Whse-Office - Demonstration Project

DETAILED ESTIMATE — Pre-Bid Design Review Cost Estimate — DETAIL PAGE 9

20. Warehouse Buildings "A" & "B"

20.01. Substructure	QUANTY UOM	Labor	Equipmnt	Material	SubContr	TOTAL COST

20. Warehouse Buildings "A" & "B"

This represents two 50' X 60' warehouses flanking the office complex on both sides. The Gross Floor Area is 6000 S. F.

SUBSTRUCTURE and SUPERSTRUCTURE

These Two Buildings Are Constructed Of Steel Framed Roofs Supported By Heavy Steel Frames founded on Spread Footings and Grade Beams.

The Slab On Grade Is Designed Using a 9" Thicknesses.

ROOFING

The Roof Is A Metal Deck on a Heavy Steel Frame. Rigid Insulation and a built-up bitiumnous roof system is also included.
There are Gutters And Downspouts.

EXTERIOR WALL

The Exterior is To Be covered with Insulated Metal Siding (PrePainted). The Exterior Doors Are Both Hollow Metal Doors And 2 Overhead Doors With Motors.

SPECILATIES

The Only Specialty Items Are Dock Levelers and Monorails for owner supplied Lifting devices.

MECHANICAL

The Mechanical estimate was complied from detail.

ELECTRICAL

The Electrical estimate was compiled from detail.

20.01. Substructure

02110 40000 Clear And Grub						
RSM GM Site Preparation		1954.19	1497.99	0.00	0.00	3452.17
	0.14 ACR	274	210	0	0	483
02220 80000 Backfill, Structural						
RSM GM Borrow; Loaded At Pit, No Haul		0.00	0.00	2.59	0.00	2.59
	334.00 CY	0	0	865	0	865
RSM GM Spread W/200Hp Dozer, Bank Yards		1.33	2.68	0.00	0.00	4.01
	334.00 CY	445	895	0	0	1,340

Building Systems Design
PROJECT WHATC2: Whatco Manufacturing-Whse-Office - Demonstration Project
DETAILED ESTIMATE Pre-Bid Design Review Cost Estimate DETAIL PAGE 10
20. Warehouse Buildings "A" & "B"

20.01. Substructure	QUANTY UOM	Labor	Equipmnt	Material	SubContr	TOTAL COST
RSM GM Borrow Delivery Charge, Min 12		2.55	4.62	0.00	0.00	7.17
Cy, 1 Hr Round Trip	334.00 CY	852	1,542	0	0	2,394
02222 60000 Compaction Vibrating						
RSM GM Compaction, Riding Vibrating		0.29	0.24	0.00	0.00	0.53
Roller, 6" Lift, 4 Pass	334.00 CY	95	81	0	0	177
02224 60000 Excavation, Bulk, Scrapers						
RSM GM Excavating Bulk - Footprint		0.67	3.04	0.00	0.00	3.72
	669.00 CY	450	2,037	0	0	2,487
02225 00000 Excavating, Structural						
RSM GM Excavating, Structural, Hand,		54.03	0.00	0.00	0.00	54.03
Spread Footings	9.00 CY	486	0	0	0	486
02226 20000 Fill						
RSM GM Spread Fill W/ Ldr, Crwlr, 300'		0.62	1.27	0.00	0.00	1.89
Haul 130 Hp	39.00 CY	24	50	0	0	74

Description	Quantity					
03210 70000 Reinforcing In Place						
RSM CW Reinforcing - Footings		0.32	0.00	0.23	0.00	0.55
	1018.00 LB	375	0	268	0	643
		----------	----------	----------	----------	----------
Substructure		3,002	4,814	1,133	0	8,950
20.02. Structural Frame						
03210 70000 Reinforcing In Place						
RSM CW Reinforcing In Place Slab On		0.32	0.00	0.25	0.00	0.57
Grade, & Load Deck	5002.00 LB	1,842	0	1,432	0	3,274
03313 00000 Concrete In Place						
RSM CW Concrete In Place,Spread Footing		36.40	5.44	64.00	0.00	105.83
& Pile Caps	9.00 CY	375	56	659	0	1,090
RSM CW Concrete In Place, Grade Beam		184.51	9.33	138.00	0.00	331.84
	40.00 CY	8,449	427	6,319	0	15,196
M RSM CW Conc In Place,Grnd Slab, 9"Thick		0.51	0.17	1.46	0.00	2.13
Trowel Finish	6000.00 SF	3,489	1,138	10,028	0	14,655
Trowel Finish						
03521 20000 Insulating						

Building Systems Design

PROJECT WHATC2: Whatco Manufacturing-Whse-Office - Demonstration Project

DETAILED ESTIMATE | Pre-Bid Design Review Cost Estimate | DETAIL PAGE 11

20. Warehouse Buildings "A" & "B"

20.02. Structural Frame	QUANTY UOM	Labor	Equipmnt	Material	SubContr	TOTAL COST
RSM CW Insul Lightweight Fill 3" Thick		0.22	0.07	0.75	0.00	1.04
	6000.00 SF	1,484	484	5,152	0	7,120
05125 50000 Structural Steel Projects						
RSM SS Structural Steel Base Plates		0.34	0.00	0.49	0.00	0.83
	504.00 LB	195	0	283	0	478
RSM SS Structural Steel Beams Light		205.65	92.90	1300.00	0.00	1598.55
	13.50 TON	3,179	1,436	20,093	0	24,707
RSM SS Structural Steel Beams Heavy		188.08	84.96	1125.00	0.00	1398.04
	20.00 TON	4,307	1,945	25,760	0	32,012
RSM PS Sand Blast		0.43	0.06	0.14	0.00	0.63
	6519.00 SF	3,143	448	1,025	0	4,616
RSM PS Paints & Protect Coats, Alkyds,		0.14	0.01	0.06	0.00	0.21
Primer	6519.00 SF	1,033	56	439	0	1,528
RSM PS Paints & Protect Coats, Epoxy,		0.18	0.01	0.14	0.00	0.33
Intermediate Coat	6519.00 SF	1,328	72	1,025	0	2,425

Item	Quantity					
RSM PS Top Coat (Elevated)		0.28	0.02	0.18	0.00	0.48
	6519.00 SF	2,066	111	1,318	0	3,495
05211 00000 Open Web Joists						
RSM SS Open Web Joists, H Series,		257.55	95.11	750.00	0.00	1102.66
	13.00 TON	3,833	1,416	11,163	0	16,412
05310 40000 Metal Decking						
RSM SS Metal Deck Galvanized 2" 20-20Ga		0.94	0.05	2.40	0.00	3.39
	6400.00 SF	6,861	369	17,586	0	24,815
		-----------	---------	-----------	----------	-----------
Structural Frame		41,582	7,957	102,283	0	151,823
20.03. Roofing						
07220 30000 Roof Deck Insulation						
RSM MR Roof Deck Insulation, Fiberboard		0.36	0.00	0.60	0.00	0.96
Mineral, 2" Thick R5.26	6000.00 SF	2,590	0	4,355	0	6,945
07510 20000 Built-Up Roofing						
RSM MR Built Up Roofing		91.56	6.75	33.00	0.00	131.31
	60.00 SQ	6,645	490	2,395	0	9,530
07620 10000 Downspouts						

Building Systems Design
PROJECT WHATC2: Whatco Manufacturing-Whse-Office - Demonstration Project
DETAILED ESTIMATE — Pre-Bid Design Review Cost Estimate — DETAIL PAGE 12
20. Warehouse Buildings "A" & "B"

20.03. Roofing	QUANTY UOM	Labor	Equipmnt	Material	SubContr	TOTAL COST
RSM MR Downspout, Galvanized, 3" X 4"		2.16	0.00	1.25	0.00	3.41
28 Ga	70.00 LF	183	0	106	0	289
07620 40000 Flashing						
RSM MR Flashing		2.11	0.00	3.15	0.00	5.26
	440.00 LF	1,125	0	1,677	0	2,802
07620 50000 Gutters						
RSM MR Gutter, 5" X 6"		5.22	0.00	2.50	0.00	7.72
	100.00 LF	632	0	302	0	934
Roofing		11,176	490	8,834	0	20,500
20.04. Exterior Closure						
07420 20000 Metal Facing Panels						
RSM MW MFG Wall Panels-18 Ga Galvanized		5.43	0.13	4.85	0.00	10.41
Steel Interior	6000.00 SF	38,365	896	34,257	0	73,517

08110 30000 Commercial Steel Doors						
RSM GM Steel Doors & Frames 3' X 7'		32.87	0.97	146.00	0.00	179.84
	2.00 EA	66	2	292	0	360
08360 40000 Overhead, Commercial						
RSM SD Steel Rollup Doors		371.49	0.00	715.00	0.00	1086.49
	2.00 EA	899	0	1,730	0	2,629
08712 50000 Mortise Lockset						
RSM GM Door Hardware		31.04	0.00	263.00	0.00	294.04
	2.00 EA	62	0	526	0	588
09910 60000 Siding						
RSM PS Painting Exterior		0.29	0.00	0.14	0.00	0.43
	2216.00 SF	711	0	349	0	1,060
		----------	--------	----------	----------	----------
Exterior Closure		40,103	898	37,153	0	78,154

20.06. Interior	QUANTY UOM	Labor	Equipmnt	Material	SubContr	TOTAL COST
20.06. Interior						
09910 60000 Siding						
RSM PS Interior Painting		0.29	0.00	0.14	0.00	0.43
	3600.00 SF	1,156	0	566	0	1,722
Interior		1,156	0	566	0	1,722
20.07. Specialties						
11160 10000 Loading Dock						
RSM SD Dock Platform Levelers, 7'X 8',		553.94	68.00	2900.00	0.00	3521.94
10 Tons, Hinged	4.00 EA	2,681	329	14,036	0	17,046
Specialties		2,681	329	14,036	0	17,046
20.09. Plumbing Systems						

Item	Quantity					
RSM PL Watercooler Wheelchar Type 8 GPH		133.37	0.00	920.00	0.00	1053.37
	2.00 EA	340	0	2,344	0	2,684
Plumbing Systems		340	0	2,344	0	2,684
20.10. Mechanical Systems						
15170 10000 Pipe, Steel						
RSM ME Heating Pipe (HW) 2" Steel		8.75	0.75	3.69	0.00	13.19
	400.00 LF	4,069	350	1,717	0	6,136
RSM PL Water Pipe, Steel Sch 40		12.41	1.07	5.70	0.00	19.17
on Hangers	200.00 LF	3,161	272	1,452	0	4,885
15198 00000 Valves, Steel						
RSM ME Gate Valves, 2" Flanged		37.05	0.00	1275.00	0.00	1312.05
	4.00 EA	172	0	5,932	0	6,104
RSM PL Gate Valves, 3" Steel, Flanged		118.55	0.00	1325.00	0.00	1443.55
Flanged	4.00 EA	604	0	6,752	0	7,357
15563 00000 Hydronic Heating						

Building Systems Design
PROJECT WHATC2: Whatco Manufacturing-Whse-Office - Demonstration Project
Pre-Bid Design Review Cost Estimate
20. Warehouse Buildings "A" & "B"

20.10. Mechanical Systems	QUANTY UOM	Labor	Equipmnt	Material	SubContr	TOTAL COST
RSM ME Unit Heaters, 140MBH, Vertical		133.37	0.00	860.00	0.00	993.37
Flow	8.00 EA	1,241	0	8,003	0	9,244
15565 10000 Insulation						
RSM IN Pipe Covering 1"Thick (2" Pipe)		2.89	0.00	2.38	0.00	5.27
Urthane Ultravoilet Resistant	400.00 LF	1,486	0	1,224	0	2,710
15729 00000 Fans						
RSM ME Air Handling Units 18,620 CFM		282.13	0.00	4250.00	0.00	4532.13
15Hp No Motor	4.00 EA	1,313	0	19,774	0	21,086
16352 00000 Motors						
RSM ME Motor 230/460V 1800 Rpm 15Hp		99.29	0.00	515.00	0.00	614.29
	4.00 EA	462	0	2,396	0	2,858
Mechanical Systems		12,509	621	47,250	0	60,380

20.11. Electrical

16020 50000 Conduit						
RSM EL Rigid Galv. Steel Conduit, 2"GRC		7.06	0.00	4.00	0.00	11.06
	1000.00 LF	8,197	0	4,644	0	12,841
16027 50000 Motor Connections						
RSM EL Motor Conn Flex Cond & Ftg 15 Hp		96.16	0.00	10.55	0.00	106.71
230 Volts 3 Ph	4.00 EA	447	0	49	0	496
OBK EL Grounding Allowance		0.00	0.00	0.00	20000.00	20000.00
120 Volt Grounded 20 Amp	1.00 LS	0	0	0	23,220	23,220
16114 00000 Mineral Insulated Cable						
RSM EL Wire, 2 Conductor, 600V #12		226.95	0.00	360.00	0.00	586.95
	15.00 CLF	3,952	0	6,269	0	10,222
16115 50000 Special Wires & Fittings						
RSM EL Wire Thermostat No Jac Twist		39.72	0.00	8.00	0.00	47.72
#18-2 Conductor	2.00 CLF	92	0	19	0	111
RSM EL Wire Fire Alrm Fep Teflon #22		39.72	0.00	93.00	0.00	132.72
2Pr	2.00 CLF	92	0	216	0	308
16116 00000 Undercarpet						

Building Systems Design

PROJECT WHATC2: Whatco Manufacturing-Whse-Office - Demonstration Project

DETAILED ESTIMATE — Pre-Bid Design Review Cost Estimate — DETAIL PAGE 15

20. Warehouse Buildings "A" & "B"

20.11. Electrical	QUANTY UOM	Labor	Equipmnt	Material	SubContr	TOTAL COST
RSM EL Transition Fitting Wall Box		13.24	0.00	19.00	0.00	32.24
Surface	10.00 EA	154	0	221	0	374
16152 00000 Cable Terminations						
RSM EL Terminations, 600V # 12		6.35	0.00	0.41	0.00	6.76
	120.00 EA	885	0	57	0	942
16211 00000 Outlet Boxes						
RSM EL Box, 4" Octagon Pressed Steel		15.89	0.00	1.15	0.00	17.04
	30.00 EA	553	0	40	0	593
16213 00000 Pull Boxes & Cabinets						
RSM EL Pull Box Sc Nema 1 12"W-12"H-8"D		63.55	0.00	26.00	0.00	89.55
	4.00 EA	295	0	121	0	416
16232 00000 Wiring Devices						
RSM EL Wiring Device Receptacle Duplex		11.77	0.00	5.00	0.00	16.77
120 Volt Grounded 20 Amp	30.00 EA	410	0	174	0	584

Item	Quantity					
16311 00000 Motor Control Center						
RSM EL MCC Starter FVNR 25 Hp Type B		158.86	0.00	760.00	0.00	918.86
	4.00 EA	738	0	3,529	0	4,267
16320 50000 Circuit Breakers						
RSM EL Crkt Brkr Disco 240V 3Pl 15Hp		113.47	0.00	150.00	0.00	263.47
Motor	4.00 EA	527	0	697	0	1,224
16611 00000 Exit And Emergency Lighting						
RSM EL Exit Lights		39.72	0.00	50.00	0.00	89.72
	6.00 EA	277	0	348	0	625
16611 50000 Exterior Fixtures						
RSM EL Lights, Metal Halide, 1500W		171.93	0.00	595.00	0.00	766.93
	12.00 EA	2,395	0	8,290	0	10,685
16612 50000 Fixture Whips						
RSM EL Fixture Whips 6'Long, Greenfield		9.93	0.00	8.35	0.00	18.28
THHN 3 Conductors #14	12.00 EA	138	0	116	0	255
Electrical		19,153	0	24,790	23,220	67,163

Building Systems Design
PROJECT WHATC2: Whatco Manufacturing-Whse-Office - Demonstration Project
DETAILED ESTIMATE Pre-Bid Design Review Cost Estimate DETAIL PAGE 16
20. Warehouse Buildings "A" & "B"

20.13. Equipment	QUANTY UOM	Labor	Equipmnt	Material	SubContr	TOTAL COST
20.13. Equipment						
B OBK GM Fork Lifts		0.00	0.00	20000.00	0.00	20000.00
	2.00 EA	0	0	40,000	0	40,000

*** MATERIAL BACKUP FOR 11000 Fork Lifts ***

ID	Name	City/State	Date	Telephone	Quote
6091234567	Fork Lifts Inc.	Philladelphia, Pa. 15678	02/02/93	(215)452-4566	20000

	Labor	Equipmnt	Material	SubContr	TOTAL COST
Equipment	0	0	40,000	0	40,000
Warehouse Buildings "A" & "B"	131,703	15,109	278,390	23,220	448,422

Building Systems Design
PROJECT WHATC2: Whatco Manufacturing-Whse-Office - Demonstration Project
DETAILED ESTIMATE Pre-Bid Design Review Cost Estimate DETAIL PAGE 17
30. Office Building

30.01. Substructure	QUANTY	UOM	Labor	Equipmnt	Material	SubContr	TOTAL COST

30. Office Building
This Building is an office building is a 70 ft x 100 ft square yeilding a gross Square Footage of 7000 sf.

STRUCTURE and SUBSTRUCTURE
The Foundation For This Facility Consists Of Spread Footings And Continous Grade Beams. A Floor Slab Is Poured Monolithically. The Floor Slab Is 6 and 9 inches thick Of 3000Psi Concrete, over a Vapor Barrier The Structural Frame Consists Of Structural Steel Columns and Roof Framing.

ROOFING
Roofing is a Built Up Roof on a Mineral Fiber Insulation Board set on a composite metal deck.

DOORS and WINDOWS
Entrance Exterior Doors Are Double Plate Glass and Hollow
Metal Door Units. Exterior Windows Are Non-Operable, With 1 in Insulated Glass.

INTERIOR PARTITIONS
Interior Walls Are Painted Gypsum Board on Metal Studs, and Furring over CMU Walls.

INTERIOR DOORS
Interior Doors Are Prehung Wood in Hollow Metal Frames.

INTERIOR FINISHES
Ceiling Construction Consists Of Suspended Acoustical Tile Concrete Sealer Is Applied To Storage Area Floors. Ceramic Tile Is Used In Toilet Areas. Office And Halls have Carpet and Resilient Tile.

SPECIALITIES
Specialties Include Toilet Partitions, Lockers, Aluminum Pipe Railings And Loading Dock Accessories.

MECHANICAL

The Mechanical Estimate was compiled from detail.

ELECTRICAL

The Electrical Estimate was compiled from detail.

30.01. Substructure

02110 40000 Clear And Grub

RSM GM Site Preparation		1954.19	1497.99	0.00	0.00	3452.17
	0.16 ACR	313	240	0	0	552

02220 80000 Backfill, Structural

30.01. Substructure	QUANTY UOM	Labor	Equipmnt	Material	SubContr	TOTAL COST
RSM GM Borrow; Loaded At Pit, No Haul		0.00	0.00	2.59	0.00	2.59
	382.00 CY	0	0	989	0	989
RSM GM Spread W/200Hp Dozer, Bank Yards		1.33	2.68	0.00	0.00	4.01
	382.00 CY	510	1,023	0	0	1,533
RSM GM Borrow Delivery Charge, Min 12		2.55	4.62	0.00	0.00	7.17
Cy, 1 Hr Round Trip	382.00 CY	975	1,764	0	0	2,738
02222 60000 Compaction Vibrating						
RSM GM Compaction, Riding Vibrating		0.29	0.24	0.00	0.00	0.53
Roller, 6" Lift, 4 Pass	382.00 CY	109	93	0	0	202
02224 60000 Excavation, Bulk, Scrapers						
RSM GM Excavating Bulk - Footprint		0.67	3.04	0.00	0.00	3.72
	764.00 CY	514	2,326	0	0	2,840
02225 00000 Excavating, Structural						
RSM GM Excavating, Structural, Hand,		54.03	0.00	0.00	0.00	54.03
Pile Caps/Spread Footings	6.00 CY	324	0	0	0	324

Item	Quantity					
02225 40000 Excavating, Trench						
RSM GM Excav Grade Beam, Common Earth, 3/8 Cy Track/Loader/Backhoe	45.00 CY	3.20 144	1.34 60	0.00 0	0.00 0	4.54 204
02226 20000 Fill						
RSM GM Spread Fill W/ Ldr, Crwlr, 300' Haul 130 Hp	22.00 CY	0.62 14	1.27 28	0.00 0	0.00 0	1.89 42
03210 70000 Reinforcing In Place						
RSM CW Reinforcing - Footings	710.00 LB	0.32 261	0.00 0	0.23 187	0.00 0	0.55 448
RSM CW Reinforcing - Grade Beam	1248.00 LB	0.19 268	0.00 0	0.25 357	0.00 0	0.44 625
07160 20000 Bituminous Asphalt Coating						
RSM WP Bituminous Dampproof Grade Beam Asphalt Waterproof Sprayed On Below Grade 25.6 Sf/Gal	367.00 SF	0.34 170	0.00 0	0.15 74	0.00 0	0.49 244
07210 90000 Perimeter Insulation						

Building Systems Design

PROJECT WHATC2: Whatco Manufacturing-Whse-Office - Demonstration Project

DETAILED ESTIMATE — Pre-Bid Design Review Cost Estimate — DETAIL PAGE 19

30. Office Building

30.01. Substructure	QUANTY UOM	Labor	Equipmnt	Material	SubContr	TOTAL COST
RSM CW Insulation Grade Beam,		0.41	0.00	0.20	0.00	0.61
Polystyrene Bead Board, 1" Thick	367.00 SF	173	0	84	0	257
Substructure		3,774	5,534	1,692	0	11,000
30.02. Structural Frame						
03210 70000 Reinforcing In Place						
RSM CW Reinforcing In Place Slab On		0.32	0.00	0.25	0.00	0.57
Grade, & Load Deck	5170.00 LB	1,904	0	1,480	0	3,384
RSM CW Concrete Reinforcing 2nd Floor		0.62	0.20	0.70	0.00	1.52
	10528 SF	7,466	2,436	8,437	0	18,339
03313 00000 Concrete In Place						
RSM CW Concrete In Place,Spread Footing		36.40	5.44	64.00	0.00	105.83
& Pile Caps	6.00 CY	250	37	440	0	727

Description	Quantity					
RSM CW Concrete In Place, Grade Beam		184.51	9.33	138.00	0.00	331.84
	23.00 CY	4,858	246	3,634	0	8,737
RSM CW Slab On Grade 6" Thick,		0.46	0.15	1.04	0.00	1.65
Trowel Finish	7000.00 SF	3,693	1,204	8,334	0	13,231
03521 20000 Insulating						
RSM CW Insul Lightweight Fill 3" Thick		0.22	0.07	0.75	0.00	1.04
	7000.00 SF	1,731	565	6,010	0	8,306
05125 50000 Structural Steel Projects						
RSM SS Structural Steel Base Plates		0.34	0.00	0.49	0.00	0.83
	336.00 LB	130	0	188	0	318
RSM SS Structural Steel Beams Heavy		188.08	84.96	1125.00	0.00	1398.04
	64.00 TON	13,781	6,225	82,433	0	102,439
RSM PS Sand Blast		0.43	0.06	0.14	0.00	0.63
	14175 SF	6,834	973	2,230	0	10,037
RSM PS Paints & Protect Coats, Alkyds,		0.14	0.01	0.06	0.00	0.21
Primer	14175 SF	2,246	121	956	0	3,322
RSM PS Paints & Protect Coats, Epoxy,		0.18	0.01	0.14	0.00	0.33
Intermediate Coat	14175 SF	2,888	156	2,230	0	5,273

Building Systems Design
PROJECT WHATC2: Whatco Manufacturing-Whse-Office - Demonstration Project
DETAILED ESTIMATE — Pre-Bid Design Review Cost Estimate — DETAIL PAGE 20
30. Office Building

30.02. Structural Frame	QUANTY UOM	Labor	Equipmnt	Material	SubContr	TOTAL COST
RSM PS Top Coat (Elevated)		0.28	0.02	0.18	0.00	0.48
	14175 SF	4,491	242	2,867	0	7,600
05211 00000 Open Web Joists						
RSM SS Open Web Joists, H Series,		257.55	95.11	750.00	0.00	1102.66
	13.00 TON	3,833	1,416	11,163	0	16,412
05310 40000 Metal Decking						
RSM SS Metal Deck Galvanized 2" 20-20Ga		0.94	0.05	2.40	0.00	3.39
	7000.00 SF	7,505	403	19,234	0	27,142
Structural Frame		61,610	14,025	149,634	0	225,268
30.03. Roofing						
05310 40000 Metal Decking						
RSM SS Metal Deck		0.99	0.05	2.50	0.00	3.55
	7000.00 SF	7,969	428	20,036	0	28,432

Item	Quantity					
07220 30000 Roof Deck Insulation						
RSM MR Roof Deck Insulation, Fiberboard Mineral, 2" Thick R5.26	7000.00 SF	0.36 3,022	0.00 0	0.60 5,080	0.00 0	0.96 8,102
07510 20000 Built-Up Roofing						
RSM MR Built Up Roofing,Asphalt Base Sheet,3 Plies #15 Asphalt Felt,Mopped	70.00 SQ	91.56 7,753	6.75 571	33.00 2,794	0.00 0	131.31 11,118
07620 10000 Downspouts						
RSM MR Downspout, Steel, Galvanized, Corrugated Rectangular, 3" X 4" 28 Ga	70.00 LF	2.16 183	0.00 0	1.25 106	0.00 0	3.41 289
07620 40000 Flashing						
RSM MR Flashing	140.00 LF	2.11 358	0.00 0	3.15 533	0.00 0	5.26 892
07620 50000 Gutters						
RSM MR Gutter, Aluminum, 5" X 6" Combination Fascia, .032" Thick, Enameled	100.00 LF	5.22 632	0.00 0	2.50 302	0.00 0	7.72 934

Building Systems Design
PROJECT WHATC2: Whatco Manufacturing-Whse-Office - Demonstration Project
DETAILED ESTIMATE Pre-Bid Design Review Cost Estimate DETAIL PAGE 21
30. Office Building

30.03. Roofing	QUANTY UOM	Labor	Equipmnt	Material	SubContr	TOTAL COST
Roofing		19,916	999	28,852	0	49,767
30.04. Exterior Closure						
01525 40000 Scaffold						
RSM MA Scaffolding,Steel Tubular;Rented		47.30	0.00	15.15	0.00	62.45
No Plank,1 Use/Mo,Buildg Ext 2St ories	36.00 CSF	1,949	0	624	0	2,574
04211 60000 Coping						
RSM MA Coping Precast Concr Stock 14"		6.39	0.00	12.50	0.00	18.89
Wide	120.00 LF	878	0	1,717	0	2,596
04218 40000 Walls						
RSM MA Walls, Common 8"X2-2/3"X4" 4"		6.02	0.00	1.79	0.00	7.81
Wall, As Face Brick	2700.00 SF	18,613	0	5,533	0	24,145

04221 60000 Concrete Block, Back-Up,						
RSM MA Concrete Block, Back-Up, Reinf.,		3.28	0.00	1.40	0.00	4.68
Alt/Course, 8"X16" 8" Thick	2700.00 SF	10,131	0	4,327	0	14,458
04223 20000 Concrete Block, Partitions						
RSM MA Concrete Block Partitions,		3.41	0.00	1.77	0.00	5.18
Tooled 2 Sides, Reinf., 8	12670 SF	49,417	0	25,673	0	75,090
"X16"X8" Thick						
07211 80000 Wall Or Ceiling Insul., Non-Rigid						
RSM MA Non-Rigid Insulation, Fiberglass		0.17	0.00	0.22	0.00	0.39
Kraft Face, 3-1/2" Thk, 15"	8850.00 SF	1,769	0	2,229	0	3,998
Wide R11						
08110 30000 Commercial Steel Doors						
RSM GM Steel Doors & Frame 1-3/8"		32.87	0.97	146.00	0.00	179.84
Tk. Full Panel, 20 Ga. 3' X 7'	8.00 EA	263	8	1,168	0	1,439
08111 80000 Steel Frames, Knock Down						
RSM GM Steel Frames, Knock Down, 6'-8"		34.92	1.03	64.50	0.00	100.45
3'-0" Wide, Single, for Wood Drs	42.00 EA	1,467	43	2,709	0	4,219
08207 00000 Wood Fire Doors						

Building Systems Design

PROJECT WHATC2: Whatco Manufacturing-Whse-Office - Demonstration Project

DETAILED ESTIMATE Pre-Bid Design Review Cost Estimate DETAIL PAGE 22

30. Office Building

30.04. Exterior Closure	QUANTY UOM	Labor	Equipmnt	Material	SubContr	TOTAL COST
RSM GM Wood.Door Birch Face, 3'X 6'-8"		42.98	1.27	169.00	0.00	213.25
	36.00 EA	1,547	46	6,084	0	7,677
08410 50000 Storefront Systems						
RSM GL Storefront System,Al & Stl Frame		4.92	0.00	11.25	0.00	16.17
3/8" Glass Plate ,6'X 7' Incl	42.00 SF	239	0	546	0	785
Hdwre,Commercial Grade						
08712 50000 Mortise Lockset						
RSM GM Door Hardware		31.04	0.00	263.00	0.00	294.04
	45.00 EA	1,397	0	11,835	0	13,232
08920 40000 Tube Framing						
M RSM GL Glass Window		9.63	0.00	9.85	0.00	19.48
	900.00 LF	10,014	0	10,244	0	20,258
Exterior Closure		97,683	97	72,690	0	170,470

30.06. Interior

Item	Quantity					
09130 40000 Suspension Systems						
RSM AT Acoustical Suspension System		0.59	0.00	0.15	0.00	0.74
	12300 SF	8,370	0	2,112	0	10,482
09260 80000 Drywall						
RSM GW Gypsum On Block		0.28	0.00	0.15	0.00	0.43
	7950.00 SF	2,496	0	1,340	0	3,836
RSM GW Drywall, Gypsum, 5/8" Thick		0.28	0.00	0.19	0.00	0.47
	17800 SF	5,588	0	3,800	0	9,388
09261 20000 Metal Studs, Drywall						
RSM GW Metal Studs 25 Ga. N.L.B.,		0.65	0.00	0.30	0.00	0.95
3-5/8" Wide, 16" O.C. Galv	8900.00 SF	6,497	0	3,000	0	9,497
09310 20000 Ceramic Tile						
RSM HT Ceramic Tile		2.85	0.00	2.96	0.00	5.81
	900.00 SF	2,883	0	2,993	0	5,876
09510 60000 Suspended Ceilings, Complete						

Building Systems Design
PROJECT WHATC2: Whatco Manufacturing-Whse-Office - Demonstration Project
DETAILED ESTIMATE Pre-Bid Design Review Cost Estimate DETAIL PAGE 23
30. Office Building

30.06. Interior	QUANTY UOM	Labor	Equipmnt	Material	SubContr	TOTAL COST
RSM AT Acoustical Tile Sus 2'X 4'X 5/8"		0.74	0.00	1.10	0.00	1.84
	12300 SF	10,352	0	15,489	0	25,842
09660 10000 Resilient						
RSM CA Resilient Tile		0.70	0.00	0.85	0.00	1.55
1/8" Thick, Color Group C & D	3825.00 SF	3,002	0	3,653	0	6,655
09685 20000 Carpet						
RSM CA Carpet		4.90	0.00	19.45	0.00	24.35
	875.00 SY	4,819	0	19,122	0	23,941
09910 60000 Siding						
RSM PS Exterior Paint		0.29	0.00	0.14	0.00	0.43
1-11/Clapboard Oil Base Ptd 2	42.00 SF	13	0	7	0	20
Coat Brushwork						
RSM PS Interior Painting		0.29	0.00	0.14	0.00	0.43
	10640 SF	3,415	0	1,674	0	5,089

09920 40000 Cabinets And Casework						
RSM PS Interior Paint, Doors		0.60	0.00	0.11	0.00	0.71
	3444.00 SF	2,325	0	426	0	2,751
09922 40000 Walls And Ceilings						
RSM PS Wall & Ceiling, Conc Drywall Or		0.43	0.00	0.11	0.00	0.54
Plaster 2 Coat Finish Brushwork	51500 SF	24,592	0	6,365	0	30,957
		----------	--------	----------	----------	----------
Interior		74,352	0	59,981	0	134,333
30.07. Specialties						
05520 30000 Railing, Pipe						
RSM SS Alum Pipe Railing, 3 Rail,		9.98	0.54	22.00	0.00	32.52
Dark Anodized	75.00 LF	857	46	1,889	0	2,792
10110 40000 Chalkboards						
RSM GM Partitions, Toilet, Ceiling Hung		139.68	0.00	700.00	0.00	839.68
Porcelained Enamel.	12.00 EA	1,676	0	8,400	0	10,076
10505 40000 Lockers						

Building Systems Design

PROJECT WHATC2: Whatco Manufacturing-Whse-Office - Demonstration Project

DETAILED ESTIMATE | Pre-Bid Design Review Cost Estimate |

30. Office Building

30.07. Specialties	QUANTY UOM	Labor	Equipmnt	Material	SubContr	TOTAL COST
RSM GM Lockers		12.06	0.00	62.50	0.00	74.56
	20.00 OPN	241	0	1,250	0	1,491
11160 10000 Loading Dock						
RSM SD Dock Platform Levelers, Hinged		553.94	68.00	2900.00	0.00	3521.94
7'X 8', 10 Tons	4.00 EA	2,681	329	14,036	0	17,046
Specialties		5,455	375	25,575	0	31,405
30.08. Fire Protection						
15411 50000 Fire Equipment Cabinets						
RSM FP Fire Ex Cab Glassdoor SS Frame		66.69	0.00	120.00	0.00	186.69
Sngl	8.00 EA	640	0	1,151	0	1,791
15412 50000 Fire Extinguishers						
RSM FP Fire Extingshr Co2/Hose 20#		0.00	0.00	202.00	0.00	202.00
	8.00 EA	0	0	1,938	0	1,938

15760 60000 Exhaust Systems

Item	Quantity					
OBK FP Wet Pipe Sprinkler System	14000 SF	0.86 14,436	0.00 0	0.43 7,218	0.00 0	1.29 21,654
Fire Protection		15,076	0	10,307	0	25,382

30.09. Plumbing Systems

15117 00000 Supports/Carriers

Item	Quantity					
RSM PL Support Drinking Fountain Top Front And Back Plate Wall Mou nted	2.00 EA	42.34 108	0.00 0	18.30 47	0.00 0	60.64 155
RSM PL Carrier,Lavatory Concealed Arm Floor Mount Flat Slab Fixture	12.00 EA	49.40 755	0.00 0	88.00 1,345	0.00 0	137.40 2,101
RSM PL Carrier,Exposed Arm Floor Mount Back To Back,Hi Back Or Flat Slab Fixt	2.00 EA	59.28 151	0.00 0	170.00 433	0.00 0	229.28 584
RSM PL Carrier Sink Wall Mount Exposed Arms, Heavy Fixture	9.00 EA	59.28 680	0.00 0	62.00 711	0.00 0	121.28 1,391

15118 10000 Traps

Building Systems Design
PROJECT WHATC2: Whatco Manufacturing-Whse-Office - Demonstration Project

DETAILED ESTIMATE

Pre-Bid Design Review Cost Estimate
30. Office Building

30.09. Plumbing Systems	QUANTY UOM	Labor	Equipmnt	Material	SubContr	TOTAL COST
RSM PL Pipe Trap Ci P 2" Diam Service		33.34	0.00	7.05	0.00	40.39
Weight	9.00 EA	382	0	81	0	463
15119 50000 Vent Flashing						
RSM PL Vent Flashing Copper 1-1/2" Pipe		14.82	0.00	14.05	0.00	28.87
	4.00 EA	76	0	72	0	147
15130 10000 Pipe, Cast Iron						
RSM PL Pipe Ci Soil 1-Hub Serv Wt Hngr		8.47	0.00	3.90	0.00	12.37
5'Oc L&O Jnt 10'Oc 2" Diam	40.00 LF	432	0	199	0	630
RSM PL Pipe Ci Soil 1-Hub Serv Wt Hngr		19.18	0.00	15.85	0.00	35.03
5'Oc L&O Jnts 10'Oc 8" Diam	100.00 LF	2,443	0	2,019	0	4,463
15132 00000 Pipe, Cast Iron, Fittings,						
RSM PL Closet Bend 3"Diam 16"X16"		44.46	0.00	35.00	0.00	79.46
W/Ring L&O Joint	12.00 EA	680	0	535	0	1,215

Item	Quantity					
15155 10000 Pipe, Plastic						
RSM PL Pipe Plastic Hi Press Pvc 40 1-1/2" Diam W/Couplings & Hangers	150.00 LF	8.23 1,573	0.00 0	1.96 375	0.00 0	10.19 1,948
15170 10000 Pipe, Steel						
RSM PL Black Steel Pipe, Schedule 40, Threaded, 1/2" Diam W/Cplgs&H ngrs 10'Oc	100.00 LF	4.70 599	0.00 0	1.24 158	0.00 0	5.94 757
RSM PL Pipe Steel Threaded Schd 40 Black 1"Diam W/Cplgs & Hngrs 10'Oc	250.00 LF	5.59 1,781	0.00 0	1.96 624	0.00 0	7.55 2,405
15213 60000 Lavatories						
RSM PL Lavatory W/Ftngs, White, Vanity Top, Pe On Ci, 20" X 18" Oval	9.00 EA	83.36 956	0.00 0	145.00 1,663	0.00 0	228.36 2,618
15215 20000 Sinks						
RSM PL Sink,Kitchn Countertop,Peci 32"X20"Double Bowl	1.00 EA	111.14 142	0.00 0	252.00 321	0.00 0	363.14 463
15216 40000 Toilet Seats						

Building Systems Design
PROJECT WHATC2: Whatco Manufacturing-Whse-Office - Demonstration Project

DETAILED ESTIMATE | Pre-Bid Design Review Cost Estimate | DETAIL PAGE 26

30. Office Building

30.09. Plumbing Systems	QUANTY UOM	Labor	Equipmnt	Material	SubContr	TOTAL COST
RSM PL Toilet Seats Industrial, W/O		12.35	0.00	19.20	0.00	31.55
Cover, Open Front, Regular Bowl	12.00 EA	189	0	294	0	482
15216 80000 Urinals						
RSM PL Wall Hung Urinal, Vitreous China		177.83	0.00	365.00	0.00	542.83
Incl. Hanger Slfcls Handle	2.00 EA	453	0	930	0	1,383
RSM PL Urinal,Wall Hung Rough-In Waste		188.49	0.00	72.50	0.00	260.99
& Vent	2.00 EA	480	0	185	0	665
15217 60000 Wash Fountains						
RSM PL Wshftn,Group,Ftcntrl,St St,54"		296.38	0.00	2125.00	0.00	2421.38
Diam	1.00 EA	378	0	2,707	0	3,085
RSM PL Washftn,Group,Rough-In Supply,		293.13	0.00	100.00	0.00	393.13
Waste & Vent	1.00 EA	373	0	127	0	501
15218 00000 Water Closets						
RSM PL Water Closet, Tank Type,		100.66	0.00	480.00	0.00	580.66
Vitreous China, Floor Mounted,	12.00 EA	1,539	0	7,339	0	8,877
1 Piece						

Description	Quantity					
RSM PL Water Closet, Tank Typ, Floor Mntd, For Rough-In Waste & Vent	12.00 EA	174.89 2,674	0.00 0	91.50 1,399	0.00 0	266.39 4,073
15311 00000 Water Heaters						
RSM PL Wtrhtr Resdtl Gas Glass Tank100 Gal	1.00 EA	227.99 290	0.00 0	855.00 1,089	0.00 0	1082.99 1,380
Plumbing Systems		17,134	0	22,652	0	39,786
30.10. Mechanical Systems						
15170 10000 Pipe, Steel						
RSM ME Pipe Steel Sch 40 Weld Jt Roll Hanger For Cvrg 10'Oc Blk 2" Diam	300.00 LF	8.75 3,052	0.75 262	3.69 1,288	0.00 0	13.19 4,602
15511 50000 Boilers, Gas Fired						
RSM PL Htg Br Gas Ci Std Htwr Gop 4488 Mbh	1.00 EA	5051.49 6,436	0.00 0	23100.00 29,431	0.00 0	28151.49 35,866
15563 00000 Hydronic Heating						

Building Systems Design
PROJECT WHATC2: Whatco Manufacturing-Whse-Office - Demonstration Project
DETAILED ESTIMATE Pre-Bid Design Review Cost Estimate DETAIL PAGE 27
30. Office Building

30.10. Mechanical Systems	QUANTY UOM	Labor	Equipmnt	Material	SubContr	TOTAL COST
RSM ME Fin Tube Rad Comml Cop/Alum		14.04	0.00	32.00	0.00	46.04
1-1/4" X 4-1/4"	600.00 LF	9,798	0	22,333	0	32,131
15565 10000 Insulation						
RSM IN Pipe Cvr Urthan Ultrav 1"W 2"Ips		2.89	0.00	2.38	0.00	5.27
	300.00 LF	1,115	0	918	0	2,032
15568 00000 Vent Chimney						
RSM SM Vent Chmy Ul Gas 2 Wall Galv		14.85	0.00	61.00	0.00	75.85
18"Dia	30.00 VLF	578	0	2,375	0	2,953
RSM SM Gasvent Dblwall Galv 45< Ell 18"		29.70	0.00	134.00	0.00	163.70
Di	1.00 EA	39	0	174	0	212
15712 50000 Central Station Air-Handling Unit						
RSM ME Ac&V Air Hndlg Chw8070-14800Cfm3		1646.58	0.00	5250.00	0.00	6896.58
Tn	1.00 EA	1,915	0	6,107	0	8,022

Item	Quantity					Total
15725 00000 Ductwork						
RSM SM Alum Ductwrk Ftgs Etc		6.75	0.00	1.24	0.00	7.99
2000-4000Lbs	8920.00 LB	78,152	0	14,353	0	92,505
15729 00000 Fans						
RSM ME Ac&V Fan Blt 18620Cfm 15Hp No		282.13	0.00	4250.00	0.00	4532.13
Motor	4.00 EA	1,313	0	19,774	0	21,086
15747 00000 Registers						
RSM SM Ac&V Register Supply Cl&Wall		16.50	0.00	21.00	0.00	37.50
10X10"	30.00 EA	642	0	818	0	1,460
OBK TB Air Conditioning & Ventilation		0.00	0.00	0.00	30000.00	30000.00
Balancing	1.00 LS	0	0	0	44,123	44,123
Mechanical Systems		103,039	262	97,568	44,123	244,992

30.11. Electrical

Item	Quantity					Total
16020 50000 Conduit						
RSM GM Rigid Galv.Steel Conduit 1-1/2"		5.78	0.00	3.15	0.00	8.93
GRC	2000.00 LF	11,554	0	6,300	0	17,854

Building Systems Design
PROJECT WHATC2: Whatco Manufacturing-Whse-Office - Demonstration Project
DETAILED ESTIMATE Pre-Bid Design Review Cost Estimate DETAIL PAGE 28
30. Office Building

30.11. Electrical	QUANTY UOM	Labor	Equipmnt	Material	SubContr	TOTAL COST
RSM GM Rigid Galv Conduit 4" GRC		15.89	0.00	13.05	0.00	28.94
	100.00 LF	1,589	0	1,305	0	2,894
16027 50000 Motor Connections						
RSM GM Motor Conn Flex Cond & Ftg 15 Hp		96.28	0.00	10.55	0.00	106.83
230 Volts 3 Ph	4.00 EA	385	0	42	0	427
RSM GM Motor Conn. 3 Pole 50Hp Motor		63.55	0.00	11.35	0.00	74.90
460V, 3 Ph	1.00 EA	64	0	11	0	75
OBK GM Grounding Allowance		0.00	0.00	0.00	10000.00	10000.00
120 Volt Grounded 20 Amp	1.00 LS	0	0	0	10,000	10,000
16114 00000 Mineral Insulated Cable						
RSM GM Wire, 3 Conductor, 600V, #12		264.77	0.00	415.00	0.00	679.77
	40.00 CLF	10,591	0	16,600	0	27,191
RSM GM Wire, # Conductor, 600V, #4		353.03	0.00	1075.00	0.00	1428.03
	2.00 CLF	706	0	2,150	0	2,856

Item	Quantity					
16115 50000 Special Wires & Fittings						
RSM GM Wire Thermostat No Jac Twist		39.72	0.00	8.00	0.00	47.72
#18-2 Conductor	2.00 CLF	79	0	16	0	95
RSM GM Wire Fire Alrm Fep Teflon #22		39.72	0.00	93.00	0.00	132.72
2Pr	2.00 CLF	79	0	186	0	265
16116 00000 Undercarpet						
RSM GM Transition Fitting Wall Box		13.24	0.00	19.00	0.00	32.24
Surface	40.00 EA	530	0	760	0	1,290
16152 00000 Cable Terminations						
RSM GM Terminations, 600V # 12		6.35	0.00	0.41	0.00	6.76
	200.00 EA	1,271	0	82	0	1,353
RSM GM Cable Term Crimp 2 Way Conn #4		13.81	0.00	2.40	0.00	16.21
	4.00 EA	55	0	10	0	65
16211 00000 Outlet Boxes						
RSM GM Box, 4" Octagon Pressed Steel		15.89	0.00	1.15	0.00	17.04
	60.00 EA	953	0	69	0	1,022
16213 00000 Pull Boxes & Cabinets						

Building Systems Design
PROJECT WHATC2: Whatco Manufacturing-Whse-Office - Demonstration Project
DETAILED ESTIMATE Pre-Bid Design Review Cost Estimate DETAIL PAGE 29
30. Office Building

30.11. Electrical	QUANTY UOM	Labor	Equipmnt	Material	SubContr	TOTAL COST
RSM GM Pull Box Sc Nema 1 12"W-12"H-8"D		63.55	0.00	26.00	0.00	89.55
	8.00 EA	508	0	208	0	716
16232 00000 Wiring Devices						
RSM GM Wiring Device Receptacle Duplex		7.94	0.00	1.90	0.00	9.84
120 Volt Grounded 15 Amp	200.00 EA	1,589	0	380	0	1,969
16311 00000 Motor Control Center						
RSM GM MCC Starter FVNR 25 Hp Type B		158.86	0.00	760.00	0.00	918.86
	4.00 EA	635	0	3,040	0	3,675
RSM GM MCC Starter FVNR 100 Hp Type B		453.90	0.00	2275.00	0.00	2728.90
	1.00 EA	454	0	2,275	0	2,729
OBK GM Main Service Switchgear (Quote)		0.00	0.00	0.00	15000.00	15000.00
	1.00 EA	0	0	0	15,000	15,000
M OBK GM Switchboard (Quote)		0.00	0.00	0.00	12500.00	12500.00
	1.00 EA	0	0	0	12,500	12,500

Item	Quantity					
16320 50000 Circuit Breakers						
RSM GM Circuit Breaker Nema 1 600V 3P		211.82	0.00	575.00	0.00	786.82
225A	1.00 EA	212	0	575	0	787
RSM GM Crkt Brkr Disco 240V 3Pl 15Hp		113.47	0.00	150.00	0.00	263.47
Motor	4.00 EA	454	0	600	0	1,054
16323 00000 Load Centers						
RSM GM Load Center 200A 3W 120/240V		423.41	0.00	250.00	0.00	673.41
Main Lugs Indoor Inc 20A Sp	4.00 EA	1,694	0	1,000	0	2,694
Bkrs 20 Circ						
16416 00000 Oil Filled Transformer						
RSM GM Transformer 1500 KVA, Oil Filled		3396.89	488.72	21000.00	0.00	24885.60
Pad Mounted, 15KV 480V 3 Phase	1.00 EA	3,397	489	21,000	0	24,886
16612 50000 Fixture Whips						
RSM GM Fix Whip 6' Greenfld W Thhn 3		9.93	0.00	8.35	0.00	18.28
#14	150.00 EA	1,489	0	1,253	0	2,742
16613 00000 Interior Lighting Fixtures						
RSM GM Rec Fluor 2X4' W 3 40W Lamp &		63.55	0.00	68.00	0.00	131.55
Lens	150.00 EA	9,532	0	10,200	0	19,732

Building Systems Design

PROJECT WHATC2: Whatco Manufacturing-Whse-Office - Demonstration Project

DETAILED ESTIMATE — Pre-Bid Design Review Cost Estimate — DETAIL PAGE 30

30. Office Building

30.11. Electrical	QUANTY UOM	Labor	Equipmnt	Material	SubContr	TOTAL COST
RSM GM Sur Incan Metal Cylinder 150W		31.77	0.00	48.00	0.00	79.77
	12.00 EA	381	0	576	0	957
Electrical		48,201	489	68,638	37,500	154,827
30.13. Conveying						
14201 10000 Elevators						
RSM EV Freight Elev,2 Story,Hydraulic		18086.31	1111.11	32100.00	0.00	51297.42
10,000 Lb Cap,Max See Sec Circl #125	1.00 EA	21,487	1,320	38,135	0	60,941
Conveying		21,487	1,320	38,135	0	60,941
Office Building		467,727	23,100	575,723	81,623	1,148,173
Whatco Manufacturing-Whse-Office		1,128,137	110,543	2,065,423	116,453	3,420,556

Building Systems Design

PROJECT WHATC2: Whatco Manufacturing-Whse-Office - Demonstration Project

Pre-Bid Design Review Cost Estimate

BACKUP PAGE 1

** LABOR BACKUP **

SRC	LABOR ID	DESCRIPTION	BASE	OVERTM	TXS/INS	FRNG	TRVL	RATE	UOM	UPDATE	**** TOTAL **** DEFAULT	HOURS
RSM	ASBE	Asbestos Workers	21.67	0.0%	31.5%	7.63	0.00	36.13	HR	12/23/92	26.15	169
RSM	BRHE	Bricklayer Helpers	16.41	0.0%	31.5%	8.42	0.00	30.00	HR	12/23/92	18.85	856
RSM	BRIC	Bricklayers	19.41	0.0%	31.5%	8.42	0.00	33.94	HR	12/23/92	24.05	1278
RSM	CARP	Carpenters	19.62	0.0%	31.5%	9.12	0.00	34.92	HR	12/23/92	23.35	1196
RSM	CEFI	Cement Finishers	17.20	0.0%	31.5%	9.84	0.00	32.46	HR	12/23/92	23.05	403
RSM	CLAB	Common Building Laborers	15.60	0.0%	31.5%	6.50	0.00	27.01	HR	12/23/92	18.55	3246
RSM	ELEC	Electricians	24.59	0.0%	31.5%	7.38	0.00	39.72	HR	12/23/92	26.70	3242
RSM	ELEF	Electrician Foreman (Inside)	24.59	2.0%	31.5%	7.38	0.00	40.37	HR	12/23/92	27.20	35
RSM	ELEV	Elevator Constructors	24.40	0.0%	31.5%	10.75	0.00	42.84	HR	12/23/92	27.25	444
RSM	EQHV	Equip.Operator, Crane Or Shovel	20.61	0.0%	31.5%	8.38	0.00	35.48	HR	12/23/92	24.95	268
RSM	EQLT	Equipment Operators, Light	16.61	0.0%	31.5%	8.38	0.00	30.22	HR	12/23/92	23.10	154
RSM	EQMD	Equipment Operators, Medium	18.68	0.0%	31.5%	8.38	0.00	32.94	HR	12/23/92	24.10	338
RSM	EQOL	Equipment Operators, Oilers	14.38	0.0%	31.5%	7.65	0.00	26.56	HR	12/23/92	20.25	242
RSM	GLAZ	Glaziers	23.40	0.0%	31.5%	10.75	0.00	41.52	HR	12/23/92	23.50	214
RSM	PLUF	Plumber Foreman (Inside)	22.47	2.2%	31.5%	7.50	0.00	37.71	HR	12/23/92	27.90	14
RSM	PLUM	Plumbers	22.47	0.0%	31.5%	7.50	0.00	37.05	HR	12/23/92	27.40	1256
RSM	PORD	Painters, Ordinary	18.64	0.0%	31.5%	7.63	0.00	32.14	HR	12/23/92	21.75	985
RSM	PSST	Painters, Structural Steel	18.32	0.0%	31.5%	7.63	0.00	31.72	HR	12/23/92	22.55	1129
RSM	RODM	Rodmen (Reinforcing)	23.40	0.0%	34.5%	10.75	0.00	42.22	HR	12/23/92	25.85	427

RSM ROFC	Roofers, Composition	19.62	0.0%	35.4%	9.12	0.00	35.69	HR	12/23/92	21.25	1297
RSM ROHE	Roofers, Helpers (Composition)	19.62	0.0%	35.4%	9.12	0.00	35.69	HR	12/23/92	15.25	313
RSM SHEE	Sheet Metal Workers	22.30	0.0%	31.5%	9.86	0.00	39.18	HR	12/23/92	26.65	3924
RSM SKWK	Skilled Worker Ave. (35 Trades)	18.10	0.0%	31.5%	6.50	0.00	30.30	HR	12/23/92	24.10	488
RSM SPRI	Sprinkler Installers	22.47	0.0%	31.5%	7.50	0.00	37.05	HR	12/23/92	29.00	16
RSM SSWK	Structural Steel Workers	23.40	0.0%	34.5%	10.75	0.00	42.22	HR	12/23/92	26.00	2400
RSM SSWL	Welders, Structural Steel	23.40	0.0%	31.5%	5.62	0.00	36.39	HR	12/23/92	26.00	374
RSM STPF	Steamfitter Foremen (Inside)	22.47	2.2%	31.5%	7.50	0.00	37.71	HR	12/23/92	27.95	36
RSM STPI	Steamfitters Or Pipefitters	22.47	0.0%	31.5%	7.50	0.00	37.05	HR	12/23/92	27.45	519
RSM TILF	Tile Layers (Floor)	19.62	0.0%	31.5%	9.12	0.00	34.92	HR	12/23/92	23.50	239
RSM TILH	Tile Layers Helpers	16.10	0.0%	31.5%	9.12	0.00	30.29	HR	12/23/92	18.70	39
RSM TRHV	Truck Drivers, Heavy	16.10	0.0%	31.5%	5.62	0.00	26.79	HR	12/23/92	19.45	291

Computerized construction document estimate (I.C.E.). A computerized construction document cost estimate for the WHATCO project was run using the I.C.E. system by Management Computer Controls, Inc. (MC^2), under the direction of Kevitt Adler, president of MC^2.

The estimate includes the entire WHATCO project except the special site work. The reports are as follows:

Standard page format (without CSI numbers)

1. Summary of results, plant-warehouse-office
2. List of cost by CSI division by plant, including cost and cost/sf
3. Detailed cost by item (CSI order but named), plant including cost and cost/sf
4. List of cost by CSI division by warehouses, including cost and cost/sf
5. Detailed cost by item (CSI order but named), warehouses, including cost and cost/sf
6. List of cost by CSI division by office building, including cost and cost/sf
7. Detailed cost by item (CSI order but named), office building, including cost and cost/sf

Horizontal page format

1. Summary page with breakdown
   - Labor
   - Material
   - Equipment
   - Subs
   - Tax and Insurance
   - Total
2. Same breakdown as 1 by CSI division (not numbered), plant
3. Detailed cost by item (CSI order, but not numbered) with same breakdown as 1, plant
4. Same breakdown as 1 by CSI division (not numbered), warehouses
5. Detailed cost by item (CSI order, but not numbered) with same breakdown as 1, warehouses
6. Same breakdown as 1 by CSI division (not numbered), office building
7. Detailed cost by item (CSI order, but not numbered) with same breakdown as 1, office building

Recap of up to four estimates, summary

1. Summary by numbered CSI division, showing: $/sf, % of total and cost, plant
2. Summary by numbered CSI division, showing: $/sf, % of total and cost, warehouses
3. Summary by numbered CSI division, showing: $/sf, % of total and cost, office building

Recap of up to four estimates, detailed

1. Summary by CSI (four digit) showing: $/sf, % of total and cost, plant
2. Summary by CSI (four digit) showing: $/sf, % of total and cost, warehouses
3. Summary by CSI (four digit) showing: $/sf, % of total and cost, office building

File ID. 2568
Project No.

MANAGEMENT COMPUTER CONTROLS, INC.
2881 DIRECTORS COVE
MEMPHIS, TENNESSEE 38131
WHATCO PROJECT

Page

Description	Qty	Unit	Unit Price	Total Price
PLANT	30,000	SQFT	81.19	2,435,823
WAREHOUSE	6,000	SQFT	91.41	548,513
OFFICE	14,000	SQFT	100.51	1,407,242
***** ESTIMATE TOTAL				$4,391,578

File ID. 2568
Project No.
30,000 SQFT

MANAGEMENT COMPUTER CONTROLS, INC.
2881 DIRECTORS COVE
MEMPHIS, TENNESSEE 38131
WHATCO PROJECT
PLANT

Page 1

Description	Qty	Unit	Unit Price	Total Price	30,000 $/SQFT
GENERAL REQUIREMENTS				$431,030	14.36
CLEARING OF SITE				$2,875	.09
EXCAV, GRADING & BACKFILL				$64,712	2.15
CONCRETE FINISHING				$20,421	.68
FORM WORK				$3,591	.12
REINFORCING STEEL				$32,163	1.07
CAST-IN-PLACE CONCRETE				$101,567	3.38
CEMENTITIOUS DECKS				$35,700	1.19
STRUCTURAL METALS				$508,061	16.93
WATERPROOF & DAMPPROOF				$5,910	.19
PREFORMED ROOFING & SIDING				$229,590	7.65
ROOFING,SHEETMETAL&ACCESS				$113,387	3.78
METAL DOORS & FRAMES				$1,837	.06
SPECIAL DOORS				$15,550	.51
HARDWARE				$1,937	.06
PAINTING & WALL COVERING				$10,632	.35
EQUIPMENT				$18,593	.62
CONVEYING SYSTEMS				$511,866	17.06
HVAC SYSTEMS & EQUIPMENT				$226,438	7.54
ELEC SYSTEMS & EQUIPMENT				$99,963	3.33
**** TOTAL PLANT				$2,435,823	81.19

File ID. 2568
Project No.
30,000 SQFT

MANAGEMENT COMPUTER CONTROLS, INC.
2881 DIRECTORS COVE
MEMPHIS, TENNESSEE 38131
WHATCO PROJECT
PLANT

Page 1

Description	Qty	Unit	Unit Price	Total Price	30,000 $/SQFT
GENERAL REQUIREMENTS					
OTHER GENERAL REQUIREMENTS					
OH/PROFIT/GC/BOND @ 21.5%	1	LS	431,030.00	431,030	14.36
** TOTAL OTHER GENERAL REQUIREMENTS				431,030	14.36
*** TOTAL GENERAL REQUIREMENTS				431,030	14.36
CONTINGENCY					
CONTINGENCY @ 0%					
*** TOTAL CONTINGENCY					
CLEARING OF SITE					
SITE PREP (CLEAR)	1	ACRE	2,875.00	2,875	.09
*** TOTAL CLEARING OF SITE				2,875	.09
EXCAV, GRADING & BACKFILL					
SITE GRADING					
EXCAV-LOAD-HAUL SURP 3-4MILES	3,345	CUYD	7.51	25,131	.83
** TOTAL SITE GRADING				25,131	.83
EXCAVATION & BACKFILL					
EXCAVATE COL FTG	84	CUYD	12.63	1,061	.03
FINE GRADE COL FTG	900	SQFT	.67	609	.02
FINE GRADE SLAB ON GRADE	30,000	SQFT	.47	14,220	.47
STRUCTURAL FILL	1,672	CUYD	14.16	23,691	.79
** TOTAL EXCAVATION & BACKFILL				39,581	1.31
*** TOTAL EXCAV, GRADING & BACKFILL				64,712	2.15

CONCRETE FINISHING				
CONCRETE FINISH				
TROWEL CEMENT FINISH	30,000 SQFT	.49	14,820	.49
POINT & PATCH	800 SQFT	.17	141	
PROTECT & CURE	30,000 SQFT	.18	5,460	.18
** TOTAL CONCRETE FINISH			20,421	.68
*** TOTAL CONCRETE FINISHING			20,421	.68
FORM WORK				
FORM WORK				
SLAB ON GRADE EDGE FORMS	800 SQFT	4.48	3,591	.12
** TOTAL FORM WORK			3,591	.12
*** TOTAL FORM WORK			3,591	.12
REINFORCING STEEL				
RE-BARS				
RE-STEEL @ SLAB ON GRADE	387 CWT	71.58	27,703	.92
RE-STEEL @ COL FTG	63 CWT	70.79	4,460	.14
** TOTAL RE-BARS			32,163	1.07
*** TOTAL REINFORCING STEEL			32,163	1.07
CAST-IN-PLACE CONCRETE				
CAST-IN-PLACE CONCRETE				
CONCRETE @ COL FTG	88 CUYD	71.46	6,289	.21
CONCRETE @ SLAB ON GRADE	1,145 CUYD	71.76	82,166	2.73
LOADING DOCK SLAB	4,000 SQFT	3.27	13,112	.43
** TOTAL CAST-IN-PLACE CONCRETE			101,567	3.38
*** TOTAL CAST-IN-PLACE CONCRETE			101,567	3.38
CEMENTITIOUS DECKS				

File ID. 2568
Project No.
30,000 SQFT

MANAGEMENT COMPUTER CONTROLS, INC.
2881 DIRECTORS COVE
MEMPHIS, TENNESSEE 38131
WHATCO PROJECT
PLANT

Description	Qty Unit	Unit Price	Total Price	30,000 $/SQFT
CEMENTITIOUS DECKS				
LTWT CONC ROOF FILL	30,000 SQFT	1.19	35,700	1.19
*** TOTAL CEMENTITIOUS DECKS			35,700	1.19
STRUCTURAL METALS				
STRUCTURAL METALS				
STRUCTURAL STEEL FRAME	3,060 CWT	93.05	284,742	9.49
MISC. SUPPORT FRAMING	122 CWT	150.76	18,393	.61
BASE PLATE	3,708 LBS	1.24	4,605	.15
SANDBLAST	22,387 SQFT	.54	12,089	.40
PRIMER	22,387 SQFT	.18	4,119	.13
INTERMEDIATE COAT	22,387 SQFT	.30	6,917	.23
TOP COAT (ELEVATED)	22,387 SQFT	.54	12,156	.40
** TOTAL STRUCTURAL METALS			343,021	11.43
METAL JOIST				
STEEL JOIST	1,440 CWT	81.48	117,340	3.91
** TOTAL METAL JOIST			117,340	3.91
METAL DECKING				
1-1/2" METAL DECK	30,000 SQFT	1.59	47,700	1.59
** TOTAL METAL DECKING			47,700	1.59
*** TOTAL STRUCTURAL METALS			508,061	16.93
WATERPROOF & DAMPPROOF				
VAPOR BARRIER @ SLAB	30,000 SQFT	.19	5,910	.19
*** TOTAL WATERPROOF & DAMPPROOF			5,910	.19
PREFORMED ROOFING & SIDING				
INSULATED METAL SIDING	18,000 SQFT	12.75	229,590	7.65
*** TOTAL PREFORMED ROOFING & SIDING			229,590	7.65

ROOFING,SHEETMETAL&ACCESS				
ROOFING & ROOF INSULATION				
3 PLY TAR& GRAVEL ROOFING	300 SQS	130.39	39,118	1.30
2" FIBER BD ROOF INSULATION	30,000 SQFT	2.04	61,200	2.04
** TOTAL ROOFING & ROOF INSULATION			100,318	3.34
FLASHING & SHEETMETAL				
24 GA GALV IRON SHEETMETAL	800 SQFT	10.96	8,771	.29
6" GALV DOWNSPOUT	245 LNFT	7.19	1,762	.05
6" GALV GUTTER	400 LNFT	6.34	2,536	.08
** TOTAL FLASHING & SHEETMETAL			13,069	.43
*** TOTAL ROOFING,SHEETMETAL&ACCESS			113,387	3.78
METAL DOORS & FRAMES				
HM FRAME	5 EACH	137.60	688	.02
HM DOOR	5 EACH	229.80	1,149	.03
*** TOTAL METAL DOORS & FRAMES			1,837	.06
SPECIAL DOORS				
STEEL ROLL-UP DOOR	4 EACH	3,887.50	15,550	.51
*** TOTAL SPECIAL DOORS			15,550	.51
HARDWARE				
FINISH HARDWARE ALLOWANCE	5 OPNG	387.40	1,937	.06
*** TOTAL HARDWARE			1,937	.06
PAINTING & WALL COVERING				
PAINTING				
PAINT EXTERIOR DOOR	10 SIDE	40.30	403	.01
PAINT DOOR FRAME	5 EACH	33.00	165	
PAINT MAS-CONC 3 CTS	60 SQS	109.91	6,595	.22
EXTERIOR PAINTING	46 SQS	75.41	3,469	.11
** TOTAL PAINTING			10,632	.35

File ID. 2568
Project No.
30,000 SQFT

MANAGEMENT COMPUTER CONTROLS, INC.
2881 DIRECTORS COVE
MEMPHIS, TENNESSEE 38131
WHATCO PROJECT
PLANT

Description	Qty	Unit	Unit Price	Total Price	30,000 $/SQFT
PAINTING & WALL COVERING					
PAINTING					
*** TOTAL PAINTING & WALL COVERING				10,632	.35
EQUIPMENT					
LOADING DOCK EQUIPMENT					
DOCK LEVELER	4	EACH	4,648.25	18,593	.62
** TOTAL LOADING DOCK EQUIPMENT				18,593	.62
*** TOTAL EQUIPMENT				18,593	.62
CONVEYING SYSTEMS					
HOISTS & CRANES					
BRIDGE CRANE	1	EACH	511,866.00	511,866	17.06
** TOTAL HOISTS & CRANES				511,866	17.06
*** TOTAL CONVEYING SYSTEMS				511,866	17.06
HVAC SYSTEMS & EQUIPMENT					
PIPE & FITTINGS					
SCH 40 STEEL PIPE		****			
BLK T&C		****			
PIPE,2"	1,200	LNFT	7.50	9,009	.30
BLACK PIPE,P.E.		****			
PIPE,3"	700	LNFT	14.10	9,876	.32
STEEL FLANGE, 150#		****			
WELD-NECK,3"	12	EACH	124.91	1,499	.05
THREADED,2"	8	EACH	98.12	785	.02
BOLT & GASKET SET		****			
FLANGE PACK,2"	4	EACH	11.00	44	
FLANGE PACK,3"	6	EACH	13.83	83	
** TOTAL PIPE & FITTINGS				21,296	.71

Item	Quantity	Unit Cost	Total	
VALVES				
IRON BODY VALVES	****			
GATE, FLANGED	****			
125#,2"	4 EACH	286.25	1,145	.03
125#,3"	4 EACH	339.75	1,359	.04
SWING CHECK, FLANGED	****			
125#,3"	2 EACH	297.00	594	.02
** TOTAL VALVES			3,098	.10
PIPE HANGERS & SUPPORTS				
PIPE HANGERS, STEEL	****			
WITH 3' ROD & C-CLAMP	****			
CLEVIS,3"	70 EACH	39.01	2,731	.09
CLEVIS,6"	120 EACH	45.11	5,414	.18
INSULATION SHIELD,6"	120 EACH	21.67	2,601	.08
** TOTAL PIPE HANGERS & SUPPORTS			10,746	.35
PIPE INSULATION				
FIBERGLASS INSULATION	****			
ASJ, 2" THICK	****			
PIPE,2"	1,200 LNFT	7.52	9,034	.30
FLANGES,2"	8 EACH	23.87	191	
** TOTAL PIPE INSULATION			9,225	.30
MAJOR HEATING EQUIPMENT				
SPACE HEATERS	****			
PROPELLER,STEAM&HW,HORIZ	****			
UH#1 3300 CFM	20 EACH	1,422.55	28,451	.94
** TOTAL MAJOR HEATING EQUIPMENT			28,451	.94
MAJOR COOLING EQUIPMENT				
AIR HANDLING UNITS	****			
LOW-PRESS,HORIZ,FLR MTD	****			
AHU#3 18620 CFM	10 EACH	15,362.20	153,622	5.12

File ID. 2568
Project No.
30,000 SQFT

MANAGEMENT COMPUTER CONTROLS, INC.
2881 DIRECTORS COVE
MEMPHIS, TENNESSEE 38131
WHATCO PROJECT
PLANT

Description	Qty	Unit	Unit Price	Total Price	30,000 $/SQFT
HVAC SYSTEMS & EQUIPMENT					
MAJOR COOLING EQUIPMENT					
** TOTAL MAJOR COOLING EQUIPMENT				153,622	5.12
*** TOTAL HVAC SYSTEMS & EQUIPMENT				226,438	7.54
ELEC SYSTEMS & EQUIPMENT					
LOW VOLTAGE CONDUCTORS					
CU WIRE THHN-THWN-12 AWG	15,600	LNFT	.39	6,209	.20
CU WIRE THHN-THWN-4 AWG	618	LNFT	1.08	669	.02
22/2 AWG	600	LNFT	.78	473	.01
TERMIN LUGS-CU WIRE 12 AWG	20	EACH	5.65	113	
TERMIN LUGS-CU WIRE 4 AWG	40	EACH	24.90	996	.03
** TOTAL LOW VOLTAGE CONDUCTORS				8,460	.28
CONDUIT, FITTINGS & ACCESS					
GALVANIZED RIGID STEEL-GRS		****			
COND-O'HEAD/WALL 2"	1,500	LNFT	9.49	14,246	.47
COND-O'HEAD/WALL 3"	1,800	LNFT	14.92	26,873	.89
4" OCTAGON BOX 1-1/2" DEEP	60	EACH	23.31	1,399	.04
4" SQUARE BOX 1-1/2"DEEP	10	EACH	23.60	236	
4" SQ BOX, 1G PLASTER RING	10	EACH	9.40	94	
OUTLET BOX CLAMP	10	EACH	.70	7	
12"X12"X8" PULL BOX	12	EACH	115.75	1,389	.04
** TOTAL CONDUIT, FITTINGS & ACCESS				44,244	1.47
GROUNDING ACCESSORIES					
GROUNDING	1	LS	3,750.00	3,750	.12
** TOTAL GROUNDING ACCESSORIES				3,750	.12

Item	Quantity	Unit Price	Amount	
LIGHTING FIXTURES				
FIXTURE EXIT ,VAR 120 V	8 EACH	105.12	841	.02
FIXTURE A ,VAR 277 V	36 EACH	471.36	16,969	.56
** TOTAL LIGHTING FIXTURES			17,810	.59
LAMPS FOR LIGHTING FIXT.				
HIGH PRESSURE SODIUM	****			
E-25 MOG LU1000	36 EACH	226.25	8,145	.27
6'--3/8" W/2-#16 TFF	36 SET	17.16	618	.02
** TOTAL LAMPS FOR LIGHTING FIXT.			8,763	.29
WIRING DEVICES				
NEMA 5-20R,DUPLEX,IVORY	60 EACH	19.85	1,191	.04
3/8" PHONE PLATE,PLAST,IV	10 EACH	8.00	80	
** TOTAL WIRING DEVICES			1,271	.04
DISCONNECT DEVICES				
ENCL CB,600/3/50AF,NEMA-1	10 EACH	473.90	4,739	.15
ENCL CB,600/3/100AF,NEMA-1	1 EACH	605.00	605	.02
** TOTAL DISCONNECT DEVICES			5,344	.17
STARTERS				
3P,POLYPH,NEMA-1,SIZE 2	10 EACH	797.20	7,972	.26
3P,POLYPH,NEMA-1,SIZE 4	1 EACH	2,349.00	2,349	.07
** TOTAL STARTERS			10,321	.34
*** TOTAL ELEC SYSTEMS & EQUIPMENT			99,963	3.33
**** TOTAL PLANT			2,435,823	81.19

File ID. 2568
Project No.
6,000 SQFT

MANAGEMENT COMPUTER CONTROLS, INC.
2881 DIRECTORS COVE
MEMPHIS, TENNESSEE 38131
WHATCO PROJECT
WAREHOUSE

Page 1

Description	Qty	Unit	Unit Price	Total Price	6,000 $/SQFT
GENERAL REQUIREMENTS				$97,062	16.17
EXCAV, GRADING & BACKFILL				$12,846	2.14
CONCRETE FINISHING				$4,116	.68
FORM WORK				$1,526	.25
REINFORCING STEEL				$4,501	.75
CAST-IN-PLACE CONCRETE				$14,369	2.39
CEMENTITIOUS DECKS				$7,140	1.19
STRUCTURAL METALS				$119,553	19.92
WATERPROOF & DAMPPROOF				$1,182	.19
PREFORMED ROOFING & SIDING				$76,530	12.75
ROOFING,SHEETMETAL&ACCESS				$26,025	4.33
METAL DOORS & FRAMES				$735	.12
SPECIAL DOORS				$7,775	1.29
HARDWARE				$775	.12
PAINTING & WALL COVERING				$5,919	.98
EQUIPMENT				$49,297	8.21
HVAC SYSTEMS & EQUIPMENT				$87,844	14.64
ELEC SYSTEMS & EQUIPMENT				$31,318	5.22
**** TOTAL WAREHOUSE				$548,513	91.41

File ID. 2568
Project No.
6,000 SQFT

MANAGEMENT COMPUTER CONTROLS, INC.
2881 DIRECTORS COVE
MEMPHIS, TENNESSEE 38131
WHATCO PROJECT
WAREHOUSE

Page 1

Description	Qty	Unit	Unit Price	Total Price	6,000 $/SQFT
GENERAL REQUIREMENTS					
OTHER GENERAL REQUIREMENTS					
OH/PROFIT/GC/BOND @ 21.5%	1	LS	97,062.00	97,062	16.17
** TOTAL OTHER GENERAL REQUIREMENTS				97,062	16.17
*** TOTAL GENERAL REQUIREMENTS				97,062	16.17
CONTINGENCY					
CONTINGENCY @ 0%					
*** TOTAL CONTINGENCY					
EXCAV, GRADING & BACKFILL					
SITE GRADING					
EXCAV-LOAD-HAUL SURP 3-4MILES	669	CUYD	7.51	5,026	.83
** TOTAL SITE GRADING				5,026	.83
EXCAVATION & BACKFILL					
EXCAVATE COLUMN FOOTING	9	CUYD	12.55	113	.01
FINE GRADE COLUMN FOOTING	192	SQFT	.67	130	.02
FINE GRADE SLAB ON GRADE	6,000	SQFT	.47	2,844	.47
STRUCTURAL FILL	334	CUYD	14.17	4,733	.78
** TOTAL EXCAVATION & BACKFILL				7,820	1.30
*** TOTAL EXCAV, GRADING & BACKFILL				12,846	2.14

Item	Quantity	Unit Cost	Total	Per SF
CONCRETE FINISHING				
CONCRETE FINISH				
TROWEL CEMENT FINISH	6,000 SQFT	.49	2,964	.49
POINT & PATCH	340 SQFT	.17	60	.01
PROTECT & CURE	6,000 SQFT	.18	1,092	.18
** TOTAL CONCRETE FINISH			4,116	.68
*** TOTAL CONCRETE FINISHING			4,116	.68
FORM WORK				
FORM WORK				
SLAB ON GRADE EDGE FORMS	340 LNFT	4.48	1,526	.25
** TOTAL FORM WORK			1,526	.25
*** TOTAL FORM WORK			1,526	.25
REINFORCING STEEL				
RE-BARS				
RE-STEEL @ SLAB ON GRADE	52 CWT	71.57	3,722	.62
RE-STEEL @ COLUMN FOOTING	11 CWT	70.81	779	.13
** TOTAL RE-BARS			4,501	.75
*** TOTAL REINFORCING STEEL			4,501	.75
CAST-IN-PLACE CONCRETE				
CAST-IN-PLACE CONCRETE				
CONCRETE @ COLUMN FTG	10 CUYD	71.50	715	.11
CONCRETE @ SLAB ON GRADE	172 CUYD	71.76	12,343	2.05
LOADING DOCK SLAB	400 SQFT	3.27	1,311	.21
** TOTAL CAST-IN-PLACE CONCRETE			14,369	2.39
*** TOTAL CAST-IN-PLACE CONCRETE			14,369	2.39
CEMENTITIOUS DECKS				
LTWT CONC ROOF FILL	6,000 SQFT	1.19	7,140	1.19
*** TOTAL CEMENTITIOUS DECKS			7,140	1.19

File ID. 2568
Project No.
6,000 SQFT

MANAGEMENT COMPUTER CONTROLS, INC.
2881 DIRECTORS COVE
MEMPHIS, TENNESSEE 38131
WHATCO PROJECT
WAREHOUSE

Description	Qty Unit	Unit Price	Total Price	6,000 $/SQFT
STRUCTURAL METALS				
STRUCTURAL METALS				
STRUCTURAL STEEL FRAME	400 CWT	93.05	37,221	6.20
MISC. SUPPORT FRAMING	270 CWT	150.75	40,705	6.78
BASE PLATE	504 LBS	1.24	626	.10
SANDBLAST	6,519 SQFT	.54	3,520	.58
PRIMER	6,519 SQFT	.18	1,200	.20
INTERMEDIATE COAT	6,519 SQFT	.30	2,014	.33
TOP COAT (ELEVATED)	6,519 SQFT	.54	3,540	.59
** TOTAL STRUCTURAL METALS			88,826	14.80
METAL JOIST				
STEEL JOIST	260 CWT	81.48	21,187	3.53
** TOTAL METAL JOIST			21,187	3.53
METAL DECKING				
1-1/2" METAL DECK	6,000 SQFT	1.59	9,540	1.59
** TOTAL METAL DECKING			9,540	1.59
*** TOTAL STRUCTURAL METALS			119,553	19.92
WATERPROOF & DAMPPROOF				
VAPOR BARRIER @ SLAB	6,000 SQFT	.19	1,182	.19
*** TOTAL WATERPROOF & DAMPPROOF			1,182	.19
PREFORMED ROOFING & SIDING				
INSULATED METAL SIDING	6,000 SQFT	12.75	76,530	12.75
*** TOTAL PREFORMED ROOFING & SIDING			76,530	12.75

Item	Quantity	Unit Cost	Total	Per Unit
ROOFING,SHEETMETAL&ACCESS				
ROOFING & ROOF INSULATION				
3 PLY TAR& GRAVEL ROOFING	60 SQS	130.40	7,824	1.30
2" FIBER BD ROOF INSULATION	6,000 SQFT	2.04	12,240	2.04
** TOTAL ROOFING & ROOF INSULATION			20,064	3.34
FLASHING & SHEETMETAL				
24 GA GALV IRON SHEETMETAL	440 SQFT	10.96	4,823	.80
6" GALV DOWNSPOUT	70 LNFT	7.18	503	.08
6" GALV GUTTER	100 LNFT	6.35	635	.10
** TOTAL FLASHING & SHEETMETAL			5,961	.99
*** TOTAL ROOFING,SHEETMETAL&ACCESS			26,025	4.33
METAL DOORS & FRAMES				
HM FRAME	2 EACH	137.50	275	.04
HM DOOR	2 EACH	230.00	460	.07
*** TOTAL METAL DOORS & FRAMES			735	.12
SPECIAL DOORS				
STEEL ROLL-UP DOOR	2 EACH	3,887.50	7,775	1.29
*** TOTAL SPECIAL DOORS			7,775	1.29
HARDWARE				
FINISH HARDWARE ALLOWANCE	2 OPNG	387.50	775	.12
*** TOTAL HARDWARE			775	.12
PAINTING & WALL COVERING				
PAINTING				
PAINT EXTERIOR DOOR	4 SIDE	40.50	162	.02
PAINT DOOR FRAME	2 EACH	33.00	66	.01
PAINT MAS-CONC 3 CTS	36 SQS	109.88	3,956	.65
EXTERIOR PAINTING	23 SQS	75.43	1,735	.28
** TOTAL PAINTING			5,919	.98
*** TOTAL PAINTING & WALL COVERING			5,919	.98

File ID. 2568
Project No.
6,000 SQFT

MANAGEMENT COMPUTER CONTROLS, INC.
2881 DIRECTORS COVE
MEMPHIS, TENNESSEE 38131
WHATCO PROJECT
WAREHOUSE

Description	Qty Unit	Unit Price	Total Price	6,000 $/SQFT
EQUIPMENT				
MAINTENANCE EQUIPMENT				
FORK LIFTS	2 EACH	20,000.00	40,000	6.66
** TOTAL MAINTENANCE EQUIPMENT			40,000	6.66
LOADING DOCK EQUIPMENT				
DOCK LEVELER	2 EACH	4,648.50	9,297	1.55
** TOTAL LOADING DOCK EQUIPMENT			9,297	1.55
*** TOTAL EQUIPMENT			49,297	8.21
HVAC SYSTEMS & EQUIPMENT				
PIPE & FITTINGS				
SCH 40 STEEL PIPE	****			
BLK T&C	****			
PIPE,2"	600 LNFT	7.50	4,504	.75
STEEL FLANGE, 150#	****			
THREADED,2"	16 EACH	98.12	1,570	.26
BOLT & GASKET SET	****			
FLANGE PACK,2"	8 EACH	11.00	88	.01
** TOTAL PIPE & FITTINGS			6,162	1.02
VALVES				
IRON BODY VALVES	****			
GATE, FLANGED	****			
125#,2"	8 EACH	286.37	2,291	.38
** TOTAL VALVES			2,291	.38

Item	Quantity	Unit Cost	Total	Per SF
PIPE HANGERS & SUPPORTS				
PIPE HANGERS, STEEL	****			
WITH 3' ROD & C-CLAMP	****			
CLEVIS,2"	20 EACH	34.40	688	.11
CLEVIS,6"	40 EACH	45.12	1,805	.30
INSULATION SHIELD,6"	40 EACH	21.67	867	.14
** TOTAL PIPE HANGERS & SUPPORTS			3,360	.56
PIPE INSULATION				
FIBERGLASS INSULATION	****			
ASJ, 2" THICK	****			
PIPE,2"	400 LNFT	7.52	3,011	.50
FLANGES,2"	8 EACH	23.87	191	.03
** TOTAL PIPE INSULATION			3,202	.53
MAJOR HEATING EQUIPMENT				
SPACE HEATERS	****			
PROPELLER,STEAM&HW,HORIZ	****			
UH#2 3300 CFM	8 EACH	1,422.50	11,380	1.89
** TOTAL MAJOR HEATING EQUIPMENT			11,380	1.89
MAJOR COOLING EQUIPMENT				
AIR HANDLING UNITS	****			
LOW-PRESS,HORIZ,FLR MTD	****			
AHU#3 18620 CFM	4 EACH	15,362.25	61,449	10.24
** TOTAL MAJOR COOLING EQUIPMENT			61,449	10.24
*** TOTAL HVAC SYSTEMS & EQUIPMENT			87,844	14.64
ELEC SYSTEMS & EQUIPMENT				
LOW VOLTAGE CONDUCTORS				
CU WIRE THHN-THWN-12 AWG	3,000 LNFT	.39	1,194	.19
22/2 AWG	200 LNFT	.78	157	.02
TERMIN LUGS-CU WIRE 12 AWG	120 EACH	5.64	677	.11
** TOTAL LOW VOLTAGE CONDUCTORS			2,028	.33
CONDUIT,FITTINGS & ACCESS				
GALVANIZED RIGID STEEL-GRS	****			

File ID. 2568
Project No.
6,000 SQFT

MANAGEMENT COMPUTER CONTROLS, INC.
2881 DIRECTORS COVE
MEMPHIS, TENNESSEE 38131
WHATCO PROJECT
WAREHOUSE

Page 4

Description	Qty	Unit	Unit Price	Total Price	6,000 $/SQFT
ELEC SYSTEMS & EQUIPMENT					
CONDUIT,FITTINGS & ACCESS					
COND-O'HEAD/WALL 2"	1,000	LNFT	9.49	9,497	1.58
4" OCTAGON BOX 1-1/2" DEEP	30	EACH	23.30	699	.11
4" SQUARE BOX 1-1/2"DEEP	10	EACH	23.60	236	.03
4" SQ BOX, 1G PLASTER RING	10	EACH	9.40	94	.01
OUTLET BOX CLAMP	10	EACH	.70	7	
12"X12"X8" PULL BOX	4	EACH	115.75	463	.07
** TOTAL CONDUIT,FITTINGS & ACCESS				10,996	1.83
GROUNDING ACCESSORIES					
GROUNDING	1	LS	750.00	750	.12
** TOTAL GROUNDING ACCESSORIES				750	.12
LIGHTING FIXTURES					
FIXTURE EXIT ,VAR 120 V	6	EACH	105.00	630	.10
FIXTURE A ,VAR 277 V	12	EACH	471.41	5,657	.94
** TOTAL LIGHTING FIXTURES				6,287	1.04
LAMPS FOR LIGHTING FIXT.					
HIGH PRESSURE SODIUM		****			
E-25 MOG LU1000	12	EACH	226.25	2,715	.45
6'--3/8" W/2-#16 TFF	162	SET	17.17	2,782	.46
** TOTAL LAMPS FOR LIGHTING FIXT.				5,497	.91

WIRING DEVICES				
NEMA 5-20R,DUPLEX,IVORY	30 EACH	19.86	596	.09
3/8" PHONE PLATE,PLAST,IV	10 EACH	8.00	80	.01
** TOTAL WIRING DEVICES			676	.11
DISCONNECT DEVICES				
ENCL CB,600/3/50AF,NEMA-1	4 EACH	473.75	1,895	.31
** TOTAL DISCONNECT DEVICES			1,895	.31
STARTERS				
3P,POLYPH,NEMA-1,SIZE 2	4 EACH	797.25	3,189	.53
** TOTAL STARTERS			3,189	.53
*** TOTAL ELEC SYSTEMS & EQUIPMENT			31,318	5.22
**** TOTAL WAREHOUSE			548,513	91.41

File ID. 2568 MANAGEMENT COMPUTER CONTROLS, INC. Page 1
Project No. 2881 DIRECTORS COVE
14,000 SQFT MEMPHIS, TENNESSEE 38131
WHATCO PROJECT
OFFICE

Description	Qty	Unit	Unit Price	Total Price	14,000 $/SQFT
GENERAL REQUIREMENTS				$249,018	17.78
EXCAV, GRADING & BACKFILL				$16,050	1.14
CONCRETE FINISHING				$9,464	.67
FORM WORK				$4,633	.33
REINFORCING STEEL				$13,226	.94
CAST-IN-PLACE CONCRETE				$21,227	1.51
CEMENTITIOUS DECKS				$8,330	.59
MASONRY				$144,038	10.28
STONE				$2,000	.14
STRUCTURAL METALS				$186,538	13.32
MISC & ORNAMENTAL METAL				$4,092	.29
WATERPROOF & DAMPPROOF				$1,639	.11
INSULATION				$10,826	.77
ROOFING,SHEETMETAL&ACCESS				$26,080	1.86
METAL DOORS & FRAMES				$7,893	.56
WOOD & PLASTIC DOORS				$6,773	.48
METAL WINDOWS				$14,884	1.06
HARDWARE				$17,047	1.21
GLASS & GLAZING				$2,867	.20
GYPSUM DRYWALL				$53,005	3.78
TILE & TERRAZZO				$7,808	.55
ACOUSTICAL TREATMENT				$20,910	1.49
FLOORING				$27,748	1.98
PAINTING & WALL COVERING				$56,363	4.02
SPECIALTIES				$8,016	.57
CONVEYING SYSTEMS				$38,588	2.75
PLUMBING SYSTEMS & EQUIP				$22,872	1.63
FIRE PROT SYSTEMS & EQUIP				$26,468	1.89
HVAC SYSTEMS & EQUIPMENT				$158,253	11.30
ELEC SYSTEMS & EQUIPMENT				$240,586	17.18
**** TOTAL OFFICE				$1,407,242	100.51

File ID. 2568
Project No.
14,000 SQFT

MANAGEMENT COMPUTER CONTROLS, INC.
2881 DIRECTORS COVE
MEMPHIS, TENNESSEE 38131
WHATCO PROJECT
OFFICE

Page 1

Description	Qty Unit	Unit Price	Total Price	14,000 $/SQFT
GENERAL REQUIREMENTS				
OTHER GENERAL REQUIREMENTS				
OH/PROFIT/GC/BOND @ 21.5%	1 LS	249,018.00	249,018	17.78
** TOTAL OTHER GENERAL REQUIREMENTS			249,018	17.78
*** TOTAL GENERAL REQUIREMENTS			249,018	17.78
CONTINGENCY				
CONTINGENCY @ 0%				
*** TOTAL CONTINGENCY				
EXCAV, GRADING & BACKFILL				
SITE GRADING				
EXCAV-LOAD-HAUL SURP 3-4MILES	764 CUYD	7.51	5,740	.41
** TOTAL SITE GRADING			5,740	.41
EXCAVATION & BACKFILL				
EXCAVATE COLUMN FOOTING	8 CUYD	12.62	101	
EXCAVATE GRADE BEAM	50 CUYD	12.46	623	.04
BACKFILL COLUMN FOOTING	2 CUYD	23.50	47	
BACKFILL GRADE BEAM	30 CUYD	23.73	712	.05
FINE GRADE COLUMN FOOTING	144 SQFT	.67	97	
FINE GRADE SLAB ON GRADE	7,000 SQFT	.47	3,318	.23
STRUCTURAL FILL	382 CUYD	14.16	5,412	.38
** TOTAL EXCAVATION & BACKFILL			10,310	.73
*** TOTAL EXCAV, GRADING & BACKFILL			16,050	1.14

CONCRETE FINISHING				
CONCRETE FINISH				
TROWEL CEMENT FINISH	14,000 SQFT	.49	6,916	.49
PROTECT & CURE	14,000 SQFT	.18	2,548	.18
** TOTAL CONCRETE FINISH			9,464	.67
*** TOTAL CONCRETE FINISHING			9,464	.67
FORM WORK				
FORM WORK				
GRADE BEAM FORMS	720 SQFT	6.43	4,633	.33
** TOTAL FORM WORK			4,633	.33
*** TOTAL FORM WORK			4,633	.33
REINFORCING STEEL				
RE-BARS				
RE-STEEL @ SLAB ON GRADE	55 CWT	71.58	3,937	.28
RE-STEEL @ SLAB ON MTL DECK	109 CWT	71.58	7,803	.55
RE-STEEL @ COLUMN FOOTING	8 CWT	70.75	566	.04
RE-STEEL @ GRADE BEAM	13 CWT	70.76	920	.06
** TOTAL RE-BARS			13,226	.94
*** TOTAL REINFORCING STEEL			13,226	.94
CAST-IN-PLACE CONCRETE				
CAST-IN-PLACE CONCRETE				
CONCRETE @ COLUMN FTG	7 CUYD	71.42	500	.03
CONCRETE @ GRADE BEAM	25 CUYD	74.80	1,870	.13
CONCRETE @ SLAB ON GRADE	134 CUYD	71.76	9,616	.68
CONCRETE OVER METAL DECK	115 CUYD	80.35	9,241	.66
** TOTAL CAST-IN-PLACE CONCRETE			21,227	1.51
*** TOTAL CAST-IN-PLACE CONCRETE			21,227	1.51

File ID. 2568
Project No.
14,000 SQFT

MANAGEMENT COMPUTER CONTROLS, INC.
2881 DIRECTORS COVE
MEMPHIS, TENNESSEE 38131
WHATCO PROJECT
OFFICE

Page 2

Description	Qty	Unit	Unit Price	Total Price	14,000 $/SQFT
CEMENTITIOUS DECKS					
LTWT CONC ROOF FILL	7,000	SQFT	1.19	8,330	.59
*** TOTAL CEMENTITIOUS DECKS				8,330	.59
MASONRY					
8" CONC BLOCK	15,370	SQFT	7.22	111,002	7.92
FACE BRICK	2,700	SQFT	10.95	29,584	2.11
EXT TUBULAR SCAFFOLDING	3,600	SQFT	.95	3,452	.24
*** TOTAL MASONRY				144,038	10.28
STONE					
PRECAST COPING	120	LNFT	16.66	2,000	.14
*** TOTAL STONE				2,000	.14
STRUCTURAL METALS					
STRUCTURAL METALS					
STRUCTURAL STEEL FRAME	1,280	CWT	93.05	119,108	8.50
BASE PLATE	336	LBS	1.24	417	.03
SANDBLAST	14,175	SQFT	.54	7,655	.54
PRIMER	14,175	SQFT	.18	2,609	.18
INTERMEDIATE COAT	14,175	SQFT	.30	4,380	.31
TOP COAT (ELEVATED)	14,175	SQFT	.54	7,697	.55
** TOTAL STRUCTURAL METALS				141,866	10.13
METAL JOIST					
STEEL JOIST	260	CWT	81.48	21,187	1.51
** TOTAL METAL JOIST				21,187	1.51

METAL DECKING				
1-1/2" METAL DECK	7,000 SQFT	1.59	11,130	.79
2" METAL DECK	7,000 SQFT	1.76	12,355	.88
** TOTAL METAL DECKING			23,485	1.67
*** TOTAL STRUCTURAL METALS			186,538	13.32
MISC & ORNAMENTAL METAL				
METAL FABRICATIONS				
ALUMINUM HORIZ HAND RAIL	75 LNFT	54.56	4,092	.29
** TOTAL METAL FABRICATIONS			4,092	.29
*** TOTAL MISC & ORNAMENTAL METAL			4,092	.29
WATERPROOF & DAMPPROOF				
DAMPPROOFING	367 SQFT	.70	260	.01
VAPOR BARRIER @ SLAB	7,000 SQFT	.19	1,379	.09
*** TOTAL WATERPROOF & DAMPPROOF			1,639	.11
INSULATION				
2" FOUNDATION INSULATION	367 SQFT	1.31	481	.03
2" RIGID INSULATION	8,850 SQFT	1.16	10,345	.73
*** TOTAL INSULATION			10,826	.77
ROOFING,SHEETMETAL&ACCESS				
ROOFING & ROOF INSULATION				
3 PLY TAR& GRAVEL ROOFING	70 SQS	130.40	9,128	.65
2" FIBER BD ROOF INSULATION	7,000 SQFT	2.04	14,280	1.02
** TOTAL ROOFING & ROOF INSULATION			23,408	1.67
FLASHING & SHEETMETAL				
24 GA GALV IRON SHEETMETAL	140 SQFT	10.95	1,534	.11
6" GALV DOWNSPOUT	70 LNFT	7.18	503	.03
6" GALV GUTTER	100 LNFT	6.35	635	.04
** TOTAL FLASHING & SHEETMETAL			2,672	.19
*** TOTAL ROOFING,SHEETMETAL&ACCESS			26,080	1.86

File ID. 2568
Project No.
14,000 SQFT

MANAGEMENT COMPUTER CONTROLS, INC.
2881 DIRECTORS COVE
MEMPHIS, TENNESSEE 38131
WHATCO PROJECT
OFFICE

Page 3

Description	Qty	Unit	Unit Price	Total Price	14,000 $/SQFT
METAL DOORS & FRAMES					
HM FRAME	44	EACH	137.59	6,054	.43
HM DOOR	8	EACH	229.87	1,839	.13
*** TOTAL METAL DOORS & FRAMES				7,893	.56
WOOD & PLASTIC DOORS					
SC WOOD DOOR	36	EACH	188.13	6,773	.48
*** TOTAL WOOD & PLASTIC DOORS				6,773	.48
METAL WINDOWS					
FIXED WINDOWS	900	SQFT	16.53	14,884	1.06
*** TOTAL METAL WINDOWS				14,884	1.06
HARDWARE					
FINISH HARDWARE ALLOWANCE	44	OPNG	387.43	17,047	1.21
*** TOTAL HARDWARE				17,047	1.21
GLASS & GLAZING					
ENTRANCES & STOREFRONTS					
PAIR ALUM GLASS DOOR	2	EACH	1,433.50	2,867	.20
** TOTAL ENTRANCES & STOREFRONTS				2,867	.20
*** TOTAL GLASS & GLAZING				2,867	.20
GYPSUM DRYWALL					
STANDARD DRYWALL PARTITION	8,900	SQFT	3.99	35,547	2.53
MTL FURRING W/GYPSUM BOARD	7,950	SQFT	2.19	17,458	1.24
*** TOTAL GYPSUM DRYWALL				53,005	3.78

Item	Quantity	Unit Cost	Total	Per SF
TILE & TERRAZZO				
CERAMIC TILE				
CERAMIC TILE FLOOR	900 SQFT	8.67	7,808	.55
** TOTAL CERAMIC TILE			7,808	.55
*** TOTAL TILE & TERRAZZO			7,808	.55
ACOUSTICAL TREATMENT				
ACOUST CEIL SYS-EXPOSED GRID	12,300 SQFT	1.70	20,910	1.49
*** TOTAL ACOUSTICAL TREATMENT			20,910	1.49
FLOORING				
RESILIENT FLOORING				
VINYL COMPOSITION TILE	3,825 SQFT	1.88	7,221	.51
** TOTAL RESILIENT FLOORING			7,221	.51
CARPETING				
CARPET	875 SQYD	23.45	20,527	1.46
** TOTAL CARPETING			20,527	1.46
*** TOTAL FLOORING			27,748	1.98
PAINTING & WALL COVERING				
PAINTING				
PAINT INTERIOR DOOR	88 SIDE	37.96	3,341	.23
PAINT DOOR FRAME	44 EACH	32.93	1,449	.10
PAINT PLASTER-GYP BD 3 CTS	515 SQS	77.15	39,737	2.83
PAINT MAS-CONC 3 CTS	107 SQS	109.90	11,760	.84
EXTERIOR PAINTING	1 SQS	76.00	76	
** TOTAL PAINTING			56,363	4.02
*** TOTAL PAINTING & WALL COVERING			56,363	4.02
SPECIALTIES				

File ID. 2568
Project No.
14,000 SQFT

MANAGEMENT COMPUTER CONTROLS, INC.
2881 DIRECTORS COVE
MEMPHIS, TENNESSEE 38131
WHATCO PROJECT
OFFICE

Description	Qty	Unit	Unit Price	Total Price	14,000 $/SQFT
SPECIALTIES					
COMPARTMENTS & CUBICLES					
TOILET COMPARTMENT	12	EACH	399.25	4,791	.34
** TOTAL COMPARTMENTS & CUBICLES				4,791	.34
LOCKERS					
DOUBLE TIER LOCKER	40	OPNG	80.62	3,225	.23
** TOTAL LOCKERS				3,225	.23
*** TOTAL SPECIALTIES				8,016	.57
CONVEYING SYSTEMS					
ELEVATORS					
ELEVATOR	1	EACH	38,588.00	38,588	2.75
** TOTAL ELEVATORS				38,588	2.75
*** TOTAL CONVEYING SYSTEMS				38,588	2.75
PLUMBING SYSTEMS & EQUIP					
PIPE & FITTINGS					
SCHEDULE 40 STEEL PIPE		****			
BLACK STEEL T&C		****			
PIPE,1/2"	100	LNFT	3.28	328	.02
PIPE,1"	250	LNFT	4.20	1,051	.07
PVC PIPE		****			
SCH 40 SOLVENT CEMENT		****			
PIPE,1-1/2"	150	LNFT	3.26	490	.03
CAST IRON PIPE		****			
SERVICE WEIGHT B&S		****			

Item	Quantity	Unit Cost	Total	
PIPE,2"	40 LNFT	7.32	293	.02
PIPE,8"	100 LNFT	23.67	2,367	.16
MISC.SVC B&S GASKET FTNG.	****			
"P"-TRAP,2"	9 EACH	25.44	229	.01
** TOTAL PIPE & FITTINGS			4,758	.34
PIPE HANGER & SUPPORTS				
PIPE HANGERS,STEEL	****			
W/3'ROD & C-CLAMP	****			
CLEVIS,1/2"	10 EACH	32.70	327	.02
CLEVIS,1"	25 EACH	32.88	822	.05
CLEVIS,2"	4 EACH	33.00	132	
CLEVIS,8"	10 EACH	50.60	506	.03
** TOTAL PIPE HANGER & SUPPORTS			1,787	.12
PLUMBING SPECIALTIES				
SEWAGE SPECIALTIES	****			
VENT FLASHING, 6# LEAD	****			
2"	4 EACH	112.50	450	.03
** TOTAL PLUMBING SPECIALTIES			450	.03
PLBG FIXTURE COMMERCIAL				
PLUMBING FIXT'S.COMMERCIAL	****			
WATER CLOSET	****			
FLR MTD W/FLUSH VALVE	12 EACH	379.16	4,550	.32
URINAL	****			
WALL HUNG W/ FLUSH VALVE	2 EACH	542.00	1,084	.07
LAVATORY W/TRIM	****			
CNTR-TP,ENAM CI,SHELF-RM	9 EACH	275.44	2,479	.17
DRINKING FTN-HANDICAP	****			
WALL HUNG,ELEC	4 EACH	949.75	3,799	.27
WASHFOUNTAIN ROUND	****			
54" STAINLESS STEEL	1 EACH	2,590.00	2,590	.18
MULTI-PURPOSE SINK W/TRIM	****			
CNTR-TOP,DBL COMPT,S.S.	2 EACH	687.50	1,375	.09
** TOTAL PLBG FIXTURE COMMERCIAL			15,877	1.13
*** TOTAL PLUMBING SYSTEMS & EQUIP			22,872	1.63

File ID. 2568 MANAGEMENT COMPUTER CONTROLS, INC. Page 5
Project No. 2881 DIRECTORS COVE
14,000 SQFT MEMPHIS, TENNESSEE 38131
WHATCO PROJECT
OFFICE

Description	Qty	Unit	Unit Price	Total Price	14,000 $/SQFT
FIRE PROT SYSTEMS & EQUIP					
HOSE CAB. & EXTINGUISHERS					
FIRE EXTINGUISHER CAB.		****			
RECESSED	8	EACH	179.50	1,436	.10
FIRE EXTINGUISHER		****			
DRY CHEMICAL		****			
CLASS ABC 20 LB.	8	EACH	234.50	1,876	.13
** TOTAL HOSE CAB. & EXTINGUISHERS				3,312	.23
MISC FIRE PROT SYS & EQUIP					
SPRINKLER SYSTEM	14,000	SQFT	1.65	23,156	1.65
** TOTAL MISC FIRE PROT SYS & EQUIP				23,156	1.65
*** TOTAL FIRE PROT SYSTEMS & EQUIP				26,468	1.89
HVAC SYSTEMS & EQUIPMENT					
PIPE & FITTINGS					
SCH 40 STEEL PIPE		****			
BLK T&C		****			
PIPE,2"	300	LNFT	7.50	2,252	.16
** TOTAL PIPE & FITTINGS				2,252	.16
PIPE HANGERS & SUPPORTS					
PIPE HANGERS, STEEL		****			
WITH 3' ROD & C-CLAMP		****			
CLEVIS,6"	30	EACH	45.13	1,354	.09
INSULATION SHIELD,6"	30	EACH	21.66	650	.04
** TOTAL PIPE HANGERS & SUPPORTS				2,004	.14

Item	Quantity	Unit Cost	Total	Per SF
PIPE INSULATION				
FIBERGLASS INSULATION	****			
ASJ, 2" THICK	****			
PIPE,2"	300 LNFT	7.52	2,258	.16
** TOTAL PIPE INSULATION			2,258	.16
MAJOR HEATING EQUIPMENT				
HEATING EQUIPMENT	****			
BOILER,STEEL TUBE PACKAGE	****			
B#3 136 HP	1 EACH	28,670.00	28,670	2.04
MISCELLANEOUS ACCESSORIES	****			
FINNED TUBE RADIATION	****			
STANDARD STYLE	600 LNFT	55.30	33,185	2.37
** TOTAL MAJOR HEATING EQUIPMENT			61,855	4.41
MAJOR COOLING EQUIPMENT				
COOLING EQUIPMENT	****			
CONDENSER,AIR COOLED	****			
CH#3 30 TONS	1 EACH	15,173.00	15,173	1.08
AIR HANDLING UNITS	****			
LOW-PRESS,HORIZ,FLR MTD	****			
AHU#3 18620 CFM	4 EACH	15,362.25	61,449	4.38
** TOTAL MAJOR COOLING EQUIPMENT			76,622	5.47
DUCT INSULATION				
GALV,CLASS "B" FLUE,18"	30 LNFT	66.56	1,997	.14
** TOTAL DUCT INSULATION			1,997	.14
REGISTER,GRILLES &DIFF				
RETURN AIR REGISTER	****			
CLG MTD,SQUARE PERF FACE	****			
1 SQFT SURF AREA	30 EACH	33.73	1,012	.07
** TOTAL REGISTER,GRILLES &DIFF			1,012	.07
TEST & BALANCE				
HVAC TEST & BALANCE	1 LS	10,253.00	10,253	.73
** TOTAL TEST & BALANCE			10,253	.73

File ID. 2568 MANAGEMENT COMPUTER CONTROLS, INC. Page 6
Project No. 2881 DIRECTORS COVE
14,000 SQFT MEMPHIS, TENNESSEE 38131
WHATCO PROJECT
OFFICE

Description	Qty	Unit	Unit Price	Total Price	14,000 $/SQFT
HVAC SYSTEMS & EQUIPMENT					
TEST & BALANCE					
*** TOTAL HVAC SYSTEMS & EQUIPMENT				158,253	11.30
ELEC SYSTEMS & EQUIPMENT					
LOW VOLTAGE CONDUCTORS					
CU WIRE THHN-THWN-12 AWG	8,000	LNFT	.39	3,184	.22
CU WIRE THHN-THWN-4 AWG	618	LNFT	1.08	669	.04
22/2 AWG	200	LNFT	.78	157	.01
TERMIN LUGS-CU WIRE 12 AWG	200	EACH	5.64	1,129	.08
TERMIN LUGS-CU WIRE 4 AWG	4	EACH	24.75	99	
** TOTAL LOW VOLTAGE CONDUCTORS				5,238	.37
CONDUIT,FITTINGS & ACCESS					
GALVANIZED RIGID STEEL-GRS		****			
GALVANIZED RIGID STEEL-GRS		****			
COND-O'HEAD/WALL 1-1/2"	2,000	LNFT	7.97	15,940	1.13
COND-O'HEAD/WALL 4"	100	LNFT	22.20	2,220	.15
4" OCTAGON BOX 1-1/2" DEEP	60	EACH	23.31	1,399	.10
4" SQUARE BOX 1-1/2"DEEP	40	EACH	23.65	946	.06
4" SQ BOX, 1G PLASTER RING	40	EACH	9.45	378	.02
OUTLET BOX CLAMP	40	EACH	.67	27	
12"X12"X8" PULL BOX	8	EACH	115.75	926	.06
** TOTAL CONDUIT,FITTINGS & ACCESS				21,836	1.56
FREESTANDING SWITCHBOARDS					
L. V. SWITCHBOARD MDP	1	EACH	94,010.00	94,010	6.71
L. V. SWITCHBOARD SWBD	1	EACH	44,715.00	44,715	3.19
SWITCHBOARD MDP RIGGING	1	EACH	750.00	750	.05
SWITCHBOARD SWBD RIGGING	1	EACH	750.00	750	.05
** TOTAL FREESTANDING SWITCHBOARDS				140,225	10.01

Item	Quantity	Unit cost	Total	%
PANELBOARDS & BREAKERS				
PANELBOARD LP	4 EACH	5,338.50	21,354	1.52
** TOTAL PANELBOARDS & BREAKERS			21,354	1.52
TRANSFORMERS				
TRAN PAD-MOUNTED,1500KVA	1 EACH	21,785.00	21,785	1.55
TRANSFORMER RIGGING	1 EACH	750.00	750	.05
** TOTAL TRANSFORMERS			22,535	1.61
GROUNDING ACCESSORIES				
GROUNDING	1 LS	875.00	875	.06
** TOTAL GROUNDING ACCESSORIES			875	.06
LIGHTING FIXTURES				
FIXTURE C ,VAR 120 V	12 EACH	53.83	646	.04
FIXTURE B ,VAR 277 V	150 EACH	107.73	16,160	1.15
** TOTAL LIGHTING FIXTURES			16,806	1.20
LAMPS FOR LIGHTING FIXT.				
***FLUORESCENT LAMPS ***	****			
F40T12/CW/RS/ENERGY SAV	600 EACH	5.22	3,136	.22
** TOTAL LAMPS FOR LIGHTING FIXT.			3,136	.22
WIRING DEVICES				
NEMA 5-15R,DUPLEX,IVORY	60 EACH	13.53	812	.05
3/8" PHONE PLATE,PLAST,IV	40 EACH	8.02	321	.02
** TOTAL WIRING DEVICES			1,133	.08
DISCONNECT DEVICES				
ENCL CB,600/3/50AF,NEMA-1	4 EACH	473.75	1,895	.13
ENCL CB,600/3/225AF,NEMA-1	1 EACH	1,304.00	1,304	.09
** TOTAL DISCONNECT DEVICES			3,199	.22

File ID. 2568
Project No.
14,000 SQFT

MANAGEMENT COMPUTER CONTROLS, INC.
2881 DIRECTORS COVE
MEMPHIS, TENNESSEE 38131
WHATCO PROJECT
OFFICE

Description	Qty	Unit	Unit Price	Total Price	14,000 $/SQFT
ELEC SYSTEMS & EQUIPMENT					
STARTERS					
3P,POLYPH,NEMA-1,SIZE 2	4	EACH	797.25	3,189	.22
3P,POLYPH,NEMA-1,SIZE 3	1	EACH	1,060.00	$1,060	.07
** TOTAL STARTERS				$4,249	.30
*** TOTAL ELEC SYSTEMS & EQUIPMENT				$240,586	17.18
**** TOTAL OFFICE				$1,407,242	100.51

FILE ID - 2568
PROJECT JOB NO -
PROJECT NAME - WHATCO PROJECT
PROJECT SIZE - 50,000 SQFT

E S T I M A T E   D E T A I L   R E C A P

** SECTION, ITEM CODE SEQUENCE **

PAGE - 1
DATE -
TIME -

MANAGEMENT COMPUTER CONTROLS, INC.
2881 DIRECTORS COVE
MEMPHIS, TENNESSEE 38131

DESCRIPTION	SECTION QUANTITY UM	TOTAL COSTS LABOR	MAT'L	EQPT.	SUB.	TOTAL	TAX & INS INCLUDED COST/UNIT PRICE	TOTAL PRICE
SECTION 01 PLANT	30,000 SQFT	583,061	1119,975	45,667	434,780	2183,483	81.194	2435,823
SECTION 02 WAREHOUSE	6,000 SQFT	139,152	207,545	9,078	137,812	493,587	91.419	548,513
SECTION 03 OFFICE	14,000 SQFT	374,738	530,686	14,771	341,891	1262,086	100.517	1407,242
***** ESTIMATE TOTAL		1096,951	1858,206	69,516	914,483	3939,156	87.832	4391,578

FILE ID - 2568
PROJECT JOB NO -
PROJECT NAME - WHATCO PROJECT
PROJECT SIZE - 30,000 SQFT
SECTION - 01 PLANT

ESTIMATE DETAIL RECAP

** SECTION, ITEM CODE SEQUENCE **

PAGE - 1
DATE -
TIME -

MANAGEMENT COMPUTER CONTROLS, INC.
2881 DIRECTORS COVE
MEMPHIS, TENNESSEE 38131

DESCRIPTION	TOTAL COSTS					TAX & INS INCLUDED	
	LABOR	MAT'L	EQPT.	SUB.	TOTAL	UNIT SQFT PRICE	TOTAL PRICE
GENERAL REQUIREMENTS				431,030	431,030	14.368	431,030
CLEARING OF SITE	1,213		1,323		2,536	.096	2,875
EXCAV, GRADING & BACKFILL	28,899	8,295	18,728		55,922	2.157	64,712
CONCRETE FINISHING	15,522	519			16,041	.681	20,421
FORM WORK	2,134	794			2,928	.120	3,591
REINFORCING STEEL	15,896	10,915			26,811	1.072	32,163
CAST-IN-PLACE CONCRETE	21,198	68,759			89,957	3.386	101,567
CEMENTITIOUS DECKS	13,920	16,530			30,450	1.190	35,700
STRUCTURAL METALS	193,122	217,308	25,616		436,046	16.935	508,061
WATERPROOF & DAMPPROOF	3,990	750			4,740	.197	5,910
PREFORMED ROOFING & SIDING	78,660	119,070			197,730	7.653	229,590
ROOFING,SHEETMETAL&ACCESS	57,250	37,069			94,319	3.780	113,387
METAL DOORS & FRAMES	457	1,158			1,615	.061	1,837
SPECIAL DOORS	2,000	12,000			14,000	.518	15,550
HARDWARE	348	1,378			1,726	.065	1,937
PAINTING & WALL COVERING	6,711	1,886			8,597	.354	10,632
EQUIPMENT	3,337	13,230			16,567	.620	18,593
CONVEYING SYSTEMS	64,236	396,900			461,136	17.062	511,866
HVAC SYSTEMS & EQUIPMENT	35,441	171,200			206,641	7.548	226,438
ELEC SYSTEMS & EQUIPMENT	38,727	42,214		3,750	84,691	3.332	99,963
***** SECTION TOTAL	583,061	1119,975	45,667	434,780	2183,483	81.194	2435,823

FILE ID - 2568
PROJECT JOB NO -
PROJECT NAME - WHATCO PROJECT
PROJECT SIZE - 30,000 SQFT
SECTION - 01 PLANT

E S T I M A T E   D E T A I L   R E P O R T
** SECTION, ITEM CODE SEQUENCE **

PAGE - 1
DATE -
TIME -

MANAGEMENT COMPUTER CONTROLS, INC.
2881 DIRECTORS COVE
MEMPHIS, TENNESSEE 38131

REF NO.	S D SC	ELEM	ITEM CODE	DESCRIPTION	UNIT MEAS	QUANTITY	UNIT COSTS LABOR	UNIT COSTS MAT'L.	UNIT COSTS EQPT.	UNIT COSTS SUB.	UNIT COSTS TOTAL	TOTAL COSTS LABOR	TOTAL COSTS MAT'L	TOTAL COSTS EQPT.	TOTAL COSTS SUB.	TOTAL COSTS TOTAL	TAX & INS INCLUDED TOT UNIT PRICE	TAX & INS INCLUDED TOTAL PRICE
GENERAL REQUIREMENTS																		
				OTHER GENERAL REQUIREMENTS														
1	01		185.901	OH/PROFIT/GC/BOND @ 21.5%	LS	1				431,030	431,030				431,030	431,030	431,030	431,030
				** TOTAL OTHER GENERAL REQUIREMENTS											431,030	431,030		431,030
				*** TOTAL GENERAL REQUIREMENTS											431,030	431,030		431,030
CONTINGENCY																		
4	01		199.901	CONTINGENCY @ 0%														
				*** TOTAL CONTINGENCY														
CLEARING OF SITE																		
7	01		211.900	SITE PREP (CLEAR)	ACRE	1	1212.750		1323.000		2535.750	1,213		1,323		2,536	2875.000	2,875
				*** TOTAL CLEARING OF SITE								1,213		1,323		2,536		2,875
EXCAV, GRADING & BACKFILL																		
				SITE GRADING														
8	01		221.011	EXCAV-LOAD-HAUL SURP 3-4MILES	CUYD	3,345	3.027		3.638		6.665	10,125		12,169		22,294	7.513	25,131
				** TOTAL SITE GRADING								10,125		12,169		22,294		25,131
				EXCAVATION & BACKFILL														
13	01	0112	222.302	EXCAVATE COL FTG	CUYD	84	8.875		1.268		10.143	746		107		853	12.631	1,061
23	01	0112	222.332	FINE GRADE COL FTG	SQFT	900	.529				.529	476				476	.677	609
24	01	0211	222.334	FINE GRADE SLAB ON GRADE	SQFT	30,000	.370				.370	11,100				11,100	.474	14,220
27	01		222.900	STRUCTURAL FILL	CUYD	1,672	3.859	4.961	3.859		12.679	6,452	8,295	6,452		21,199	14.169	23,691
				** TOTAL EXCAVATION & BACKFILL								18,774	8,295	6,559		33,628		39,581
				*** TOTAL EXCAV, GRADING & BACKFILL								28,899	8,295	18,728		55,922		64,712
CONCRETE FINISHING																		
				CONCRETE FINISH														
31	01	0211	301.019	TROWEL CEMENT FINISH	SQFT	30,000	.386				.386	11,580				11,580	.494	14,820
34	01	0312	301.020	POINT & PATCH	SQFT	800	.128	.011			.139	102	9			111	.176	141
36	01	0211	301.050	PROTECT & CURE	SQFT	30,000	.128	.017			.145	3,840	510			4,350	.182	5,460
				** TOTAL CONCRETE FINISH								15,522	519			16,041		20,421
				*** TOTAL CONCRETE FINISHING								15,522	519			16,041		20,421

FILE ID - 2568
PROJECT JOB NO -
PROJECT NAME - WHATCO PROJECT
PROJECT SIZE - 30,000 SQFT
SECTION - 01 PLANT

E S T I M A T E   D E T A I L   R E P O R T
** SECTION, ITEM CODE SEQUENCE **

PAGE - 2
DATE -
TIME -

MANAGEMENT COMPUTER CONTROLS, INC.
2881 DIRECTORS COVE
MEMPHIS, TENNESSEE 38131

REF NO.	S D SC ELEM	ITEM CODE	DESCRIPTION	UNIT MEAS	QUANTITY	UNIT COSTS LABOR	UNIT COSTS MAT'L.	UNIT COSTS EQPT.	UNIT COSTS SUB.	UNIT COSTS TOTAL	TOTAL COSTS LABOR	TOTAL COSTS MAT'L	TOTAL COSTS EQPT.	TOTAL COSTS SUB.	TOTAL COSTS TOTAL	TAX & INS INCLUDED TOT UNIT PRICE	TAX & INS INCLUDED TOTAL PRICE
			FORM WORK														
			FORM WORK														
43	01 0211	310.105	SLAB ON GRADE EDGE FORMS	SQFT	800	2.668	.992			3.660	2,134	794			2,928	4.489	3,591
			** TOTAL FORM WORK								2,134	794			2,928		3,591
			*** TOTAL FORM WORK								2,134	794			2,928		3,591
			REINFORCING STEEL														
			RE-BARS														
44	01 0211	321.009	RE-STEEL @ SLAB ON GRADE	CWT	387	35.412	24.255			59.667	13,704	9,387			23,091	71.584	27,703
50	01 0112	321.120	RE-STEEL @ COL FTG	CWT	63	34.798	24.255			59.053	2,192	1,528			3,720	70.794	4,460
			** TOTAL RE-BARS								15,896	10,915			26,811		32,163
			*** TOTAL REINFORCING STEEL								15,896	10,915			26,811		32,163
			CAST-IN-PLACE CONCRETE														
			CAST-IN-PLACE CONCRETE														
56	01 0112	333.010	CONCRETE @ COL FTG	CUYD	88	12.156	51.652			63.808	1,070	4,545			5,615	71.466	6,289
60	01 0211	333.035	CONCRETE @ SLAB ON GRADE	CUYD	1,145	12.381	51.652			64.033	14,176	59,142			73,318	71.761	82,166
64	01	333.910	LOADING DOCK SLAB	SQFT	4,000	1.488	1.268			2.756	5,952	5,072			11,024	3.278	13,112
			** TOTAL CAST-IN-PLACE CONCRETE								21,198	68,759			89,957		101,567
			*** TOTAL CAST-IN-PLACE CONCRETE								21,198	68,759			89,957		101,567
			CEMENTITIOUS DECKS														
66	01 0321	350.100	LTWT CONC ROOF FILL	SQFT	30,000	.464	.551			1.015	13,920	16,530			30,450	1.190	35,700
			*** TOTAL CEMENTITIOUS DECKS								13,920	16,530			30,450		35,700
			STRUCTURAL METALS														
			STRUCTURAL METALS														
73	01	501.011	STRUCTURAL STEEL FRAME	CWT	3,060	35.756	38.588	5.513		79.857	109,413	118,079	16,870		244,362	93.053	284,742
76	01	501.051	MISC. SUPPORT FRAMING	CWT	122	71.513	49.613	5.513		126.639	8,725	6,053	673		15,451	150.762	18,393
78	01	501.900	BASE PLATE	LBS	3,708	.551	.496			1.047	2,043	1,839			3,882	1.242	4,605
81	01	510.901	SANDBLAST	SQFT	22,387	.276	.132	.044		.452	6,179	2,955	985		10,119	.540	12,089
84	01	510.902	PRIMER	SQFT	22,387	.088	.055	.011		.154	1,970	1,231	246		3,447	.184	4,119
87	01	510.903	INTERMEDIATE COAT	SQFT	22,387	.121	.132	.011		.264	2,709	2,955	246		5,910	.309	6,917
90	01	510.904	TOP COAT (ELEVATED)	SQFT	22,387	.276	.165	.011		.452	6,179	3,694	246		10,119	.543	12,156
			** TOTAL STRUCTURAL METALS								137,218	136,806	19,266		293,290		34
			** TOTAL STRUCTURAL METALS								137,218	136,806	19,266		293,290		343,021

FILE ID - 2568
PROJECT JOB NO -
PROJECT NAME - WHATCO PROJECT
PROJECT SIZE - 30,000 SQFT
SECTION - 01 PLANT

ESTIMATE DETAIL REPORT

** SECTION, ITEM CODE SEQUENCE **

DATE -
TIME -

MANAGEMENT COMPUTER CONTROLS, INC.
2881 DIRECTORS COVE
MEMPHIS, TENNESSEE 38131

REF NO.	S D	SC	ELEM	ITEM CODE	DESCRIPTION	UNIT MEAS	QUANTITY	UNIT COSTS LABOR	UNIT COSTS MAT'L.	UNIT COSTS EQPT.	UNIT COSTS SUB.	UNIT COSTS TOTAL	TOTAL COSTS LABOR	TOTAL COSTS MAT'L	TOTAL COSTS EQPT.	TOTAL COSTS SUB.	TOTAL COSTS TOTAL	TAX & INS INCLUDED TOT UNIT PRICE	TAX & INS INCLUDED TOTAL PRICE
STRUCTURAL METALS																			
					METAL JOIST														
93		01	0312	521.011	STEEL JOIST	CWT	1,440	29.447	36.383	4.410		70.240	42,404	52,392	6,350		101,146	81.486	117,340
					** TOTAL METAL JOIST								42,404	52,392	6,350		101,146		117,340
					METAL DECKING														
96		01	0312	532.010	1-1/2" METAL DECK	SQFT	30,000	.450	.937			1.387	13,500	28,110			41,610	1.590	47,700
					** TOTAL METAL DECKING								13,500	28,110			41,610		47,700
					*** TOTAL STRUCTURAL METALS								193,122	217,308	25,616		436,046		508,061
WATERPROOF & DAMPPROOF																			
104		01	0211	710.000	VAPOR BARRIER @ SLAB	SQFT	30,000	.133	.025			.158	3,990	750			4,740	.197	5,910
					*** TOTAL WATERPROOF & DAMPPROOF								3,990	750			4,740		5,910
PREFORMED ROOFING & SIDING																			
111		01	0411	740.010	INSULATED METAL SIDING	SQFT	18,000	4.370	6.615			10.985	78,660	119,070			197,730	12.755	229,590
					*** TOTAL PREFORMED ROOFING & SIDING								78,660	119,070			197,730		229,590
ROOFING, SHEETMETAL&ACCESS																			
					ROOFING & ROOF INSULATION														
113		01	0501	751.002	3 PLY TAR& GRAVEL ROOFING	SQS	300	59.515	50.081			109.596	17,855	15,024			32,879	130.393	39,118
116		01	0503	755.003	2" FIBER BD ROOF INSULATION	SQFT	30,000	1.037	.659			1.696	31,110	19,770			50,880	2.040	61,200
					** TOTAL ROOFING & ROOF INSULATION								48,965	34,794			83,759		100,318
					FLASHING & SHEETMETAL														
119		01	0504	762.040	24 GA GALV IRON SHEETMETAL	SQFT	800	7.513	1.243			8.756	6,010	994			7,004	10.964	8,771
122		01	0504	762.064	6" GALV DOWNSPOUT	LNFT	245	3.807	2.143			5.950	933	525			1,458	7.192	1,762
125		01	0504	762.069	6" GALV GUTTER	LNFT	400	3.356	1.889			5.245	1,342	756			2,098	6.340	2,536
					** TOTAL FLASHING & SHEETMETAL								8,285	2,275			10,560		13,069
					*** TOTAL ROOFING, SHEETMETAL&ACCESS								57,250	37,069			94,319		113,387
METAL DOORS & FRAMES																			
128		01	0616	805.100	HM FRAME	EACH	5	51.543	66.150			117.693	258	331			589	137.600	688
131		01	0616	805.200	HM DOOR	EACH	5	39.762	165.375			205.137	199	827			1,026	229.800	1,149
					*** TOTAL METAL DOORS & FRAMES								457	1,158			1,615		1,837

FILE ID - 2568
PROJECT JOB NO -

PROJECT JOB NO - ** SECTION, ITEM CODE SEQUENCE **

ESTIMATE DETAIL REPORT

PAGE - 4

** SECTION, ITEM CODE SEQUENCE **

PROJECT JOB NO -
PROJECT NAME - WHATCO PROJECT
PROJECT SIZE - 30,000 SQFT
SECTION - 01 PLANT

DATE -
TIME -
MANAGEMENT COMPUTER CONTROLS, INC.
2881 DIRECTORS COVE
MEMPHIS, TENNESSEE 38131

REF NO.	S D	SC	ELEM	ITEM CODE	DESCRIPTION	UNIT MEAS	QUANTITY	UNIT COSTS LABOR	UNIT COSTS MAT'L.	UNIT COSTS EQPT.	UNIT COSTS SUB.	UNIT COSTS TOTAL	TOTAL COSTS LABOR	TOTAL COSTS MAT'L	TOTAL COSTS EQPT.	TOTAL COSTS SUB.	TOTAL COSTS TOTAL	TAX & INS INCLUDED TOT UNIT PRICE	TAX & INS INCLUDED TOTAL PRICE
SPECIAL DOORS																			
135		01		830.900	STEEL ROLL-UP DOOR	EACH	4	500.000	3000.000			3500.000	2,000	12,000			14,000	3887.500	15,550
					*** TOTAL SPECIAL DOORS								2,000	12,000			14,000		15,550
HARDWARE																			
138		01	0616	871.027	FINISH HARDWARE ALLOWANCE	OPNG	5	69.583	275.625			345.208	348	1,378			1,726	387.400	1,937
					*** TOTAL HARDWARE								348	1,378			1,726		1,937
PAINTING & WALL COVERING																			
					PAINTING														
148		01	0423	990.008	PAINT EXTERIOR DOOR	SIDE	10	29.607	2.205			31.812	296	22			318	40.300	403
151		01	0616	990.032	PAINT DOOR FRAME	EACH	5	23.398	2.756			26.154	117	14			131	33.000	165
155		01	0621	990.081	PAINT MAS-CONC 3 CTS	SQS	60	69.082	19.845			88.927	4,145	1,191			5,336	109.917	6,595
158		01	0411	992.000	EXTERIOR PAINTING	SQS	46	46.797	14.333			61.130	2,153	659			2,812	75.413	3,469
					** TOTAL PAINTING								6,711	1,886			8,597		10,632
					*** TOTAL PAINTING & WALL COVERING								6,711	1,886			8,597		10,632
EQUIPMENT																			
					LOADING DOCK EQUIPMENT														
163		01	1116	1116.000	DOCK LEVELER	EACH	4	834.314	3307.500			4141.814	3,337	13,230			16,567	4648.250	18,593
					** TOTAL LOADING DOCK EQUIPMENT								3,337	13,230			16,567		18,593
					*** TOTAL EQUIPMENT								3,337	13,230			16,567		18,593
CONVEYING SYSTEMS																			
					HOISTS & CRANES														
166		01		1430.900	BRIDGE CRANE	EACH	1	64,236	396,900			461,136	64,236	396,900			461,136	511,866	511,866
					** TOTAL HOISTS & CRANES								64,236	396,900			461,136		511,866
					*** TOTAL CONVEYING SYSTEMS								64,236	396,900			461,136		511,866
HVAC SYSTEMS & EQUIPMENT																			
					PIPE & FITTINGS														
211		01	0824	1560.001	SCH 40 STEEL PIPE	****													
214		01	0824	1560.002	BLK T&C	****													
217		01	0824	1560.003	PIPE,2"	LNFT	1,200	3.848	2.811			6.659	4,618	3,373			7,991	7.508	9,009
220		01	0824	1560.004	BLACK PIPE,P.E.	****													

FILE ID - 2568
PROJECT JOB NO -
PROJECT NAME - WHATCO PROJECT
PROJECT SIZE - 30,000 SQFT
SECTION - 01 PLANT

ESTIMATE DETAIL REPORT

** SECTION, ITEM CODE SEQUENCE **

DATE -
TIME -

MANAGEMENT COMPUTER CONTROLS, INC.
2881 DIRECTORS COVE
MEMPHIS, TENNESSEE 38131

REF NO.	S D SC	ELEM	ITEM CODE	DESCRIPTION	UNIT MEAS	QUANTITY	UNIT COSTS LABOR	UNIT COSTS MAT'L.	UNIT COSTS EQPT.	UNIT COSTS SUB.	UNIT COSTS TOTAL	TOTAL COSTS LABOR	TOTAL COSTS MAT'L	TOTAL COSTS EQPT.	TOTAL COSTS SUB.	TOTAL COSTS TOTAL	TAX & INS INCLUDED TOT UNIT PRICE	TAX & INS INCLUDED TOTAL PRICE
				HVAC SYSTEMS & EQUIPMENT														
				PIPE & FITTINGS														
221	01	0824	1560.005	PIPE,3"	LNFT	700	7.132	5.391			12.523	4,992	3,774			8,766	14.109	9,876
222	01	0824	1560.006	STEEL FLANGE, 150#	****													
224	01	0824	1560.007	WELD-NECK,3"	EACH	12	68.297	42.171			110.468	820	506			1,326	124.917	1,499
225	01	0824	1560.008	THREADED,2"	EACH	8	32.109	56.228			88.337	257	450			707	98.125	785
227	01	0824	1560.009	BOLT & GASKET SET	****													
229	01	0824	1560.010	FLANGE PACK,2"	EACH	4	5.772	3.980			9.752	23	16			39	11.000	44
231	01	0824	1560.011	FLANGE PACK,3"	EACH	6	6.833	5.402			12.235	41	32			73	13.833	83
				** TOTAL PIPE & FITTINGS								10,751	8,151			18,902		21,296
				VALVES														
232	01	0824	1575.001	IRON BODY VALVES	****													
234	01	0824	1575.002	GATE, FLANGED	****													
237	01	0824	1575.003	125#,2"	EACH	4	36.188	225.737			261.925	145	903			1,048	286.250	1,145
238	01	0824	1575.004	125#,3"	EACH	4	45.808	264.600			310.408	183	1,058			1,241	339.750	1,359
239	01	0824	1575.005	SWING CHECK, FLANGED	****													
240	01	0824	1575.006	125#,3"	EACH	2	40.036	231.525			271.561	80	463			543	297.000	594
				** TOTAL VALVES								408	2,424			2,832		3,098
				PIPE HANGERS & SUPPORTS														
241	01	0824	1580.001	PIPE HANGERS, STEEL	****													
244	01	0824	1580.002	WITH 3' ROD & C-CLAMP	****													
248	01	0824	1580.004	CLEVIS,3"	EACH	70	27.166	6.924			34.090	1,902	485			2,387	39.014	2,731
250	01	0824	1580.005	CLEVIS,6"	EACH	120	30.450	9.047			39.497	3,654	1,086			4,740	45.117	5,414
252	01	0824	1580.006	INSULATION SHIELD,6"	EACH	120	6.335	13.230			19.565	760	1,588			2,348	21.675	2,601
				** TOTAL PIPE HANGERS & SUPPORTS								6,316	3,159			9,475		10,746
				PIPE INSULATION														
256	01	0824	1583.001	FIBERGLASS INSULATION	****													
258	01	0824	1583.002	ASJ, 2" THICK	****													
261	01	0824	1583.003	PIPE,2"	LNFT	1,200	1.500	5.347			6.847	1,800	6,416			8,216	7.528	9,034
264	01	0824	1583.004	FLANGES,2"	EACH	8	12.034	9.096			21.130	96	73			169	23.875	191
				** TOTAL PIPE INSULATION								1,896	6,489			8,385		9,225
				MAJOR HEATING EQUIPMENT														
269	01	0822	1588.004	SPACE HEATERS	****													
271	01	0822	1588.005	PROPELLER,STEAM&HW,HORIZ	****													

FILE ID - 2568
PROJECT JOB NO -
PROJECT NAME - WHATCO PROJECT
PROJECT SIZE - 30,000 SQFT
SECTION - 01 PLANT

ESTIMATE DETAIL REPORT
** SECTION, ITEM CODE SEQUENCE **

PAGE - 6
DATE -
TIME -

MANAGEMENT COMPUTER CONTROLS, INC.
2881 DIRECTORS COVE
MEMPHIS, TENNESSEE 38131

REF NO.	S D SC ELEM	ITEM CODE	DESCRIPTION	UNIT MEAS	QUANTITY	UNIT COSTS LABOR	UNIT COSTS MAT'L.	UNIT COSTS EQPT.	UNIT COSTS SUB.	UNIT COSTS TOTAL	TOTAL COSTS LABOR	TOTAL COSTS MAT'L	TOTAL COSTS EQPT.	TOTAL COSTS SUB.	TOTAL COSTS TOTAL	TAX & INS INCLUDED TOT UNIT PRICE	TAX & INS INCLUDED TOTAL PRICE
			HVAC SYSTEMS & EQUIPMENT														
			MAJOR HEATING EQUIPMENT														
274	01 0822	1588.006	UH#1 3300 CFM	EACH	20	164.556	1137.780			1302.336	3,291	22,756			26,047	1422.550	28,451
			** TOTAL MAJOR HEATING EQUIPMENT								3,291	22,756			26,047		28,451
			MAJOR COOLING EQUIPMENT														
281	01 0824	1589.004	AIR HANDLING UNITS	****													
284	01 0824	1589.005	LOW-PRESS,HORIZ,FLR MTD	****													
287	01 0824	1589.006	AHU#3 18620 CFM	EACH	10	1277.864	12,822			14,099	12,779	128,221			141,000	15,362	153,622
			** TOTAL MAJOR COOLING EQUIPMENT								12,779	128,221			141,000		153,622
			*** TOTAL HVAC SYSTEMS & EQUIPMENT								35,441	171,200			206,641		226,438
			ELEC SYSTEMS & EQUIPMENT														
			LOW VOLTAGE CONDUCTORS														
295	01 0921	1601.029	CU WIRE THHN-THWN-12 AWG	LNFT	15,600	.226	.095			.321	3,526	1,482			5,008	.398	6,209
298	01 0912	1601.033	CU WIRE THHN-THWN-4 AWG	LNFT	618	.390	.531			.921	241	328			569	1.083	669
300	01 0921	1607.761	22/2 AWG	LNFT	600	.494	.131			.625	296	79			375	.788	473
303	01 0921	1609.002	TERMIN LUGS-CU WIRE 12 AWG	EACH	20	4.111	.240			4.351	82	5			87	5.650	113
306	01 0912	1609.006	TERMIN LUGS-CU WIRE 4 AWG	EACH	40	14.390	5.626			20.016	576	225			801	24.900	996
			** TOTAL LOW VOLTAGE CONDUCTORS								4,721	2,119			6,840		8,460
			CONDUIT,FITTINGS & ACCESS														
309	01 0912	1610.100	GALVANIZED RIGID STEEL-GRS	****													
313	01 0912	1610.130	COND-O'HEAD/WALL 2"	LNFT	1,500	5.756	1.816			7.572	8,634	2,724			11,358	9.497	14,246
315	01 0912	1610.132	COND-O'HEAD/WALL 3"	LNFT	1,800	8.223	3.858			12.081	14,801	6,944			21,745	14.929	26,873
317	01 0921	1614.520	4" OCTAGON BOX 1-1/2" DEEP	EACH	60	16.445	1.634			18.079	987	98			1,085	23.317	1,399
320	01 0921	1614.530	4" SQUARE BOX 1-1/2"DEEP	EACH	10	16.445	1.943			18.388	164	19			183	23.600	236
323	01 0921	1614.536	4" SQ BOX, 1G PLASTER RING	EACH	10	5.139	2.524			7.663	51	25			76	9.400	94
326	01 0921	1614.595	OUTLET BOX CLAMP	EACH	10		.625			.625		6			6	.700	7
329	01	1615.039	12"X12"X8" PULL BOX	EACH	12	61.669	32.500			94.169	740	390			1,130	115.750	1,389
			** TOTAL CONDUIT,FITTINGS & ACCESS								25,377	10,206			35,583		44,244
			GROUNDING ACCESSORIES														
339	01	1640.950	GROUNDING	LS	1				3750.000	3750.000				3,750	3,750	3750.000	3,750
			** TOTAL GROUNDING ACCESSORIES											3,750	3,750		3,750

FILE ID - 2568
PROJECT JOB NO -
PROJECT NAME - WHATCO PROJECT
PROJECT SIZE - 30,000 SQFT
SECTION - 01 PLANT

ESTIMATE DETAIL REPORT

** SECTION, ITEM CODE SEQUENCE **

PAGE - 7
DATE -
TIME -

MANAGEMENT COMPUTER CONTROLS, INC.
2881 DIRECTORS COVE
MEMPHIS, TENNESSEE 38131

REF NO.	S D SC ELEM	ITEM CODE	DESCRIPTION	UNIT MEAS	QUANTITY	UNIT COSTS LABOR	UNIT COSTS MAT'L.	UNIT COSTS EQPT.	UNIT COSTS SUB.	UNIT COSTS TOTAL	TOTAL COSTS LABOR	TOTAL COSTS MAT'L	TOTAL COSTS EQPT.	TOTAL COSTS SUB.	TOTAL COSTS TOTAL	TAX & INS INCLUDED TOT UNIT PRICE	TAX & INS INCLUDED TOTAL PRICE
ELEC SYSTEMS & EQUIPMENT																	
			LIGHTING FIXTURES														
342	01 0922	1643.000	FIXTURE EXIT ,VAR 120 V	EACH	8	28.779	62.500			91.279	230	500			730	105.125	841
345	01 0922	1643.000	FIXTURE A ,VAR 277 V	EACH	36	102.781	312.500			415.281	3,700	11,250			14,950	471.361	16,969
			** TOTAL LIGHTING FIXTURES								3,930	11,750			15,680		17,810
			LAMPS FOR LIGHTING FIXT.														
350	01 0922	1644.560	***HIGH PRESSURE SODIUM***	****													
352	01 0922	1644.571	E-25 MOG LU1000	EACH	36		209.988			209.988		7,560			7,560	226.250	8,145
354	01 0922	1644.801	6'--3/8" W/2-#16 TFF	SET	36	10.279	3.438			13.717	370	124			494	17.167	618
			** TOTAL LAMPS FOR LIGHTING FIXT.								370	7,684			8,054		8,763
			WIRING DEVICES														
358	01 0921	1645.024	NEMA 5-20R,DUPLEX,IVORY	EACH	60	12.334	3.439			15.773	740	206			946	19.850	1,191
360	01 0921	1646.781	3/8" PHONE PLATE,PLAST,IV	EACH	10	4.111	2.450			6.561	41	25			66	8.000	80
			** TOTAL WIRING DEVICES								781	231			1,012		1,271
			DISCONNECT DEVICES														
363	01 0923	1647.011	ENCL CB,600/3/50AF,NEMA-1	EACH	10	67.836	357.325			425.161	678	3,573			4,251	473.900	4,739
366	01 0923	1647.013	ENCL CB,600/3/100AF,NEMA-1	EACH	1	157.050	370.000			527.050	157	370			527	605.000	605
			** TOTAL DISCONNECT DEVICES								835	3,943			4,778		5,344
			STARTERS														
368	01 0923	1649.126	3P,POLYPH,NEMA-1,SIZE 2	EACH	10	226.119	464.975			691.094	2,261	4,650			6,911	797.200	7,972
372	01 0923	1649.128	3P,POLYPH,NEMA-1,SIZE 4	EACH	1	452.238	1630.900			2083.138	452	1,631			2,083	2349.000	2,349
			** TOTAL STARTERS								2,713	6,281			8,994		10,321
			*** TOTAL ELEC SYSTEMS & EQUIPMENT								38,727	42,214		3,750	84,691		99,963
			**** TOTAL PLANT								583,061	1119,975	45,667	434,780	2183,483		2435,823

FILE ID - 2568
PROJECT JOB NO -
PROJECT NAME - WHATCO PROJECT
PROJECT SIZE - 6,000 SQFT
SECTION - 02 WAREHOUSE

ESTIMATE DETAIL RECAP
** SECTION, ITEM CODE SEQUENCE **

PAGE - 1
DATE -
TIME -

MANAGEMENT COMPUTER CONTROLS, INC.
2881 DIRECTORS COVE
MEMPHIS, TENNESSEE 38131

DESCRIPTION	TOTAL COSTS LABOR	MAT'L	EQPT.	SUB.	TOTAL	TAX & INS INCLUDED UNIT SQFT PRICE	TOTAL PRICE
GENERAL REQUIREMENTS				97,062	97,062	16.177	97,062
EXCAV, GRADING & BACKFILL	5,716	1,657	3,734		11,107	2.141	12,846
CONCRETE FINISHING	3,128	106			3,234	.686	4,116
FORM WORK	907	337			1,244	.254	1,526
REINFORCING STEEL	2,224	1,528			3,752	.750	4,501
CAST-IN-PLACE CONCRETE	2,847	9,908			12,755	2.395	14,369
CEMENTITIOUS DECKS	2,784	3,306			6,090	1.190	7,140
STRUCTURAL METALS	49,206	47,320	5,344		101,870	19.926	119,553
WATERPROOF & DAMPPROOF	798	150			948	.197	1,182
PREFORMED ROOFING & SIDING	26,220	39,690			65,910	12.755	76,530
ROOFING,SHEETMETAL&ACCESS	13,701	7,845			21,546	4.338	26,025
METAL DOORS & FRAMES	183	463			646	.123	735
SPECIAL DOORS	1,000	6,000			7,000	1.296	7,775
HARDWARE	139	551			690	.129	775
PAINTING & WALL COVERING	3,728	1,059			4,787	.987	5,919
EQUIPMENT	1,669	6,615		40,000	48,284	8.216	49,297
HVAC SYSTEMS & EQUIPMENT	12,264	68,006			80,270	14.641	87,844
ELEC SYSTEMS & EQUIPMENT	12,638	13,004		750	26,392	5.220	31,318
***** SECTION TOTAL	139,152	207,545	9,078	137,812	493,587	91.419	548,513

FILE ID - 2568
PROJECT JOB NO -
PROJECT NAME - WHATCO PROJECT
PROJECT SIZE - 6,000 SQFT
SECTION - 02 WAREHOUSE

ESTIMATE DETAIL REPORT

** SECTION, ITEM CODE SEQUENCE **

PAGE - 1
DATE -
TIME -

MANAGEMENT COMPUTER CONTROLS, INC.
2881 DIRECTORS COVE
MEMPHIS, TENNESSEE 38131

REF NO.	S D	SC ELEM	ITEM CODE	DESCRIPTION	UNIT MEAS	QUANTITY	UNIT COSTS LABOR	UNIT COSTS MAT'L.	UNIT COSTS EQPT.	UNIT COSTS SUB.	UNIT COSTS TOTAL	TOTAL COSTS LABOR	TOTAL COSTS MAT'L	TOTAL COSTS EQPT.	TOTAL COSTS SUB.	TOTAL COSTS TOTAL	TAX & INS INCLUDED TOT UNIT PRICE	TAX & INS INCLUDED TOTAL PRICE
GENERAL REQUIREMENTS																		
				OTHER GENERAL REQUIREMENTS														
2	02		185.902	OH/PROFIT/GC/BOND @ 21.5%	LS	1				97,062	97,062				97,062	97,062	97,062	97,062
				** TOTAL OTHER GENERAL REQUIREMENTS											97,062	97,062		97,062
				*** TOTAL GENERAL REQUIREMENTS											97,062	97,062		97,062
CONTINGENCY																		
5	02		199.902	CONTINGENCY @ 0%														
				*** TOTAL CONTINGENCY														
EXCAV, GRADING & BACKFILL																		
				SITE GRADING														
9	02		221.011	EXCAV-LOAD-HAUL SURP 3-4MILES	CUYD	669	3.027		3.638		6.665	2,025		2,434		4,459	7.513	5,026
				** TOTAL SITE GRADING								2,025		2,434		4,459		5,026
				EXCAVATION & BACKFILL														
11	02	0112	222.301	EXCAVATE COLUMN FOOTING	CUYD	9	8.875		1.268		10.143	80		11		91	12.556	113
21	02	0112	222.331	FINE GRADE COLUMN FOOTING	SQFT	192	.529				.529	102				102	.677	130
25	02	0211	222.334	FINE GRADE SLAB ON GRADE	SQFT	6,000	.370				.370	2,220				2,220	.474	2,844
28	02		222.900	STRUCTURAL FILL	CUYD	334	3.859	4.961	3.859		12.679	1,289	1,657	1,289		4,235	14.171	4,733
				** TOTAL EXCAVATION & BACKFILL								3,691	1,657	1,300		6,648		7,820
				*** TOTAL EXCAV, GRADING & BACKFILL								5,716	1,657	3,734		11,107		12,846
CONCRETE FINISHING																		
				CONCRETE FINISH														
32	02	0211	301.019	TROWEL CEMENT FINISH	SQFT	6,000	.386				.386	2,316				2,316	.494	2,964
35	02	0312	301.020	POINT & PATCH	SQFT	340	.128	.011			.139	44	4			48	.176	60
37	02	0211	301.050	PROTECT & CURE	SQFT	6,000	.128	.017			.145	768	102			870	.182	1,092
				** TOTAL CONCRETE FINISH								3,128	106			3,234		4,116
				*** TOTAL CONCRETE FINISHING								3,128	106			3,234		4,116

FORM WORK

FILE ID - 2568
PROJECT JOB NO -
PROJECT NAME - WHATCO PROJECT
PROJECT SIZE - 6,000 SQFT
SECTION - 02 WAREHOUSE

ESTIMATE DETAIL REPORT
** SECTION, ITEM CODE SEQUENCE **

PAGE - 2
DATE -
TIME -

MANAGEMENT COMPUTER CONTROLS, INC.
2881 DIRECTORS COVE
MEMPHIS, TENNESSEE 38131

REF NO.	S D	SC ELEM	ITEM CODE	DESCRIPTION	UNIT MEAS	QUANTITY	UNIT COSTS LABOR	MAT'L.	EQPT.	SUB.	TOTAL	TOTAL COSTS LABOR	MAT'L	EQPT.	SUB.	TOTAL	TAX & INS INCLUDED TOT UNIT PRICE	TOTAL PRICE
FORM WORK																		
				FORM WORK														
42	02	0211	310.100	SLAB ON GRADE EDGE FORMS	LNFT	340	2.668	.992			3.660	907	337			1,244	4.488	1,526
				** TOTAL FORM WORK								907	337			1,244		1,526
				*** TOTAL FORM WORK								907	337			1,244		1,526
REINFORCING STEEL																		
				RE-BARS														
45	02	0211	321.009	RE-STEEL @ SLAB ON GRADE	CWT	52	35.412	24.255			59.667	1,841	1,261			3,102	71.577	3,722
48	02	0112	321.110	RE-STEEL @ COLUMN FOOTING	CWT	11	34.798	24.255			59.053	383	267			650	70.818	779
				** TOTAL RE-BARS								2,224	1,528			3,752		4,501
				*** TOTAL REINFORCING STEEL								2,224	1,528			3,752		4,501
CAST-IN-PLACE CONCRETE																		
				CAST-IN-PLACE CONCRETE														
54	02	0112	333.005	CONCRETE @ COLUMN FTG	CUYD	10	12.156	51.652			63.808	122	517			639	71.500	715
61	02	0211	333.035	CONCRETE @ SLAB ON GRADE	CUYD	172	12.381	51.652			64.033	2,130	8,884			11,014	71.762	12,343
65	02		333.910	LOADING DOCK SLAB	SQFT	400	1.488	1.268			2.756	595	507			1,102	3.278	1,311
				** TOTAL CAST-IN-PLACE CONCRETE								2,847	9,908			12,755		14,369
				*** TOTAL CAST-IN-PLACE CONCRETE								2,847	9,908			12,755		14,369
CEMENTITIOUS DECKS																		
67	02	0321	350.100	LTWT CONC ROOF FILL	SQFT	6,000	.464	.551			1.015	2,784	3,306			6,090	1.190	7,140
				*** TOTAL CEMENTITIOUS DECKS								2,784	3,306			6,090		7,140
STRUCTURAL METALS																		
				STRUCTURAL METALS														
74	02		501.011	STRUCTURAL STEEL FRAME	CWT	400	35.756	38.588	5.513		79.857	14,302	15,435	2,205		31,942	93.053	37,221
77	02		501.051	MISC. SUPPORT FRAMING	CWT	270	71.513	49.613	5.513		126.639	19,309	13,396	1,489		34,194	150.759	40,705
79	02		501.900	BASE PLATE	LBS	504	.551	.496			1.047	278	250			528	1.242	626
82	02		510.901	SANDBLAST	SQFT	6,519	.276	.132	.044		.452	1,799	861	287		2,947	.540	3,520
85	02		510.902	PRIMER	SQFT	6,519	.088	.055	.011		.154	574	359	72		1,005	.184	1,200
88	02		510.903	INTERMEDIATE COAT	SQFT	6,519	.121	.132	.011		.264	789	861	72		1,722	.309	2,014
91	02		510.904	TOP COAT (ELEVATED)	SQFT	6,519	.276	.165	.011		.452	1,799	1,076	72		2,947	.543	3,540
				** TOTAL STRUCTURAL METALS								38,850	32,238	4,197		75,285		88,826

FILE ID - 256B
PROJECT JOB NO -
PROJECT NAME - WHATCO PROJECT
PROJECT SIZE - 6,000 SQFT
SECTION - 02 WAREHOUSE

ESTIMATE DETAIL REPORT

** SECTION, ITEM CODE SEQUENCE **

DATE -
TIME -

MANAGEMENT COMPUTER CONTROLS, INC.
2881 DIRECTORS COVE
MEMPHIS, TENNESSEE 38131

REF NO.	S D SC	ELEM	ITEM CODE	DESCRIPTION	UNIT MEAS	QUANTITY	UNIT COSTS LABOR	UNIT COSTS MAT'L.	UNIT COSTS EQPT.	UNIT COSTS SUB.	UNIT COSTS TOTAL	TOTAL COSTS LABOR	TOTAL COSTS MAT'L	TOTAL COSTS EQPT.	TOTAL COSTS SUB.	TOTAL COSTS TOTAL	TAX & INS INCLUDED TOT UNIT PRICE	TAX & INS INCLUDED TOTAL PRICE
				STRUCTURAL METALS														
				METAL JOIST														
94	02	0312	521.011	STEEL JOIST	CWT	260	29.447	36.383	4.410		70.240	7,656	9,460	1,147		18,263	81.488	21,187
				** TOTAL METAL JOIST								7,656	9,460	1,147		18,263		21,187
				METAL DECKING														
97	02	0312	532.010	1-1/2" METAL DECK	SQFT	6,000	.450	.937			1.387	2,700	5,622			8,322	1.590	9,540
				** TOTAL METAL DECKING								2,700	5,622			8,322		9,540
				*** TOTAL STRUCTURAL METALS								49,206	47,320	5,344		101,870		119,553
				WATERPROOF & DAMPPROOF														
105	02	0211	710.000	VAPOR BARRIER @ SLAB	SQFT	6,000	.133	.025			.158	798	150			948	.197	1,182
				*** TOTAL WATERPROOF & DAMPPROOF								798	150			948		1,182
				PREFORMED ROOFING & SIDING														
112	02	0411	740.010	INSULATED METAL SIDING	SQFT	6,000	4.370	6.615			10.985	26,220	39,690			65,910	12.755	76,530
				*** TOTAL PREFORMED ROOFING & SIDING								26,220	39,690			65,910		76,530
				ROOFING,SHEETMETAL&ACCESS														
				ROOFING & ROOF INSULATION														
114	02	0501	751.002	3 PLY TAR& GRAVEL ROOFING	SQS	60	59.515	50.081			109.596	3,571	3,005			6,576	130.400	7,824
117	02	0503	755.003	2" FIBER BD ROOF INSULATION	SQFT	6,000	1.037	.659			1.696	6,222	3,954			10,176	2.040	12,240
				** TOTAL ROOFING & ROOF INSULATION								9,793	6,959			16,752		20,064
				FLASHING & SHEETMETAL														
120	02	0504	762.040	24 GA GALV IRON SHEETMETAL	SQFT	440	7.513	1.243			8.756	3,306	547			3,853	10.961	4,823
123	02	0504	762.064	6" GALV DOWNSPOUT	LNFT	70	3.807	2.143			5.950	266	150			416	7.186	503
126	02	0504	762.069	6" GALV GUTTER	LNFT	100	3.356	1.889			5.245	336	189			525	6.350	635
				** TOTAL FLASHING & SHEETMETAL								3,908	886			4,794		5,961
				*** TOTAL ROOFING,SHEETMETAL&ACCESS								13,701	7,845			21,546		26,025
				METAL DOORS & FRAMES														
129	02	0616	805.100	HM FRAME	EACH	2	51.543	66.150			117.693	103	132			235	137.500	275
132	02	0616	805.200	HM DOOR	EACH	2	39.762	165.375			205.137	80	331			411	230.000	460
				*** TOTAL METAL DOORS & FRAMES								183	463			646		735

FILE ID - 2568
PROJECT JOB NO -
PROJECT NAME - WHATCO PROJECT
PROJECT SIZE - 6,000 SQFT
SECTION - 02 WAREHOUSE

E S T I M A T E   D E T A I L   R E P O R T
** SECTION, ITEM CODE SEQUENCE **

DATE -
TIME -

MANAGEMENT COMPUTER CONTROLS, INC.
2881 DIRECTORS COVE
MEMPHIS, TENNESSEE 38131

REF NO.	S D SC	ELEM	ITEM CODE	DESCRIPTION	UNIT MEAS	QUANTITY	UNIT COSTS LABOR	UNIT COSTS MAT'L.	UNIT COSTS EQPT.	UNIT COSTS SUB.	UNIT COSTS TOTAL	TOTAL COSTS LABOR	TOTAL COSTS MAT'L	TOTAL COSTS EQPT.	TOTAL COSTS SUB.	TOTAL COSTS TOTAL	TAX & INS INCLUDED TOT UNIT PRICE	TAX & INS INCLUDED TOTAL PRICE
SPECIAL DOORS																		
136	02		830.900	STEEL ROLL-UP DOOR	EACH	2	500.000	3000.000			3500.000	1,000	6,000			7,000	3887.500	7,775
				*** TOTAL SPECIAL DOORS								1,000	6,000			7,000		7,775
HARDWARE																		
139	02	0616	871.027	FINISH HARDWARE ALLOWANCE	OPNG	2	69.583	275.625			345.208	139	551			690	387.500	775
				*** TOTAL HARDWARE								139	551			690		775
PAINTING & WALL COVERING																		
				PAINTING														
149	02	0423	990.008	PAINT EXTERIOR DOOR	SIDE	4	29.607	2.205			31.812	118	9			127	40.500	162
152	02	0616	990.032	PAINT DOOR FRAME	EACH	2	23.398	2.756			26.154	47	6			53	33.000	66
156	02	0621	990.081	PAINT MAS-CONC 3 CTS	SQS	36	69.082	19.845			88.927	2,487	714			3,201	109.889	3,956
159	02	0411	992.000	EXTERIOR PAINTING	SQS	23	46.797	14.333			61.130	1,076	330			1,406	75.435	1,735
				** TOTAL PAINTING								3,728	1,059			4,787		5,919
				*** TOTAL PAINTING & WALL COVERING								3,728	1,059			4,787		5,919
EQUIPMENT																		
				MAINTENANCE EQUIPMENT														
167	02		1101.900	FORK LIFTS	EACH	2				20,000	20,000				40,000	40,000	20,000	40,000
				** TOTAL MAINTENANCE EQUIPMENT											40,000	40,000		40,000
				LOADING DOCK EQUIPMENT														
164	02	1116	1116.000	DOCK LEVELER	EACH	2	834.314	3307.500			4141.814	1,669	6,615			8,284	4648.500	9,297
				** TOTAL LOADING DOCK EQUIPMENT								1,669	6,615			8,284		9,297
				*** TOTAL EQUIPMENT								1,669	6,615		40,000	48,284		49,297
HVAC SYSTEMS & EQUIPMENT																		
				PIPE & FITTINGS														
212	02	0824	1560.001	SCH 40 STEEL PIPE	****													
215	02	0824	1560.002	BLK T&C	****													
218	02	0824	1560.003	PIPE,2"	LNFT	600	3.848	2.811			6.659	2,309	1,687			3,996	7.507	4,504
223	02	0824	1560.006	STEEL FLANGE, 150#	****													
226	02	0824	1560.008	THREADED,2"	EACH	16	32.109	56.228			88.337	514	900			1,414	98.125	1,570
228	02	0824	1560.009	BOLT & GASKET SET	****													
230	02	0824	1560.010	FLANGE PACK,2"	EACH	8	5.772	3.980			9.752	46	32			78	11.000	88
				** TOTAL PIPE & FITTINGS								2,869	2,619			5,488		6,162

FILE ID - 2568
PROJECT JOB NO -
PROJECT NAME - WHATCO PROJECT
PROJECT SIZE - 6,000 SQFT
SECTION - 02 WAREHOUSE

ESTIMATE DETAIL REPORT

** SECTION, ITEM CODE SEQUENCE **

PAGE - 5
DATE -
TIME -

MANAGEMENT COMPUTER CONTROLS, INC.
2881 DIRECTORS COVE
MEMPHIS, TENNESSEE 38131

REF NO.	S D SC ELEM	ITEM CODE	DESCRIPTION	UNIT MEAS	QUANTITY	UNIT COSTS LABOR	UNIT COSTS MAT'L.	UNIT COSTS EQPT.	UNIT COSTS SUB.	UNIT COSTS TOTAL	TOTAL COSTS LABOR	TOTAL COSTS MAT'L	TOTAL COSTS EQPT.	TOTAL COSTS SUB.	TOTAL COSTS TOTAL	TAX & INS INCLUDED TOT UNIT PRICE	TAX & INS INCLUDED TOTAL PRICE
			HVAC SYSTEMS & EQUIPMENT														
			VALVES														
233	02 0824	1575.001	IRON BODY VALVES	****													
235	02 0824	1575.002	GATE, FLANGED	****													
236	02 0824	1575.003	125#,2"	EACH	8	36.188	225.737			261.925	290	1,806			2,096	286.375	2,291
			** TOTAL VALVES								290	1,806			2,096		2,291
			PIPE HANGERS & SUPPORTS														
242	02 0824	1580.001	PIPE HANGERS, STEEL	****													
245	02 0824	1580.002	WITH 3' ROD & C-CLAMP	****													
247	02 0824	1580.003	CLEVIS,2"	EACH	20	25.541	4.399			29.940	511	88			599	34.400	688
251	02 0824	1580.005	CLEVIS,6"	EACH	40	30.450	9.047			39.497	1,218	362			1,580	45.125	1,805
253	02 0824	1580.006	INSULATION SHIELD,6"	EACH	40	6.335	13.230			19.565	253	529			782	21.675	867
			** TOTAL PIPE HANGERS & SUPPORTS								1,982	979			2,961		3,360
			PIPE INSULATION														
257	02 0824	1583.001	FIBERGLASS INSULATION	****													
259	02 0824	1583.002	ASJ, 2" THICK	****													
262	02 0824	1583.003	PIPE,2"	LNFT	400	1.500	5.347			6.847	600	2,139			2,739	7.528	3,011
265	02 0824	1583.004	FLANGES,2"	EACH	8	12.034	9.096			21.130	96	73			169	23.875	191
			** TOTAL PIPE INSULATION								696	2,212			2,908		3,202
			MAJOR HEATING EQUIPMENT														
270	02 0822	1588.004	SPACE HEATERS	****													
272	02 0822	1588.005	PROPELLER,STEAM&HW,HORIZ	****													
273	02 0822	1588.006	UH#2 3300 CFM	EACH	8	164.556	1137.780			1302.336	1,316	9,102			10,418	1422.500	11,380
			** TOTAL MAJOR HEATING EQUIPMENT								1,316	9,102			10,418		11,380
			MAJOR COOLING EQUIPMENT														
282	02 0824	1589.004	AIR HANDLING UNITS	****													
285	02 0824	1589.005	LOW-PRESS,HORIZ,FLR MTD	****													
288	02 0824	1589.006	AHU#3 18620 CFM	EACH	4	1277.864	12,822			14,099	5,111	51,288			56,399	15,362	61,449
			** TOTAL MAJOR COOLING EQUIPMENT								5,111	51,288			56,399		61,449
			*** TOTAL HVAC SYSTEMS & EQUIPMENT								12,264	68,006			80,270		87,844

FILE ID - 2568
PROJECT JOB NO -
PROJECT NAME - WHATCO PROJECT
PROJECT SIZE - 6,000 SQFT
SECTION - 02 WAREHOUSE

E S T I M A T E   D E T A I L   R E P O R T
** SECTION, ITEM CODE SEQUENCE **

PAGE - 6
DATE -
TIME -

MANAGEMENT COMPUTER CONTROLS, INC.
2881 DIRECTORS COVE
MEMPHIS, TENNESSEE 38131

REF NO.	S D SC ELEM	ITEM CODE	DESCRIPTION	UNIT MEAS	QUANTITY	UNIT COSTS LABOR	MAT'L.	EQPT.	SUB.	TOTAL	TOTAL COSTS LABOR	MAT'L	EQPT.	SUB.	TOTAL	TAX & INS INCLUDED TOT UNIT PRICE	TOTAL PRICE
			ELEC SYSTEMS & EQUIPMENT														
			LOW VOLTAGE CONDUCTORS														
296	02 0921	1601.029	CU WIRE THHN-THWN-12 AWG	LNFT	3,000	.226	.095			.321	678	285			963	.398	1,194
301	02 0921	1607.761	22/2 AWG	LNFT	200	.494	.131			.625	99	26			125	.785	157
304	02 0921	1609.002	TERMIN LUGS-CU WIRE 12 AWG	EACH	120	4.111	.240			4.351	493	29			522	5.642	677
			** TOTAL LOW VOLTAGE CONDUCTORS								1,270	340			1,610		2,028
			CONDUIT, FITTINGS & ACCESS														
310	02 0912	1610.100	GALVANIZED RIGID STEEL-GRS	****													
314	02 0912	1610.130	COND-O'HEAD/WALL 2"	LNFT	1,000	5.756	1.816			7.572	5,756	1,816			7,572	9.497	9,497
318	02 0921	1614.520	4" OCTAGON BOX 1-1/2" DEEP	EACH	30	16.445	1.634			18.079	493	49			542	23.300	699
321	02 0921	1614.530	4" SQUARE BOX 1-1/2"DEEP	EACH	10	16.445	1.943			18.388	164	19			183	23.600	236
324	02 0921	1614.536	4" SQ BOX, 1G PLASTER RING	EACH	10	5.139	2.524			7.663	51	25			76	9.400	94
327	02 0921	1614.595	OUTLET BOX CLAMP	EACH	10		.625			.625		6			6	.700	7
330	02	1615.039	12"X12"X8" PULL BOX	EACH	4	61.669	32.500			94.169	247	130			377	115.750	463
			** TOTAL CONDUIT,FITTINGS & ACCESS								6,711	2,045			8,756		10,996
			GROUNDING ACCESSORIES														
340	02	1640.950	GROUNDING	LS	1				750.000	750.000				750	750	750.000	750
			** TOTAL GROUNDING ACCESSORIES											750	750		750
			LIGHTING FIXTURES														
343	02 0922	1643.000	FIXTURE EXIT ,VAR 120 V	EACH	6	28.779	62.500			91.279	173	375			548	105.000	630
346	02 0922	1643.000	FIXTURE A ,VAR 277 V	EACH	12	102.781	312.500			415.281	1,233	3,750			4,983	471.417	5,657
			** TOTAL LIGHTING FIXTURES								1,406	4,125			5,531		6,287
			LAMPS FOR LIGHTING FIXT.														
351	02 0922	1644.560	***HIGH PRESSURE SODIUM***	****													
353	02 0922	1644.571	E-25 MOG LU1000	EACH	12		209.988			209.988		2,520			2,520	226.250	2,715
355	02 0922	1644.801	6'--3/8" W/2-#16 TFF	SET	12	10.279	3.438			13.717	123	41			164	17.167	206
356	02 0922	1644.801	6'--3/8" W/2-#16 TFF	SET	150	10.279	3.438			13.717	1,542	516			2,058	17.173	2,576
			** TOTAL LAMPS FOR LIGHTING FIXT.								1,665	3,077			4,742		5,497
			WIRING DEVICES														
359	02 0921	1645.024	NEMA 5-20R,DUPLEX,IVORY	EACH	30	12.334	3.439			15.773	370	103			473	19.867	596
361	02 0921	1646.781	3/8" PHONE PLATE,PLAST,IV	EACH	10	4.111	2.450			6.561	41	25			66	8.000	80
			** TOTAL WIRING DEVICES								411	128			539		676

FILE ID - 256B
PROJECT JOB NO -
PROJECT NAME - WHATCO PROJECT
PROJECT SIZE - 6,000 SQFT
SECTION - 02 WAREHOUSE

E S T I M A T E   D E T A I L   R E P O R T

** SECTION, ITEM CODE SEQUENCE **

PAGE - 7
DATE -
TIME -

MANAGEMENT COMPUTER CONTROLS, INC.
2881 DIRECTORS COVE
MEMPHIS, TENNESSEE 38131

REF NO.	S D SC	ELEM	ITEM CODE	DESCRIPTION	UNIT MEAS	QUANTITY	UNIT COSTS LABOR	UNIT COSTS MAT'L.	UNIT COSTS EQPT.	UNIT COSTS SUB.	UNIT COSTS TOTAL	TOTAL COSTS LABOR	TOTAL COSTS MAT'L	TOTAL COSTS EQPT.	TOTAL COSTS SUB.	TOTAL COSTS TOTAL	TAX & INS INCLUDED TOT UNIT PRICE	TAX & INS INCLUDED TOTAL PRICE
ELEC SYSTEMS & EQUIPMENT																		
				DISCONNECT DEVICES														
364	02	0923	1647.011	ENCL CB,600/3/50AF,NEMA-1	EACH	4	67.836	357.325			425.161	271	1,429			1,700	473.750	1,895
				** TOTAL DISCONNECT DEVICES								271	1,429			1,700		1,895
				STARTERS														
369	02	0923	1649.126	3P,POLYPH,NEMA-1,SIZE 2	EACH	4	226.119	464.975			691.094	904	1,860			2,764	797.250	3,189
				** TOTAL STARTERS								904	1,860			2,764		3,189
				*** TOTAL ELEC SYSTEMS & EQUIPMENT								12,638	13,004		750	26,392		31,318
				**** TOTAL WAREHOUSE								139,152	207,545	9,078	137,812	495,587		548,513

FILE ID - 2568
PROJECT JOB NO -
PROJECT NAME - WHATCO PROJECT
PROJECT SIZE - 14,000 SQFT
SECTION - 03 OFFICE

E S T I M A T E   D E T A I L   R E C A P
** SECTION, ITEM CODE SEQUENCE **

PAGE - 1
DATE -
TIME -

MANAGEMENT COMPUTER CONTROLS, INC.
2881 DIRECTORS COVE
MEMPHIS, TENNESSEE 38131

DESCRIPTION	TOTAL COSTS LABOR	MAT'L	EQPT.	SUB.	TOTAL	TAX & INS INCLUDED UNIT SQFT PRICE	TOTAL PRICE
GENERAL REQUIREMENTS				249,018	249,018	17.787	249,018
EXCAV, GRADING & BACKFILL	7,561	1,895	4,318		13,774	1.146	16,050
CONCRETE FINISHING	7,196	238			7,434	.676	9,464
FORM WORK	3,049	675			3,724	.331	4,633
REINFORCING STEEL	6,538	4,487			11,025	.945	13,226
CAST-IN-PLACE CONCRETE	3,715	14,514	761		18,990	1.516	21,227
CEMENTITIOUS DECKS	3,248	3,857			7,105	.595	8,330
MASONRY	83,287	34,222	396		117,905	10.288	144,038
STONE	891	794			1,685	.143	2,000
STRUCTURAL METALS	70,933	79,846	9,296		160,075	13.324	186,538
MISC & ORNAMENTAL METAL	959	2,646			3,605	.292	4,092
WATERPROOF & DAMPPROOF	1,100	215			1,315	.117	1,639
INSULATION	4,645	4,511			9,156	.773	10,826
ROOFING,SHEETMETAL&ACCESS	13,079	8,632			21,711	1.863	26,080
METAL DOORS & FRAMES	2,586	4,234			6,820	.564	7,893
WOOD & PLASTIC DOORS	1,431	4,564			5,995	.484	6,773
METAL WINDOWS				14,884	14,884	1.063	14,884
HARDWARE	3,062	12,128			15,190	1.218	17,047
GLASS & GLAZING				2,867	2,867	.205	2,867
GYPSUM DRYWALL	31,806	11,356			43,162	3.786	53,005
TILE & TERRAZZO	4,842	1,489			6,331	.558	7,808
ACOUSTICAL TREATMENT	8,881	8,819			17,700	1.494	20,910
FLOORING	7,551	16,705			24,256	1.982	27,748
PAINTING & WALL COVERING	36,596	8,795			45,391	4.026	56,363
SPECIALTIES	2,159	4,851			7,010	.573	8,016
CONVEYING SYSTEMS				38,588	38,588	2.756	38,588
PLUMBING SYSTEMS & EQUIP	7,589	13,003			20,592	1.634	22,872
FIRE PROT SYSTEMS & EQUIP	963	2,029		23,156	26,148	1.891	26,468
HVAC SYSTEMS & EQUIPMENT	23,072	111,995		10,253	145,320	11.304	158,253
ELEC SYSTEMS & EQUIPMENT	37,999	174,186		3,125	215,310	17.185	240,586
***** SECTION TOTAL	374,738	530,686	14,771	341,891	1262,086	100.517	1407,242

FILE ID - 2568
PROJECT JOB NO -
PROJECT NAME - WHATCO PROJECT
PROJECT SIZE - 14,000 SQFT
SECTION - 03 OFFICE

E S T I M A T E   D E T A I L   R E P O R T
** SECTION, ITEM CODE SEQUENCE **

PAGE - 1
DATE -
TIME -
MANAGEMENT COMPUTER CONTROLS, INC.
2881 DIRECTORS COVE
MEMPHIS, TENNESSEE 38131

REF NO.	S D	SC	ELEM	ITEM CODE	DESCRIPTION	UNIT MEAS	QUANTITY	UNIT COSTS LABOR	UNIT COSTS MAT'L.	UNIT COSTS EQPT.	UNIT COSTS SUB.	UNIT COSTS TOTAL	TOTAL COSTS LABOR	TOTAL COSTS MAT'L	TOTAL COSTS EQPT.	TOTAL COSTS SUB.	TOTAL COSTS TOTAL	TAX & INS INCLUDED TOT UNIT PRICE	TAX & INS INCLUDED TOTAL PRICE
GENERAL REQUIREMENTS																			
					OTHER GENERAL REQUIREMENTS														
3		03		185.903	OH/PROFIT/GC/BOND @ 21.5%	LS	1				249,018	249,018				249,018	249,018	249,018	249,018
					** TOTAL OTHER GENERAL REQUIREMENTS											249,018	249,018		249,018
					*** TOTAL GENERAL REQUIREMENTS											249,018	249,018		249,018
CONTINGENCY																			
6		03		199.903	CONTINGENCY @ 0%														
					*** TOTAL CONTINGENCY														
EXCAV, GRADING & BACKFILL																			
					SITE GRADING														
10		03		221.011	EXCAV-LOAD-HAUL SURP 3-4MILES	CUYD	764	3.027		3.638		6.665	2,313		2,779		5,092	7.513	5,740
					** TOTAL SITE GRADING								2,313		2,779		5,092		5,740
					EXCAVATION & BACKFILL														
12		03	0112	222.301	EXCAVATE COLUMN FOOTING	CUYD	8	8.875		1.268		10.143	71		10		81	12.625	101
16		03	0111	222.308	EXCAVATE GRADE BEAM	CUYD	50	8.875		1.103		9.978	444		55		499	12.460	623
17		03	0112	222.321	BACKFILL COLUMN FOOTING	CUYD	2	18.529				18.529	37				37	23.500	47
20		03	0111	222.328	BACKFILL GRADE BEAM	CUYD	30	18.529				18.529	556				556	23.733	712
22		03	0112	222.331	FINE GRADE COLUMN FOOTING	SQFT	144	.529				.529	76				76	.674	97
26		03	0211	222.334	FINE GRADE SLAB ON GRADE	SQFT	7,000	.370				.370	2,590				2,590	.474	3,318
29		03		222.900	STRUCTURAL FILL	CUYD	382	3.859	4.961	3.859		12.679	1,474	1,895	1,474		4,843	14.168	5,412
					** TOTAL EXCAVATION & BACKFILL								5,248	1,895	1,539		8,682		10,310
					*** TOTAL EXCAV, GRADING & BACKFILL								7,561	1,895	4,318		13,774		16,050
CONCRETE FINISHING																			
					CONCRETE FINISH														
33		03	0211	301.019	TROWEL CEMENT FINISH	SQFT	14,000	.386				.386	5,404				5,404	.494	6,916
38		03	0211	301.050	PROTECT & CURE	SQFT	14,000	.128	.017			.145	1,792	238			2,030	.182	2,548
					** TOTAL CONCRETE FINISH								7,196	238			7,434		9,464
					*** TOTAL CONCRETE FINISHING								7,196	238			7,434		9,464

FILE ID - 2568
PROJECT JOB NO -
PROJECT NAME - WHATCO PROJECT
PROJECT SIZE - 14,000 SQFT
SECTION - 03 OFFICE

E S T I M A T E D E T A I L R E P O R T
** SECTION, ITEM CODE SEQUENCE **

PAGE - 2
DATE -
TIME -

MANAGEMENT COMPUTER CONTROLS, INC.
2881 DIRECTORS COVE
MEMPHIS, TENNESSEE 38131

REF NO.	S D SC ELEM	ITEM CODE	DESCRIPTION	UNIT MEAS	QUANTITY	UNIT COSTS LABOR	MAT'L.	EQPT.	SUB.	TOTAL	TOTAL COSTS LABOR	MAT'L	EQPT.	SUB.	TOTAL	TAX & INS INCLUDED TOT UNIT PRICE	TOTAL PRICE
FORM WORK																	
			FORM WORK														
41	03 0111	310.025	GRADE BEAM FORMS	SQFT	720	4.235	.937			5.172	3,049	675			3,724	6.435	4,633
			** TOTAL FORM WORK								3,049	675			3,724		4,633
			*** TOTAL FORM WORK								3,049	675			3,724		4,633
REINFORCING STEEL																	
			RE-BARS														
46	03 0211	321.009	RE-STEEL @ SLAB ON GRADE	CWT	55	35.412	24.255			59.667	1,948	1,334			3,282	71.582	3,937
47	03 0312	321.040	RE-STEEL @ SLAB ON MTL DECK	CWT	109	35.412	24.255			59.667	3,860	2,644			6,504	71.587	7,803
49	03 0112	321.110	RE-STEEL @ COLUMN FOOTING	CWT	8	34.798	24.255			59.053	278	194			472	70.750	566
53	03 0111	321.200	RE-STEEL @ GRADE BEAM	CWT	13	34.798	24.255			59.053	452	315			767	70.769	920
			** TOTAL RE-BARS								6,538	4,487			11,025		13,226
			*** TOTAL REINFORCING STEEL								6,538	4,487			11,025		13,226
CAST-IN-PLACE CONCRETE																	
			CAST-IN-PLACE CONCRETE														
55	03 0112	333.005	CONCRETE @ COLUMN FTG	CUYD	7	12.156	51.652			63.808	85	362			447	71.429	500
59	03 0111	333.025	CONCRETE @ GRADE BEAM	CUYD	25	14.761	51.652			66.413	369	1,291			1,660	74.800	1,870
62	03 0211	333.035	CONCRETE @ SLAB ON GRADE	CUYD	134	12.381	51.652			64.033	1,659	6,921			8,580	71.761	9,616
63	03 0312	333.080	CONCRETE OVER METAL DECK	CUYD	115	13.929	51.652	6.615		72.196	1,602	5,940	761		8,303	80.357	9,241
			** TOTAL CAST-IN-PLACE CONCRETE								3,715	14,514	761		18,990		21,227
			*** TOTAL CAST-IN-PLACE CONCRETE								3,715	14,514	761		18,990		21,227
CEMENTITIOUS DECKS																	
68	03 0321	350.100	LTWT CONC ROOF FILL	SQFT	7,000	.464	.551			1.015	3,248	3,857			7,105	1.190	8,330
			*** TOTAL CEMENTITIOUS DECKS								3,248	3,857			7,105		8,330
MASONRY																	
69	03 0611	402.102	8" CONC BLOCK	SQFT	15,370	4.030	1.907			5.937	61,941	29,311			91,252	7.222	111,002
70	03 0411	402.300	FACE BRICK	SQFT	2,700	7.022	1.819			8.841	18,959	4,911			23,870	10.957	29,584
71	03	430.101	EXT TUBULAR SCAFFOLDING	SQFT	3,600	.663		.110		.773	2,387		396		2,783	.959	3,452
			*** TOTAL MASONRY								83,287	34,222	396		117,905		144,038

FILE ID - 2568
PROJECT JOB NO -
PROJECT NAME - WHATCO PROJECT
PROJECT SIZE - 14,000 SQFT
SECTION - 03 OFFICE

E S T I M A T E   D E T A I L   R E P O R T

** SECTION, ITEM CODE SEQUENCE **

MANAGEMENT COMPUTER CONTROLS, INC.
2881 DIRECTORS COVE
MEMPHIS, TENNESSEE 38131

REF NO.	S D	SC	ELEM	ITEM CODE	DESCRIPTION	UNIT MEAS	QUANTITY	UNIT COSTS LABOR	UNIT COSTS MAT'L.	UNIT COSTS EQPT.	UNIT COSTS SUB.	UNIT COSTS TOTAL	TOTAL COSTS LABOR	TOTAL COSTS MAT'L	TOTAL COSTS EQPT.	TOTAL COSTS SUB.	TOTAL COSTS TOTAL	TAX & INS INCLUDED TOT UNIT PRICE	TAX & INS INCLUDED TOTAL PRICE
STONE																			
72		03	0411	445.200	PRECAST COPING	LNFT	120	7.426	6.615			14.041	891	794			1,685	16.667	2,000
					*** TOTAL STONE								891	794			1,685		2,000
STRUCTURAL METALS																			
					STRUCTURAL METALS														
75		03		501.011	STRUCTURAL STEEL FRAME	CWT	1,280	35.756	38.588	5.513		79.857	45,768	49,393	7,057		102,218	93.053	119,108
80		03		501.900	BASE PLATE	LBS	336	.551	.496			1.047	185	167			352	1.241	417
83		03		510.901	SANDBLAST	SQFT	14,175	.276	.132	.044		.452	3,912	1,871	624		6,407	.540	7,655
86		03		510.902	PRIMER	SQFT	14,175	.088	.055	.011		.154	1,247	780	156		2,183	.184	2,609
89		03		510.903	INTERMEDIATE COAT	SQFT	14,175	.121	.132	.011		.264	1,715	1,871	156		3,742	.309	4,380
92		03		510.904	TOP COAT (ELEVATED)	SQFT	14,175	.276	.165	.011		.452	3,912	2,339	156		6,407	.543	7,697
					** TOTAL STRUCTURAL METALS								56,739	56,421	8,149		121,309		141,866
					METAL JOIST														
95		03	0312	521.011	STEEL JOIST	CWT	260	29.447	36.383	4.410		70.240	7,656	9,460	1,147		18,263	81.488	21,187
					** TOTAL METAL JOIST								7,656	9,460	1,147		18,263		21,187
					METAL DECKING														
98		03	0312	532.010	1-1/2" METAL DECK	SQFT	7,000	.450	.937			1.387	3,150	6,559			9,709	1.590	11,130
99		03	0312	532.011	2" METAL DECK	SQFT	7,000	.484	1.058			1.542	3,388	7,406			10,794	1.765	12,355
					** TOTAL METAL DECKING								6,538	13,965			20,503		23,485
					*** TOTAL STRUCTURAL METALS								70,933	79,846	9,296		160,075		186,538
MISC & ORNAMENTAL METAL																			
					METAL FABRICATIONS														
100		03		551.013	ALUMINUM HORIZ HAND RAIL	LNFT	75	12.790	35.280			48.070	959	2,646			3,605	54.560	4,092
					** TOTAL METAL FABRICATIONS								959	2,646			3,605		4,092
					*** TOTAL MISC & ORNAMENTAL METAL								959	2,646			3,605		4,092
WATERPROOF & DAMPPROOF																			
103		03		700.110	DAMPPROOFING	SQFT	367	.460	.110			.570	169	40			209	.708	260
106		03	0211	710.000	VAPOR BARRIER @ SLAB	SQFT	7,000	.133	.025			.158	931	175			1,106	.197	1,379
					*** TOTAL WATERPROOF & DAMPPROOF								1,100	215			1,315		1,639

FILE ID - 2568
PROJECT JOB NO -
PROJECT NAME - WHATCO PROJECT
PROJECT SIZE - 14,000 SQFT
SECTION - 03 OFFICE

E S T I M A T E   D E T A I L   R E P O R T
** SECTION, ITEM CODE SEQUENCE **

PAGE - 4
DATE -
TIME -

MANAGEMENT COMPUTER CONTROLS, INC.
2881 DIRECTORS COVE
MEMPHIS, TENNESSEE 38131

REF NO.	S D SC ELEM	ITEM CODE	DESCRIPTION	UNIT MEAS	QUANTITY	UNIT COSTS LABOR	UNIT COSTS MAT'L.	UNIT COSTS EQPT.	UNIT COSTS SUB.	UNIT COSTS TOTAL	TOTAL COSTS LABOR	TOTAL COSTS MAT'L	TOTAL COSTS EQPT.	TOTAL COSTS SUB.	TOTAL COSTS TOTAL	TAX & INS INCLUDED TOT UNIT PRICE	TAX & INS INCLUDED TOTAL PRICE
INSULATION																	
109	03 0111	720.030	2" FOUNDATION INSULATION	SQFT	367	.745	.331			1.076	273	121			394	1.311	481
110	03 0411	720.042	2" RIGID INSULATION	SQFT	8,850	.494	.496			.990	4,372	4,390			8,762	1.169	10,345
			*** TOTAL INSULATION								4,645	4,511			9,156		10,826
ROOFING,SHEETMETAL&ACCESS																	
			ROOFING & ROOF INSULATION														
115	03 0501	751.002	3 PLY TAR& GRAVEL ROOFING	SQS	70	59.515	50.081			109.596	4,166	3,506			7,672	130.400	9,128
118	03 0503	755.003	2" FIBER BD ROOF INSULATION	SQFT	7,000	1.037	.659			1.696	7,259	4,613			11,872	2.040	14,280
			** TOTAL ROOFING & ROOF INSULATION								11,425	8,119			19,544		23,408
			FLASHING & SHEETMETAL														
121	03 0504	762.040	24 GA GALV IRON SHEETMETAL	SQFT	140	7.513	1.243			8.756	1,052	174			1,226	10.957	1,534
124	03 0504	762.064	6" GALV DOWNSPOUT	LNFT	70	3.807	2.143			5.950	266	150			416	7.186	503
127	03 0504	762.069	6" GALV GUTTER	LNFT	100	3.356	1.889			5.245	336	189			525	6.350	635
			** TOTAL FLASHING & SHEETMETAL								1,654	513			2,167		2,672
			*** TOTAL ROOFING,SHEETMETAL&ACCESS								13,079	8,632			21,711		26,080
METAL DOORS & FRAMES																	
130	03 0616	805.100	HM FRAME	EACH	44	51.543	66.150			117.693	2,268	2,911			5,179	137.591	6,054
133	03 0616	805.200	HM DOOR	EACH	8	39.762	165.375			205.137	318	1,323			1,641	229.875	1,839
			*** TOTAL METAL DOORS & FRAMES								2,586	4,234			6,820		7,893
WOOD & PLASTIC DOORS																	
134	03 0616	823.000	SC WOOD DOOR	EACH	36	39.762	126.788			166.550	1,431	4,564			5,995	188.139	6,773
			*** TOTAL WOOD & PLASTIC DOORS								1,431	4,564			5,995		6,773
METAL WINDOWS																	
137	03 0421	851.110	FIXED WINDOWS	SQFT	900				16.538	16.538				14,884	14,884	16.538	14,884
			*** TOTAL METAL WINDOWS											14,884	14,884		14,884
HARDWARE																	
140	03 0616	871.027	FINISH HARDWARE ALLOWANCE	OPNG	44	69.583	275.625			345.208	3,062	12,128			15,190	387.432	17,047
			*** TOTAL HARDWARE								3,062	12,128			15,190		17,047

FILE ID - 2568
PROJECT JOB NO -
PROJECT NAME - WHATCO PROJECT
PROJECT SIZE - 14,000 SQFT
SECTION - 03 OFFICE

ESTIMATE DETAIL REPORT

** SECTION, ITEM CODE SEQUENCE **

PAGE - 5
DATE -
TIME -

MANAGEMENT COMPUTER CONTROLS, INC.
2881 DIRECTORS COVE
MEMPHIS, TENNESSEE 38131

REF NO.	S D SC	ELEM	ITEM CODE	DESCRIPTION	UNIT MEAS	QUANTITY	UNIT COSTS LABOR	UNIT COSTS MAT'L.	UNIT COSTS EQPT.	UNIT COSTS SUB.	UNIT COSTS TOTAL	TOTAL COSTS LABOR	TOTAL COSTS MAT'L	TOTAL COSTS EQPT.	TOTAL COSTS SUB.	TOTAL COSTS TOTAL	TAX & INS INCLUDED TOT UNIT PRICE	TAX & INS INCLUDED TOTAL PRICE
GLASS & GLAZING																		
				ENTRANCES & STOREFRONTS														
141	03	0423	885.205	PAIR ALUM GLASS DOOR	EACH	2				1433.250	1433.250				2,867	2,867	1433.500	2,867
				** TOTAL ENTRANCES & STOREFRONTS											2,867	2,867		2,867
				*** TOTAL GLASS & GLAZING											2,867	2,867		2,867
GYPSUM DRYWALL																		
142	03	0611	926.620	STANDARD DRYWALL PARTITION	SQFT	8,900	2.374	.882			3.256	21,129	7,850			28,979	3.994	35,547
143	03	0611	926.625	MTL FURRING W/GYPSUM BOARD	SQFT	7,950	1.343	.441			1.784	10,677	3,506			14,183	2.196	17,458
				*** TOTAL GYPSUM DRYWALL								31,806	11,356			43,162		53,005
TILE & TERRAZZO																		
				CERAMIC TILE														
144	03	0622	931.010	CERAMIC TILE FLOOR	SQFT	900	5.380	1.654			7.034	4,842	1,489			6,331	8.676	7,808
				** TOTAL CERAMIC TILE								4,842	1,489			6,331		7,808
				*** TOTAL TILE & TERRAZZO								4,842	1,489			6,331		7,808
ACOUSTICAL TREATMENT																		
145	03	0623	950.400	ACOUST CEIL SYS-EXPOSED GRID	SQFT	12,300	.722	.717			1.439	8,881	8,819			17,700	1.700	20,910
				*** TOTAL ACOUSTICAL TREATMENT								8,881	8,819			17,700		20,910
FLOORING																		
				RESILIENT FLOORING														
146	03	0622	965.010	VINYL COMPOSITION TILE	SQFT	3,825	.981	.584			1.565	3,752	2,234			5,986	1.888	7,221
				** TOTAL RESILIENT FLOORING								3,752	2,234			5,986		7,221
				CARPETING														
147	03	0622	968.200	CARPET	SQYD	875	4.342	16.538			20.880	3,799	14,471			18,270	23.459	20,527
				** TOTAL CARPETING								3,799	14,471			18,270		20,527
				*** TOTAL FLOORING								7,551	16,705			24,256		27,748
PAINTING & WALL COVERING																		

FILE ID - 2568
PROJECT JOB NO -
PROJECT NAME - WHATCO PROJECT
PROJECT SIZE - 14,000 SQFT
SECTION - 03 OFFICE

ESTIMATE DETAIL REPORT

** SECTION, ITEM CODE SEQUENCE **

PAGE - 6
DATE -
TIME -

MANAGEMENT COMPUTER CONTROLS, INC.
2881 DIRECTORS COVE
MEMPHIS, TENNESSEE 38131

REF NO.	S D SC ELEM	ITEM CODE	DESCRIPTION	UNIT MEAS	QUANTITY	UNIT COSTS LABOR	UNIT COSTS MAT'L.	UNIT COSTS EQPT.	UNIT COSTS SUB.	UNIT COSTS TOTAL	TOTAL COSTS LABOR	TOTAL COSTS MAT'L	TOTAL COSTS EQPT.	TOTAL COSTS SUB.	TOTAL COSTS TOTAL	TAX & INS INCLUDED TOT UNIT PRICE	TAX & INS INCLUDED TOTAL PRICE
			PAINTING & WALL COVERING														
			PAINTING														
150	03 0616	990.031	PAINT INTERIOR DOOR	SIDE	88	26.865	3.308			30.173	2,364	291			2,655	37.966	3,341
153	03 0616	990.032	PAINT DOOR FRAME	EACH	44	23.398	2.756			26.154	1,030	121			1,151	32.932	1,449
154	03 0621	990.061	PAINT PLASTER-GYP BD 3 CTS	SQS	515	50.025	12.128			62.153	25,763	6,246			32,009	77.159	39,737
157	03 0621	990.081	PAINT MAS-CONC 3 CTS	SQS	107	69.082	19.845			88.927	7,392	2,123			9,515	109.907	11,760
160	03 0411	992.000	EXTERIOR PAINTING	SQS	1	46.797	14.333			61.130	47	14			61	76.000	76
			** TOTAL PAINTING								36,596	8,795			45,391		56,363
			*** TOTAL PAINTING & WALL COVERING								36,596	8,795			45,391		56,363
			SPECIALTIES														
			COMPARTMENTS & CUBICLES														
161	03 0631	1018.010	TOILET COMPARTMENT	EACH	12	125.435	220.500			345.935	1,505	2,646			4,151	399.250	4,791
			** TOTAL COMPARTMENTS & CUBICLES								1,505	2,646			4,151		4,791
			LOCKERS														
162	03 0631	1050.001	DOUBLE TIER LOCKER	OPNG	40	16.361	55.125			71.486	654	2,205			2,859	80.625	3,225
			** TOTAL LOCKERS								654	2,205			2,859		3,225
			*** TOTAL SPECIALTIES								2,159	4,851			7,010		8,016
			CONVEYING SYSTEMS														
			ELEVATORS														
165	03	1420.900	ELEVATOR	EACH	1				38,587	38,587				38,588	38,588	38,588	38,588
			** TOTAL ELEVATORS											38,588	38,588		38,588
			*** TOTAL CONVEYING SYSTEMS											38,588	38,588		38,588
			PLUMBING SYSTEMS & EQUIP														
			PIPE & FITTINGS														
168	03 0811	1500.001	SCHEDULE 40 STEEL PIPE	****													
169	03 0811	1500.002	BLACK STEEL T&C	****													
170	03 0811	1500.003	PIPE,1/2"	LNFT	100	2.055	.827			2.882	206	83			289	3.280	328
171	03 0811	1500.004	PIPE,1"	LNFT	250	2.339	1.378			3.717	585	345			930	4.204	1,051
172	03 0812	1500.005	PVC PIPE	****													
173	03 0812	1500.006	SCH 40 SOLVENT CEMENT	****													
174	03 0812	1500.007	PIPE,1-1/2"	LNFT	150	1.391	1.529			2.920	209	229			438	3.267	490
175	03 0812	1500.008	CAST IRON PIPE	****													

FILE ID - 2568
PROJECT JOB NO -
PROJECT NAME - WHATCO PROJECT
PROJECT SIZE - 14,000 SQFT
SECTION - 03 OFFICE

ESTIMATE DETAIL REPORT

** SECTION, ITEM CODE SEQUENCE **

PAGE - 7
DATE -
TIME -

MANAGEMENT COMPUTER CONTROLS, INC.
2881 DIRECTORS COVE
MEMPHIS, TENNESSEE 38131

REF NO.	S D SC ELEM	ITEM CODE	DESCRIPTION	UNIT MEAS	QUANTITY	UNIT COSTS LABOR	MAT'L.	EQPT.	SUB.	TOTAL	TOTAL COSTS LABOR	MAT'L	EQPT.	SUB.	TOTAL	TAX & INS INCLUDED TOT UNIT PRICE	TOTAL PRICE
			PLUMBING SYSTEMS & EQUIP														
			PIPE & FITTINGS														
176	03 0812	1500.009	SERVICE WEIGHT B&S	****													
177	03 0812	1500.010	PIPE,2"	LNFT	40	3.667	2.844			6.511	147	114			261	7.325	293
178	03 0812	1500.011	PIPE,8"	LNFT	100	11.506	9.537			21.043	1,151	954			2,105	23.670	2,367
179	03 0812	1500.012	MISC.SVC B&S GASKET FTNG.	****													
180	03 0812	1500.013	"P"-TRAP,2"	EACH	9	12.296	10.397			22.693	111	94			205	25.444	229
			** TOTAL PIPE & FITTINGS								2,409	1,819			4,228		4,758
			PIPE HANGER & SUPPORTS														
181	03 0811	1530.001	PIPE HANGERS,STEEL	****													
182	03 0812	1530.001	PIPE HANGERS,STEEL	****													
183	03 0811	1530.002	W/3"ROD & C-CLAMP	****													
184	03 0812	1530.002	W/3"ROD & C-CLAMP	****													
185	03 0811	1530.003	CLEVIS,1/2"	EACH	10	24.340	4.190			28.530	243	42			285	32.700	327
186	03 0811	1530.004	CLEVIS,1"	EACH	25	24.340	4.278			28.618	609	107			716	32.880	822
187	03 0812	1530.005	CLEVIS,2"	EACH	4	24.340	4.399			28.739	97	18			115	33.000	132
188	03 0812	1530.006	CLEVIS,8"	EACH	10	30.598	13.947			44.545	306	139			445	50.600	506
			** TOTAL PIPE HANGER & SUPPORTS								1,255	306			1,561		1,787
			PLUMBING SPECIAL PLUMBING SPECIALTIES														
			PLUMBING SPECIALTIES														
189	03 0812	1539.001	SEWAGE SPECIALTIES	****													
190	03 0812	1539.002	VENT FLASHING, 6# LEAD	****													
191	03 0812	1539.003	2"	EACH	4	91.669	5.678			97.347	367	23			390	112.500	450
			** TOTAL PLUMBING SPECIALTIES								367	23			390		450
			PLBG FIXTURE COMMERCIAL														
192	03 0814	1542.001	PLUMBING FIXT'S.COMMERCIAL	****													
193	03 0814	1542.002	WATER CLOSET	****													
194	03 0814	1542.003	FLR MTD W/FLUSH VALVE	EACH	12	101.942	241.062			343.004	1,223	2,893			4,116	379.167	4,550
195	03 0814	1542.004	URINAL	****													
196	03 0814	1542.005	WALL HUNG W/ FLUSH VALVE	EACH	2	125.460	366.471			491.931	251	733			984	542.000	1,084
197	03 0814	1542.006	LAVATORY W/TRIM	****													
198	03 0814	1542.007	CNTR-TP,ENAM CI,SHELF-RM	EACH	9	78.424	170.336			248.760	706	1,533			2,239	275.444	2,479
199	03 0814	1542.008	DRINKING FTN-HANDICAP	****													
200	03 0814	1542.009	WALL HUNG,ELEC	EACH	4	138.515	728.753			867.268	554	2,915			3,469	949.750	3,799
201	03 0814	1542.010	WASHFOUNTAIN ROUND	****													
202	03 0814	1542.011	54" STAINLESS STEEL	EACH	1	509.680	1846.688			2356.368	510	1,847			2,357	2590.000	2,590
203	03 0814	1542.012	MULTI-PURPOSE SINK W/TRIM	****													

FILE ID - 2568
PROJECT JOB NO -
PROJECT NAME - WHATCO PROJECT
PROJECT SIZE - 14,000 SQFT
SECTION - 03 OFFICE

ESTIMATE DETAIL REPORT

** SECTION, ITEM CODE SEQUENCE **

PAGE - 8
DATE -
TIME -

MANAGEMENT COMPUTER CONTROLS, INC.
2881 DIRECTORS COVE
MEMPHIS, TENNESSEE 38131

REF NO.	S D SC ELEM	ITEM CODE	DESCRIPTION	UNIT MEAS	QUANTITY	UNIT COSTS LABOR	MAT'L.	EQPT.	SUB.	TOTAL	TOTAL COSTS LABOR	MAT'L	EQPT.	SUB.	TOTAL	TAX & INS INCLUDED TOT UNIT PRICE	TOTAL PRICE
PLUMBING SYSTEMS & EQUIP																	
			PLBG FIXTURE COMMERCIAL														
204	03 0814	1542.013	CNTR-TOP,DBL COMPT,S.S.	EACH	2	156.817	467.019			623.836	314	934			1,248	687.500	1,375
			** TOTAL PLBG FIXTURE COMMERCIAL								3,558	10,855			14,413		15,877
			*** TOTAL PLUMBING SYSTEMS & EQUIP								7,589	13,003			20,592		22,872
FIRE PROT SYSTEMS & EQUIP																	
			HOSE CAB. & EXTINGUISHERS														
205	03 0833	1554.001	FIRE EXTINGUISHER CAB.	****													
206	03 0833	1554.002	RECESSED	EACH	8	77.638	82.688			160.326	621	662			1,283	179.500	1,436
207	03 0833	1554.003	FIRE EXTINGUISHER	****													
208	03 0833	1554.004	DRY CHEMICAL	****													
209	03 0833	1554.005	CLASS ABC 20 LB.	EACH	8	42.710	170.888			213.598	342	1,367			1,709	234.500	1,876
			** TOTAL HOSE CAB. & EXTINGUISHERS								963	2,029			2,992		3,312
			MISC FIRE PROT SYS & EQUIP														
210	03 8040	1558.001	SPRINKLER SYSTEM	SQFT	14,000				1.654	1.654				23,156	23,156	1.654	23,156
			** TOTAL MISC FIRE PROT SYS & EQUIP											23,156	23,156		23,156
			*** TOTAL FIRE PROT SYSTEMS & EQUIP								963	2,029		23,156	26,148		26,468
HVAC SYSTEMS & EQUIPMENT																	
			PIPE & FITTINGS														
213	03 0824	1560.001	SCH 40 STEEL PIPE	****													
216	03 0824	1560.002	BLK T&C	****													
219	03 0824	1560.003	PIPE,2"	LNFT	300	3.848	2.811			6.659	1,154	843			1,997	7.507	2,252
			** TOTAL PIPE & FITTINGS								1,154	843			1,997		2,252
			PIPE HANGERS & SUPPORTS														
243	03 0824	1580.001	PIPE HANGERS, STEEL	****													
246	03 0824	1580.002	WITH 3' ROD & C-CLAMP	****													
249	03 0824	1580.005	CLEVIS,6"	EACH	30	30.450	9.047			39.497	914	271			1,185	45.133	1,354
254	03 0824	1580.006	INSULATION SHIELD,6"	EACH	30	6.335	13.230			19.565	190	397			587	21.667	650
			** TOTAL PIPE HANGERS & SUPPORTS								1,104	668			1,772		2,004
			PIPE INSULATION														
255	03 0824	1583.001	FIBERGLASS INSULATION	****													
260	03 0824	1583.002	ASJ, 2" THICK														
260	03 0824	1583.002	ASJ, 2" THICK	****													
260	03 0824	1583.002	ASJ, 2" THICK	****													

FILE ID - 2568
PROJECT JOB NO -
PROJECT NAME - WHATCO PROJECT
PROJECT SIZE - 14,000 SQFT
SECTION - 03 OFFICE

ESTIMATE DETAIL REPORT

** SECTION, ITEM CODE SEQUENCE **

DATE -
TIME -

MANAGEMENT COMPUTER CONTROLS, INC.
2881 DIRECTORS COVE
MEMPHIS, TENNESSEE 38131

REF NO.	S D	SC	ELEM	ITEM CODE	DESCRIPTION	UNIT MEAS	QUANTITY	UNIT COSTS LABOR	UNIT COSTS MAT'L.	UNIT COSTS EQPT.	UNIT COSTS SUB.	UNIT COSTS TOTAL	TOTAL COSTS LABOR	TOTAL COSTS MAT'L	TOTAL COSTS EQPT.	TOTAL COSTS SUB.	TOTAL COSTS TOTAL	TAX & INS INCLUDED TOT UNIT PRICE	TAX & INS INCLUDED TOTAL PRICE
					HVAC SYSTEMS & EQUIPMENT														
					PIPE INSULATION														
263		03	0824	1583.003	PIPE,2"	LNFT	300	1.500	5.347			6.847	450	1,604			2,054	7.527	2,258
					** TOTAL PIPE INSULATION								450	1,604			2,054		2,258
					MAJOR HEATING EQUIPMENT														
266		03	0822	1588.001	HEATING EQUIPMENT	****													
267		03	0822	1588.002	BOILER,STEEL TUBE PACKAGE	****													
268		03	0822	1588.003	B#3 136 HP	EACH	1	877.678	25,544			26,422	878	25,545			26,423	28,670	28,670
275		03	0822	1588.007	MISCELLANEOUS ACCESSORIES	****													
276		03	0822	1588.008	FINNED TUBE RADIATION	****													
277		03	0822	1588.009	STANDARD STYLE	LNFT	600	21.959	27.563			49.522	13,175	16,538			29,713	55.308	33,185
					** TOTAL MAJOR HEATING EQUIPMENT								14,053	42,083			56,136		61,855
					MAJOR COOLING EQUIPMENT														
278		03	0823	1589.001	COOLING EQUIPMENT	****													
279		03	0823	1589.002	CONDENSER,AIR COOLED	****													
280		03	0823	1589.003	CHW3 30 TONS	EACH	1	219.420	13,781			14,000	219	13,781			14,000	15,173	15,173
283		03	0824	1589.004	AIR HANDLING UNITS	****													
286		03	0824	1589.005	LOW-PRESS,HORIZ,FLR MTD	****													
289		03	0824	1589.006	AHU#3 18620 CFM	EACH	4	1277.864	12,822			14,099	5,111	51,288			56,399	15,362	61,449
					** TOTAL MAJOR COOLING EQUIPMENT								5,330	65,069			70,399		76,622
					DUCT INSULATION														
290		03	0824	1593.001	GALV,CLASS "B" FLUE,18"	LNFT	30	13.909	46.581			60.490	417	1,397			1,814	66.567	1,997
					** TOTAL DUCT INSULATION								417	1,397			1,814		1,997
					REGISTER,GRILLES &DIFF														
291		03	0824	1594.001	RETURN AIR REGISTER	****													
292		03	0824	1594.002	CLG MTD,SQUARE PERF FACE	****													
293		03	0824	1594.003	1 SQFT SURF AREA	EACH	30	18.784	11.025			29.809	564	331			895	33.733	1,012
					** TOTAL REGISTER,GRILLES &DIFF								564	331			895		1,012
					TEST & BALANCE														
294		03	0827	1596.001	HVAC TEST & BALANCE	LS	1				10,253	10,253				10,253	10,253	10,253	10,253
					** TOTAL TEST & BALANCE											10,253	10,253		10,253
					*** TOTAL HVAC SYSTEMS & EQUIPMENT								23,072	111,995		10,253	145,320		158,253

FILE ID	- 2568
PROJECT JOB NO	-
PROJECT NAME	- WHATCO PROJECT
PROJECT SIZE	- 14,000 SQFT
SECTION	- 03 OFFICE

E S T I M A T E   D E T A I L   R E P O R T

** SECTION, ITEM CODE SEQUENCE **

PAGE - 10
DATE -
TIME -

MANAGEMENT COMPUTER CONTROLS, INC.
2881 DIRECTORS COVE
MEMPHIS, TENNESSEE 38131

REF NO.	S D	SC ELEM	ITEM CODE	DESCRIPTION	UNIT MEAS	QUANTITY	UNIT COSTS LABOR	UNIT COSTS MAT'L.	UNIT COSTS EQPT.	UNIT COSTS SUB.	UNIT COSTS TOTAL	TOTAL COSTS LABOR	TOTAL COSTS MAT'L	TOTAL COSTS EQPT.	TOTAL COSTS SUB.	TOTAL COSTS TOTAL	TAX & INS INCLUDED TOT UNIT PRICE	TAX & INS INCLUDED TOTAL PRICE
ELEC SYSTEMS & EQUIPMENT																		
				LOW VOLTAGE CONDUCTORS														
297	03	0921	1601.029	CU WIRE THHN-THWN-12 AWG	LNFT	8,000	.226	.095			.321	1,808	760			2,568	.398	3,184
299	03	0912	1601.033	CU WIRE THHN-THWN-4 AWG	LNFT	618	.390	.531			.921	241	328			569	1.083	669
302	03	0921	1607.761	22/2 AWG	LNFT	200	.494	.131			.625	99	26			125	.785	157
305	03	0921	1609.002	TERMIN LUGS-CU WIRE 12 AWG	EACH	200	4.111	.240			4.351	822	48			870	5.645	1,129
307	03	0912	1609.006	TERMIN LUGS-CU WIRE 4 AWG	EACH	4	14.390	5.626			20.016	58	23			81	24.750	99
				** TOTAL LOW VOLTAGE CONDUCTORS								3,028	1,185			4,213		5,238
				CONDUIT,FITTINGS & ACCESS														
308	03	0921	1610.099	GALVANIZED RIGID STEEL-GRS	****													
311	03	0912	1610.100	GALVANIZED RIGID STEEL-GRS	****													
312	03	0921	1610.129	COND-O'HEAD/WALL 1-1/2"	LNFT	2,000	4.934	1.398			6.332	9,868	2,796			12,664	7.970	15,940
316	03	0912	1610.134	COND-O'HEAD/WALL 4"	LNFT	100	12.334	5.603			17.937	1,233	560			1,793	22.200	2,220
319	03	0921	1614.520	4" OCTAGON BOX 1-1/2" DEEP	EACH	60	16.445	1.634			18.079	987	98			1,085	23.317	1,399
322	03	0921	1614.530	4" SQUARE BOX 1-1/2"DEEP	EACH	40	16.445	1.943			18.388	658	78			736	23.650	946
325	03	0921	1614.536	4" SQ BOX, 1G PLASTER RING	EACH	40	5.139	2.524			7.663	206	101			307	9.450	378
328	03	0921	1614.595	OUTLET BOX CLAMP	EACH	40		.625			.625		25			25	.675	27
331	03		1615.039	12"X12"X8" PULL BOX	EACH	8	61.669	32.500			94.169	493	260			753	115.750	926
				** TOTAL CONDUIT,FITTINGS & ACCESS								13,445	3,918			17,363		21,836
				FREESTANDING SWITCHBOARDS														
332	03	0912	1633.000	L. V. SWITCHBOARD MDP	EACH	1	4933.500	81,250			86,183	4,934	81,250			86,184	94,010	94,010
333	03	0912	1633.001	L. V. SWITCHBOARD SWBD	EACH	1	3289.000	37,500			40,789	3,289	37,500			40,789	44,715	44,715
334	03	0912	1633.500	SWITCHBOARD MDP RIGGING	EACH	1				750.000	750.000				750	750	750.000	750
335	03	0912	1633.501	SWITCHBOARD SWBD RIGGING	EACH	1				750.000	750.000				750	750	750.000	750
				** TOTAL FREESTANDING SWITCHBOARDS								8,223	118,750		1,500	128,473		140,225
				PANELBOARDS & BREAKERS														
336	03	0912	1634.000	PANELBOARD LP	EACH	4	888.030	3875.000			4763.030	3,552	15,500			19,052	5338.500	21,354
				** TOTAL PANELBOARDS & BREAKERS								3,552	15,500			19,052		21,354
				TRANSFORMERS														
337	03	0912	1638.088	TRAN PAD-MOUNTED,1500KVA	EACH	1	1068.925	18,918			19,987	1,069	18,919			19,988	21,785	21,785
338	03	0912	1638.099	TRANSFORMER RIGGING	EACH	1				750.000	750.000				750	750	750.000	750
				** TOTAL TRANSFORMERS								1,069	18,919		750	20,738		22,535

FILE ID - 2568
PROJECT JOB NO -
PROJECT NAME - WHATCO PROJECT
PROJECT SIZE - 14,000 SQFT
SECTION - 03 OFFICE

ESTIMATE DETAIL REPORT

** SECTION, ITEM CODE SEQUENCE **

MANAGEMENT COMPUTER CONTROLS, INC.
2881 DIRECTORS COVE
MEMPHIS, TENNESSEE 38131

REF NO.	S D	SC ELEM	ITEM CODE	DESCRIPTION	UNIT MEAS	QUANTITY	UNIT COSTS LABOR	UNIT COSTS MAT'L.	UNIT COSTS EQPT.	UNIT COSTS SUB.	UNIT COSTS TOTAL	TOTAL COSTS LABOR	TOTAL COSTS MAT'L	TOTAL COSTS EQPT.	TOTAL COSTS SUB.	TOTAL COSTS TOTAL	TAX & INS INCLUDED TOT UNIT PRICE	TAX & INS INCLUDED TOTAL PRICE
				ELEC SYSTEMS & EQUIPMENT														
				GROUNDING ACCESSORIES														
341	03		1640.950	GROUNDING	LS	1				875.000	875.000				875	875	875.000	875
				** TOTAL GROUNDING ACCESSORIES											875	875		875
				LIGHTING FIXTURES														
344	03	0922	1643.000	FIXTURE C ,VAR 120 V	EACH	12	20.556	25.000			45.556	247	300			547	53.833	646
347	03	0922	1643.001	FIXTURE B ,VAR 277 V	EACH	150	41.113	50.000			91.113	6,167	7,500			13,667	107.733	16,160
				** TOTAL LIGHTING FIXTURES								6,414	7,800			14,214		16,806
				LAMPS FOR LIGHTING FIXT.														
348	03	0922	1644.100	***FLUORESCENT LAMPS ***	****													
349	03	0922	1644.241	F40T12/CW/RS/ENERGY SAV	EACH	600		4.850			4.850		2,910			2,910	5.227	3,136
				** TOTAL LAMPS FOR LIGHTING FIXT.									2,910			2,910		3,136
				WIRING DEVICES														
357	03	0921	1645.022	NEMA 5-15R,DUPLEX,IVORY	EACH	60	8.223	2.573			10.796	493	154			647	13.533	812
362	03	0921	1646.781	3/8" PHONE PLATE,PLAST,IV	EACH	40	4.111	2.450			6.561	164	98			262	8.025	321
				** TOTAL WIRING DEVICES								657	252			909		1,133
				DISCONNECT DEVICES														
365	03	0923	1647.011	ENCL CB,600/3/50AF,NEMA-1	EACH	4	67.836	357.325			425.161	271	1,429			1,700	473.750	1,895
367	03	0923	1647.014	ENCL CB,600/3/225AF,NEMA-1	EACH	1	240.919	916.975			1157.894	241	917			1,158	1304.000	1,304
				** TOTAL DISCONNECT DEVICES								512	2,346			2,858		3,199
				STARTERS														
370	03	0923	1649.126	3P,POLYPH,NEMA-1,SIZE 2	EACH	4	226.119	464.975			691.094	904	1,860			2,764	797.250	3,189
371	03	0923	1649.127	3P,POLYPH,NEMA-1,SIZE 3	EACH	1	195.285	746.050			941.335	195	746			941	1060.000	1,060
				** TOTAL STARTERS								1,099	2,606			3,705		4,249
				*** TOTAL ELEC SYSTEMS & EQUIPMENT								37,999	174,186		3,125	215,310		240,586
				**** TOTAL OFFICE								374,738	530,686	14,771	341,891	1262,086		1407,242
				***** GRAND TOTAL								1096,951	1858,206	69,516	914,483	3939,156		4391,578

RECAP OF COMPARISON OF FOUR ESTIMATES (A B C D) IN ITEM (CSI DIVISION) CODE SEQUENCE FOR A SECTION OF PROJECT	ESTIMATE (A) WHATCO PROJECT SECTION-01:PLANT ONLY File ID.:3256 Job Type:SCHEMATIC Job Loc.: Est Date:11/11/92			ESTIMATE (B) WHATCO PROJECT SECTION-01:PLANT ONLY File ID.:256A Job Type:DESIGN DEVELOP Job Loc.: Est Date: 1/05/93			ESTIMATE (C) WHATCO PROJECT SECTION-01:PLANT ONLY File ID.:2568 Job Type:CONST. DOC. Job Loc.: Est Date: 2/02/93			Page: 1 Date: at:
Description	30,000 $/SQFT	% of Sec.	TOTAL COST	30,000 $/SQFT	% of Sec.	TOTAL COST	30,000 $/SQFT	% of Sec.	TOTAL COST	
DIV-01 GENERAL REQUIREMENTS	$12.78	19.5%	$383,322	$28.23	21.6%	$846,783	$14.37	17.7%	$431,030	
DIV-02 SITEWORK	$1.83	2.8%	$54,879	$5.35	4.1%	$160,647	$2.25	2.7%	$67,587	
DIV-03 CONCRETE	$9.34	14.3%	$280,310	$7.92	6.0%	$237,614	$6.45	7.9%	$193,442	
DIV-05 METALS	$6.92	10.5%	$207,503	$31.42	24.0%	$942,456	$16.94	20.8%	$508,061	
DIV-06 WOOD & PLASTICS	$.11	.1%	$3,339							
DIV-07 THERMAL & MOISTURE PROT.	$8.51	13.0%	$255,252	$11.75	9.0%	$352,533	$11.63	14.3%	$348,887	
DIV-08 DOORS & WINDOWS	$.93	1.4%	$27,816	$2.26	1.7%	$67,834	$.64	.7%	$19,324	
DIV-09 FINISHES	$.02		$597	$.35	.2%	$10,632	$.35	.4%	$10,632	
DIV-11 EQUIPMENT	$.38	.5%	$11,285	$.62	.4%	$18,593	$.62	.7%	$18,593	
DIV-14 CONVEYING SYSTEMS	$11.67	17.8%	$350,000	$17.06	13.0%	$511,866	$17.06	21.0%	$511,866	
DIV-15 MECHANICAL	$6.38	9.7%	$191,475	$14.00	10.7%	$419,910	$7.55	9.3%	$226,438	
DIV-16 ELECTRICAL	$6.50	9.9%	$195,000	$11.63	8.9%	$348,754	$3.33	4.1%	$99,963	
** SECTION TOTAL	$65.36	100.0%	$1,960,778	$130.59	100.0%	$3,917,622	$81.19	100.0%	$2,435,823	

Page: 1
Date:
at:

RECAP OF COMPARISON OF FOUR ESTIMATES (A B C D) IN ITEM (CSI DIVISION) CODE SEQUENCE FOR A SECTION OF PROJECT	ESTIMATE (A) WHATCO PROJECT SECTION-02:WAREHOUSE ONLY File ID.:3256 Job Type:SCHEMATIC Job Loc.: Est Date:11/11/92			ESTIMATE (B) WHATCO PROJECT SECTION-02:WAREHOUSE ONLY File ID.:256A Job Type:DESIGN DEVELOP Job Loc.: Est Date: 1/05/93			ESTIMATE (C) WHATCO PROJECT SECTION-02:WAREHOUSE ONLY File ID.:256B Job Type:CONST. DOC. Job Loc.: Est Date: 2/02/93			
	9,600			6,000			6,000			
Description	$/SQFT	% of Sec.	TOTAL COST	$/SQFT	% of Sec.	TOTAL COST	$/SQFT	% of Sec.	TOTAL COST	
DIV-01 GENERAL REQUIREMENTS	$10.09	19.5%	$96,911	$37.06	21.6%	$222,360	$16.18	17.7%	$97,062	
DIV-02 SITEWORK	$.69	1.3%	$6,579	$2.71	1.5%	$16,259	$2.14	2.3%	$12,846	
DIV-03 CONCRETE	$9.06	17.5%	$86,992	$8.23	4.8%	$49,401	$5.28	5.7%	$31,652	
DIV-04 MASONRY	$.76	1.4%	$7,329							
DIV-05 METALS	$5.51	10.6%	$52,923	$32.70	19.0%	$196,226	$19.93	21.8%	$119,553	
DIV-06 WOOD & PLASTICS	$.42	.8%	$4,051							
DIV-07 THERMAL & MOISTURE PROT.	$11.33	21.9%	$108,726	$17.51	10.2%	$105,047	$17.29	18.9%	$103,737	
DIV-08 DOORS & WINDOWS	$.72	1.3%	$6,899	$5.59	3.2%	$33,540	$1.55	1.6%	$9,285	
DIV-09 FINISHES	$.29	.5%	$2,809	$.99	.5%	$5,919	$.99	1.0%	$5,919	
DIV-10 SPECIALTIES	$.68	1.3%	$6,500							
DIV-11 EQUIPMENT	$.78	1.5%	$7,523	$1.55	.9%	$9,297	$8.22	8.9%	$49,297	
DIV-14 CONVEYING SYSTEMS				$41.19	24.0%	$247,121				
DIV-15 MECHANICAL	$7.80	15.1%	$74,880	$16.92	9.8%	$101,510	$14.64	16.0%	$87,844	
DIV-16 ELECTRICAL	$3.50	6.7%	$33,600	$7.01	4.0%	$42,062	$5.22	5.7%	$31,318	
** SECTION TOTAL	$51.64	100.0%	$495,722	$171.46	100.0%	$1,028,742	$91.42	100.0%	$548,513	

RECAP OF COMPARISON OF FOUR ESTIMATES (A B C D) IN ITEM (CSI DIVISION) CODE SEQUENCE FOR A SECTION OF PROJECT	ESTIMATE (A) WHATCO PROJECT SECTION-03:OFFICE ONLY File ID.:3256 Job Type:SCHEMATIC Job Loc.: Est Date:11/11/92			ESTIMATE (B) WHATCO PROJECT SECTION-03:OFFICE ONLY File ID.:256A Job Type:DESIGN DEVELOP Job Loc.: Est Date: 1/05/93			ESTIMATE (C) WHATCO PROJECT SECTION-03:OFFICE ONLY File ID.:256B Job Type:CONST. DOC. Job Loc.: Est Date: 2/02/93				Page: Date: at:
Description	6,400 $/SQFT	% of Sec.	TOTAL COST	14,000 $/SQFT	% of Sec.	TOTAL COST	14,000 $/SQFT	% of Sec.	TOTAL COST		
DIV-01 GENERAL REQUIREMENTS	$18.21	19.5%	$116,538	$17.11	21.6%	$239,538	$17.79	17.7%	$249,018		
DIV-02 SITEWORK	$.57	.6%	$3,663	$1.15	1.4%	$16,050	$1.15	1.1%	$16,050		
DIV-03 CONCRETE	$33.23	35.6%	$212,665	$4.06	5.1%	$56,880	$4.06	4.0%	$56,880		
DIV-04 MASONRY				$10.43	13.1%	$146,038	$10.43	10.3%	$146,038		
DIV-05 METALS				$16.92	21.3%	$236,828	$13.62	13.5%	$190,630		
DIV-06 WOOD & PLASTICS	$1.00	1.0%	$6,371								
DIV-07 THERMAL & MOISTURE PROT.	$4.14	4.4%	$26,468	$2.75	3.4%	$38,545	$2.75	2.7%	$38,545		
DIV-08 DOORS & WINDOWS	$3.52	3.7%	$22,515	$3.53	4.4%	$49,464	$3.53	3.5%	$49,464		
DIV-09 FINISHES	$13.42	14.4%	$85,898	$11.85	14.9%	$165,834	$11.85	11.7%	$165,834		
DIV-10 SPECIALTIES	$.78	.8%	$5,000	$.57	.7%	$8,016	$.57	.5%	$8,016		
DIV-11 EQUIPMENT	$.28	.3%	$1,800								
DIV-14 CONVEYING SYSTEMS				$2.76	3.4%	$38,588	$2.76	2.7%	$38,588		
DIV-15 MECHANICAL	$12.25	13.1%	$78,400	$5.11	6.4%	$71,542	$14.83	14.7%	$207,593		
DIV-16 ELECTRICAL	$5.75	6.1%	$36,800	$2.92	3.6%	$40,892	$17.18	17.1%	$240,586		
** SECTION TOTAL	$93.14	100.0%	$596,118	$79.16	100.0%	$1,108,215	$100.52	100.0%	$1,407,242		

C O M P A R I S O N OF FOUR ESTIMATES (A B C D) IN MAJOR ITEM (CSI) CODE SEQUENCE FOR A SECTION OF PROJECT	ESTIMATE (A) WHATCO PROJECT SECTION-01:PLANT ONLY File ID.:3256 Job Type:SCHEMATIC Job Loc.: Est Date:11/11/92			ESTIMATE (B) WHATCO PROJECT SECTION-01:PLANT ONLY File ID.:256A Job Type:DESIGN DEVELOP Job Loc.: Est Date: 1/05/93			ESTIMATE (C) WHATCO PROJECT SECTION-01:PLANT ONLY File ID.:256B Job Type:CONST. DOC. Job Loc.: Est Date: 2/02/93		
D e s c r i p t i o n	30,000 $/SQFT	% of Sec.	TOTAL COST	30,000 $/SQFT	% of Sec.	TOTAL COST	30,000 $/SQFT	% of Sec.	TOTAL COST
0100-GENERAL REQUIREMENTS				$22.01	16.8%	$660,230	$14.37	17.7%	$431,030
0198-ESCALATION	$7.52	11.5%	$225,576						
0199-CONTINGENCY	$5.26	8.0%	$157,746	$6.22	4.7%	$186,553			
0210-CLEARING OF SITE				$.10		$2,875	$.10	.1%	$2,875
0220-EXCAV, GRADING & BACKFILL	$.58	.8%	$17,545	$2.43	1.8%	$72,963	$2.16	2.6%	$64,712
0230-PILING & CAISSONS	$1.24	1.9%	$37,334	$2.83	2.1%	$84,809			
0300-CONCRETE FINISHING	$.67	1.0%	$20,054	$.68	.5%	$20,421	$.68	.8%	$20,421
0310-FORM WORK	$.82	1.2%	$24,527	$1.15	.8%	$34,479	$.12	.1%	$3,591
0320-REINFORCING STEEL	$.44	.6%	$13,303	$1.22	.9%	$36,694	$1.07	1.3%	$32,163
0330-CAST-IN-PLACE CONCRETE	$2.41	3.6%	$72,306	$3.68	2.8%	$110,320	$3.39	4.1%	$101,567
0340-PRECAST CONCRETE	$5.00	7.6%	$150,120						
0350-CEMENTITIOUS DECKS				$1.19	.9%	$35,700	$1.19	1.4%	$35,700
0500-STRUCTURAL METALS	$6.68	10.2%	$200,254	$31.42	24.0%	$942,456	$16.94	20.8%	$508,061
0550-MISC & ORNAMENTAL METAL	$.24	.3%	$7,249						
0600-ROUGH CARPENTRY	$.11	.1%	$3,339						
0700-WATERPROOF & DAMPPROOF				$.24	.1%	$7,188	$.20	.2%	$5,910
0720-INSULATION	$.08	.1%	$2,272	$.08		$2,368			
0740-PREFORMED ROOFING & SIDIN	$4.83	7.3%	$144,980	$7.65	5.8%	$229,590	$7.65	9.4%	$229,590
0750-ROOFING,SHEETMETAL&ACCESS	$3.50	5.3%	$105,000	$3.78	2.8%	$113,387	$3.78	4.6%	$113,387
0790-CAULKING & SEALANTS	$.10	.1%	$3,000						
0800-METAL DOORS & FRAMES	$.09	.1%	$2,837	$.06		$1,837	$.06		$1,837

C O M P A R I S O N OF FOUR ESTIMATES (A B C D) IN MAJOR ITEM (CSI) CODE SEQUENCE FOR A SECTION OF PROJECT	ESTIMATE (A) WHATCO PROJECT SECTION-01:PLANT ONLY File ID.:3256 Job Type:SCHEMATIC Job Loc.: Est Date:11/11/92			ESTIMATE (B) WHATCO PROJECT SECTION-01:PLANT ONLY File ID.:256A Job Type:DESIGN DEVELOP Job Loc.: Est Date: 1/05/93			ESTIMATE (C) WHATCO PROJECT SECTION-01:PLANT ONLY File ID.:256B Job Type:CONST. DOC. Job Loc.: Est Date: 2/02/93		
Description	30,000 $/SQFT	% of Sec.	TOTAL COST	30,000 $/SQFT	% of Sec.	TOTAL COST	30,000 $/SQFT	% of Sec.	TOTAL COST
0830-SPECIAL DOORS	$.73	1.1%	$21,844	$2.14	1.6%	$64,060	$.52	.6%	$15,550
0870-HARDWARE	$.10	.1%	$3,135	$.06		$1,937	$.06		$1,937
0990-PAINTING & WALL COVERING	$.02		$597	$.35	.2%	$10,632	$.35	.4%	$10,632
1100-EQUIPMENT	$.38	.5%	$11,285	$.62	.4%	$18,593	$.62	.7%	$18,593
1400-CONVEYING SYSTEMS	$11.67	17.8%	$350,000	$17.06	13.0%	$511,866	$17.06	21.0%	$511,866
1500-PLUMBING SYSTEMS & EQUIP	$.75	1.1%	$22,500						
1550-FIRE PROT SYSTEMS & EQUIP	$.33	.5%	$9,975						
1560-HVAC SYSTEMS & EQUIPMENT	$5.30	8.1%	$159,000	$14.00	10.7%	$419,910	$7.55	9.3%	$226,438
1600-ELEC SYSTEMS & EQUIPMENT	$6.50	9.9%	$195,000	$11.63	8.9%	$348,754	$3.33	4.1%	$99,963
** SECTION TOTAL	$65.36	100.0%	$1,960,778	$130.59	100.0%	$3,917,622	$81.19	100.0%	$2,435,823

COMPARISON OF FOUR ESTIMATES (A B C D) IN MAJOR ITEM (CSI) CODE SEQUENCE FOR A SECTION OF PROJECT	ESTIMATE (A) WHATCO PROJECT SECTION-02:WAREHOUSE ONLY File ID.:3256 Job Type:SCHEMATIC Job Loc.: Est Date:11/11/92			ESTIMATE (B) WHATCO PROJECT SECTION-02:WAREHOUSE ONLY File ID.:256A Job Type:DESIGN DEVELOP Job Loc.: Est Date: 1/05/93			ESTIMATE (C) WHATCO PROJECT SECTION-02:WAREHOUSE ONLY File ID.:2568 Job Type:CONST. DOC. Job Loc.: Est Date: 2/02/93				Page: 1 Date: at:
Description	9,600 $/SQFT	% of Sec.	TOTAL COST	6,000 $/SQFT	% of Sec.	TOTAL COST	6,000 $/SQFT	% of Sec.	TOTAL COST		
0100-GENERAL REQUIREMENTS				$28.90	16.8%	$173,372	$16.18	17.7%	$97,062		
0198-ESCALATION	$5.94	11.5%	$57,030								
0199-CONTINGENCY	$4.15	8.0%	$39,881	$8.16	4.7%	$48,988					
0220-EXCAV, GRADING & BACKFILL	$.69	1.3%	$6,579	$2.71	1.5%	$16,259	$2.14	2.3%	$12,846		
0300-CONCRETE FINISHING	$.67	1.3%	$6,434	$.69	.4%	$4,116	$.69	.7%	$4,116		
0310-FORM WORK	$1.37	2.6%	$13,127	$2.44	1.4%	$14,654	$.25	.2%	$1,526		
0320-REINFORCING STEEL	$.53	1.0%	$5,042	$1.02	.6%	$6,129	$.75	.8%	$4,501		
0330-CAST-IN-PLACE CONCRETE	$1.49	2.8%	$14,351	$2.89	1.6%	$17,362	$2.39	2.6%	$14,369		
0340-PRECAST CONCRETE	$5.00	9.6%	$48,038								
0350-CEMENTITIOUS DECKS				$1.19	.6%	$7,140	$1.19	1.3%	$7,140		
0400-MASONRY	$.76	1.4%	$7,329								
0500-STRUCTURAL METALS	$5.20	10.0%	$49,923	$32.70	19.0%	$196,226	$19.93	21.8%	$119,553		
0550-MISC & ORNAMENTAL METAL	$.31	.6%	$3,000								
0600-ROUGH CARPENTRY	$.42	.8%	$4,051								
0700-WATERPROOF & DAMPPROOF				$.27	.1%	$1,641	$.20	.2%	$1,182		
0720-INSULATION	$.12	.2%	$1,136	$.14		$851					
0740-PREFORMED ROOFING & SIDIN	$7.55	14.6%	$72,490	$12.76	7.4%	$76,530	$12.76	13.9%	$76,530		
0750-ROOFING,SHEETMETAL&ACCESS	$3.50	6.7%	$33,600	$4.34	2.5%	$26,025	$4.34	4.7%	$26,025		
0790-CAULKING & SEALANTS	$.16	.3%	$1,500								
0800-METAL DOORS & FRAMES	$.24	.4%	$2,270	$.12		$735	$.12	.1%	$735		
0830-SPECIAL DOORS	$.22	.4%	$2,121	$5.34	3.1%	$32,030	$1.30	1.4%	$7,775		

Page: 2
Date:
at:

COMPARISON OF FOUR ESTIMATES (A B C D) IN MAJOR ITEM (CSI) CODE SEQUENCE FOR A SECTION OF PROJECT	ESTIMATE (A) WHATCO PROJECT SECTION-02:WAREHOUSE ONLY File ID.:3256 Job Type:SCHEMATIC Job Loc.: Est Date:11/11/92			ESTIMATE (B) WHATCO PROJECT SECTION-02:WAREHOUSE ONLY File ID.:256A Job Type:DESIGN DEVELOP Job Loc.: Est Date: 1/05/93			ESTIMATE (C) WHATCO PROJECT SECTION-02:WAREHOUSE ONLY File ID.:256B Job Type:CONST. DOC. Job Loc.: Est Date: 2/02/93			
Description	9,600 $/SQFT	% of Sec.	TOTAL COST	6,000 $/SQFT	% of Sec.	TOTAL COST	6,000 $/SQFT	% of Sec.	TOTAL COST	
0870-HARDWARE	$.26	.5%	$2,508	$.13		$775	$.13	.1%	$775	
0925-GYPSUM DRYWALL	$.05	.1%	$487							
0990-PAINTING & WALL COVERING	$.24	.4%	$2,322	$.99	.5%	$5,919	$.99	1.0%	$5,919	
1000-SPECIALTIES	$.68	1.3%	$6,500							
1100-EQUIPMENT	$.78	1.5%	$7,523	$1.55	.9%	$9,297	$8.22	8.9%	$49,297	
1400-CONVEYING SYSTEMS				$41.19	24.0%	$247,121				
1500-PLUMBING SYSTEMS & EQUIP	$2.00	3.8%	$19,200							
1550-FIRE PROT SYSTEMS & EQUIP	$1.30	2.5%	$12,480							
1560-HVAC SYSTEMS & EQUIPMENT	$4.50	8.7%	$43,200	$16.92	9.8%	$101,510	$14.64	16.0%	$87,844	
1600-ELEC SYSTEMS & EQUIPMENT	$3.50	6.7%	$33,600	$7.01	4.0%	$42,062	$5.22	5.7%	$31,318	
** SECTION TOTAL	$51.64	100.0%	$495,722	$171.46	100.0%	$1,028,742	$91.42	100.0%	$548,513	

Page: 1
Date:
at:

COMPARISON OF FOUR ESTIMATES (A B C D) IN MAJOR ITEM (CSI) CODE SEQUENCE FOR A SECTION OF PROJECT	ESTIMATE (A) WHATCO PROJECT SECTION-03:OFFICE ONLY File ID.:3256 Job Type:SCHEMATIC Job Loc.: Est Date:11/11/92			ESTIMATE (B) WHATCO PROJECT SECTION-03:OFFICE ONLY File ID.:256A Job Type:DESIGN DEVELOP Job Loc.: Est Date: 1/05/93			ESTIMATE (C) WHATCO PROJECT SECTION-03:OFFICE ONLY File ID.:256B Job Type:CONST. DOC. Job Loc.: Est Date: 2/02/93		
Description	6,400 $/SQFT	% of Sec.	TOTAL COST	14,000 $/SQFT	% of Sec.	TOTAL COST	14,000 $/SQFT	% of Sec.	TOTAL COST
0100-GENERAL REQUIREMENTS				$13.34	16.8%	$186,766	$17.79	17.7%	$249,018
0198-ESCALATION	$7.49	8.0%	$47,958						
0199-CONTINGENCY	$10.72	11.5%	$68,580	$3.77	4.7%	$52,772			
0220-EXCAV, GRADING & BACKFILL	$.57	.6%	$3,663	$1.15	1.4%	$16,050	$1.15	1.1%	$16,050
0300-CONCRETE FINISHING	$.42	.4%	$2,693	$.68	.8%	$9,464	$.68	.6%	$9,464
0310-FORM WORK	$.47	.5%	$2,987	$.33	.4%	$4,633	$.33	.3%	$4,633
0320-REINFORCING STEEL	$.41	.4%	$2,635	$.94	1.1%	$13,226	$.94	.9%	$13,226
0330-CAST-IN-PLACE CONCRETE	$1.24	1.3%	$7,934	$1.52	1.9%	$21,227	$1.52	1.5%	$21,227
0340-PRECAST CONCRETE	$30.69	32.9%	$196,416						
0350-CEMENTITIOUS DECKS				$.60	.7%	$8,330	$.60	.5%	$8,330
0400-MASONRY				$10.29	13.0%	$144,038	$10.29	10.2%	$144,038
0440-STONE				$.14	.1%	$2,000	$.14	.1%	$2,000
0500-STRUCTURAL METALS				$16.62	21.0%	$232,736	$13.32	13.2%	$186,538
0550-MISC & ORNAMENTAL METAL				$.29	.3%	$4,092	$.29	.2%	$4,092
0600-ROUGH CARPENTRY	$.50	.5%	$3,187						
0620-FINISH CARPENTRY	$.50	.5%	$3,184						
0700-WATERPROOF & DAMPPROOF				$.12	.1%	$1,639	$.12	.1%	$1,639
0720-INSULATION	$.45	.4%	$2,868	$.77	.9%	$10,826	$.77	.7%	$10,826
0750-ROOFING,SHEETMETAL&ACCESS	$3.50	3.7%	$22,400	$1.86	2.3%	$26,080	$1.86	1.8%	$26,080
0790-CAULKING & SEALANTS	$.19	.2%	$1,200						
0800-METAL DOORS & FRAMES	$.09	.1%	$568	$.56	.7%	$7,893	$.56	.5%	$7,893

COMPARISON OF FOUR ESTIMATES (A B C D) IN MAJOR ITEM (CSI) CODE SEQUENCE FOR A SECTION OF PROJECT	ESTIMATE (A) WHATCO PROJECT SECTION-03:OFFICE ONLY File ID.:3256 Job Type:SCHEMATIC Job Loc.: Est Date:11/11/92			ESTIMATE (B) WHATCO PROJECT SECTION-03:OFFICE ONLY File ID.:256A Job Type:DESIGN DEVELOP Job Loc.: Est Date: 1/05/93			ESTIMATE (C) WHATCO PROJECT SECTION-03:OFFICE ONLY File ID.:256B Job Type:CONST. DOC. Job Loc.: Est Date: 2/02/93		
Description	6,400 $/SQFT	% of Sec.	TOTAL COST	14,000 $/SQFT	% of Sec.	TOTAL COST	14,000 $/SQFT	% of Sec.	TOTAL COST
0820-WOOD & PLASTIC DOORS	$.81	.8%	$5,176	$.48	.6%	$6,773	$.48	.4%	$6,773
0850-METAL WINDOWS	$1.69	1.8%	$10,800	$1.06	1.3%	$14,884	$1.06	1.0%	$14,884
0870-HARDWARE	$.56	.6%	$3,574	$1.22	1.5%	$17,047	$1.22	1.2%	$17,047
0880-GLASS & GLAZING	$.37	.4%	$2,397	$.20	.2%	$2,867	$.20	.2%	$2,867
0925-GYPSUM DRYWALL	$6.81	7.3%	$43,569	$3.79	4.7%	$53,005	$3.79	3.7%	$53,005
0930-TILE & TERRAZZO	$.57	.6%	$3,654	$.56	.7%	$7,808	$.56	.5%	$7,808
0950-ACOUSTICAL TREATMENT	$1.14	1.2%	$7,309	$1.49	1.8%	$20,910	$1.49	1.4%	$20,910
0955-FLOORING	$2.62	2.8%	$16,773	$1.98	2.5%	$27,748	$1.98	1.9%	$27,748
0990-PAINTING & WALL COVERING	$2.28	2.4%	$14,593	$4.03	5.0%	$56,363	$4.03	4.0%	$56,363
1000-SPECIALTIES	$.78	.8%	$5,000	$.57	.7%	$8,016	$.57	.5%	$8,016
1100-EQUIPMENT	$.28	.3%	$1,800						
1400-CONVEYING SYSTEMS				$2.76	3.4%	$38,588	$2.76	2.7%	$38,588
1500-PLUMBING SYSTEMS & EQUIP	$2.75	2.9%	$17,600				$1.63	1.6%	$22,872
1550-FIRE PROT SYSTEMS & EQUIP	$1.50	1.6%	$9,600				$1.89	1.8%	$26,468
1560-HVAC SYSTEMS & EQUIPMENT	$8.00	8.5%	$51,200	$5.11	6.4%	$71,542	$11.30	11.2%	$158,253
1600-ELEC SYSTEMS & EQUIPMENT	$5.75	6.1%	$36,800	$2.92	3.6%	$40,892	$17.18	17.1%	$240,586
** SECTION TOTAL	$93.14	100.0%	$596,118	$79.16	100.0%	$1,108,215	$100.52	100.0%	$1,407,242

TABLE 10.8 Comparison of Construction Document Estimates (without Site) in Dollars

	Manual (from Table 10.5)	Composer Gold	I.C.E.
Plant	$2,758,559	$2,535,360	$2,435,823
Warehouses	643,309	622,720	548,513
Office building	1,591,958	1,595,720	1,407,242
Total	$4,993,826	$4,755,800	$4,391,578

Tables 10.8, 10.9, and 10.10 compare the three construction document estimates:

- Table 10.8 is a comparison in dollar estimates (total, plus subsets for plant, warehouses, and office buildings).
- Table 10.9 is a comparison using the manual estimate as the base reference (i.e., 100%).
- Table 10.10 is a comparison by cost per square foot (total, plus subsets).

TABLE 10.9 Comparison of Construction Document Estimates (without Site) in Percent

	Manual	Composer Gold	I.C.E.
Plant	100%	92%	88%
Warehouses	100%	97%	85%
Office building	100%	100%	88%

TABLE 10.10 Comparison of Construction Document Estimates (without Site) in Cost/sf

	Manual	Composer Gold	I.C.E.
Plant	91.97	84.50	81.20
Warehouses	107.17	103.83	91.50
Office building	113.71	114.00	100.50

Chapter

11

Reconciliation

The conceptual estimate is the basis for the budget. In some companies, the concept estimate is the basis for an appropriation request (AR) or similar term. Once the budget is approved, each level of estimate (schematic, design development, construction documents) should be reconciled with the budget. This chapter describes a typical reconciliation.

The reconciliation should follow the format of the budget for a one-to-one check:

- Base construction cost
  - Subsets
- Escalation to construction midpoint
- Contingency
- Design fees
- Construction management (CM) fees
- Owner's project management oversight (PMO)
- Land cost

Base Construction Cost

This is the category that estimators are most involved in. At each chronological phase more is known about the specifics of the construction scope. This is evidenced by evolution of the drawings and development of the specifications.

A second estimate may be available at some point (or points). An example of such a point would be at value engineering. The value team is often authorized to prepare an independent estimate.

If a second estimate is available, the estimators cross-check by CSI division and by major equipment item. The purpose of the cross-check is not to determine which estimate is the "right" one. An estimate is never more than a best evaluation of anticipated cost based upon various factors, some identified, others assumed. The estimator's goal in cross-checking is to identify either omissions or specific instances where one estimate appears more probable than the others. The optimum approach is to have the two estimating teams conduct a resolution of the two base cost estimates into a single baseline. This optimum approach is not used in the majority of opportunities.

Some differences between estimates cannot be reconciled because they are due to selection of different approaches to markups. Examples of this include:

- *Profit.* Prebid percentage is a judgment call. In periods of high construction activity, 10% profit factor is the norm. In slow times, 5% is more the norm.
- *General conditions.* Some estimators use a factor of direct labor, while others use a percentage of base construction labor and materials.
- *Subcontracts.* Usual practice is for estimators to add a factor for profit to be taken by subcontractors.

The markups above may be constrained by the format of the estimate. For instance, an estimator may prefer to use a factor based upon direct labor. However, if the estimate provides only a total cost per item, the factor must then be adjusted accordingly.

Escalation to Construction Midpoint

The base construction cost should be a "today" cost. The escalation to construction midpoint extrapolates the "today" cost to the midpoint of construction. The variables are projected rate of inflation and the projected duration of the project.

The rate of inflation can be an average of one to three prior years, with cognizance of the trend (i.e., were prior years trending up or down).

Contingency

This factor will vary with the type of estimate. The early (i.e., conceptual and schematic) estimates should carry the higher contingency factor. A subjective evaluation of the specific project will decide whether the factor will be low or high in the range. Typical ranges of contingencies would be:

Estimate	Contingency range, %
Concept	15–20
Schematic	10–15
Design development	5–10
Construction document	0–5

Design Fees

These fees should follow professional (AIA, ASCE) guidelines. A typical design fee would be 6% of base design cost. Current practice is to put a cap on the fee—that is, the designer is not rewarded for a cost overrun. Another current practice requires the designer to design-to-cost, that is, redesign so that the project can be successfully rebid.

If properly structured, the design fee should stay within budget.

Construction Management Fee

These fees should follow professional (CMAA) guidelines. Typical construction management (CM) fees are in the range of 3 to 5% of base construction and can increase with increase in inspection (QC) scope. A major portion of CM fees is linear with time, so that time overruns in construction will require change in scope for the CM.

Project Management Oversight

This cost can be in-house staff where the owner has technical staff or can be on consulting basis where the owner either does not have in-house technical staff or does not want to increase staff. A reasonable range for budget costs would be 1 to 2%.

Land Cost

This category is not usually in the estimator's purview. If the site is not selected but the acreage is known, real estate appraisers would be a good source of a reasonable budget.

Change Orders

For budget purposes, it is reasonable to add a reserve of 5% for change orders.

WHATCO Project

The principal part of the WHATCO project still unestimated is the site work. Figure 11.1 is the WHATCO site plan. Table 11.1 is a sum-

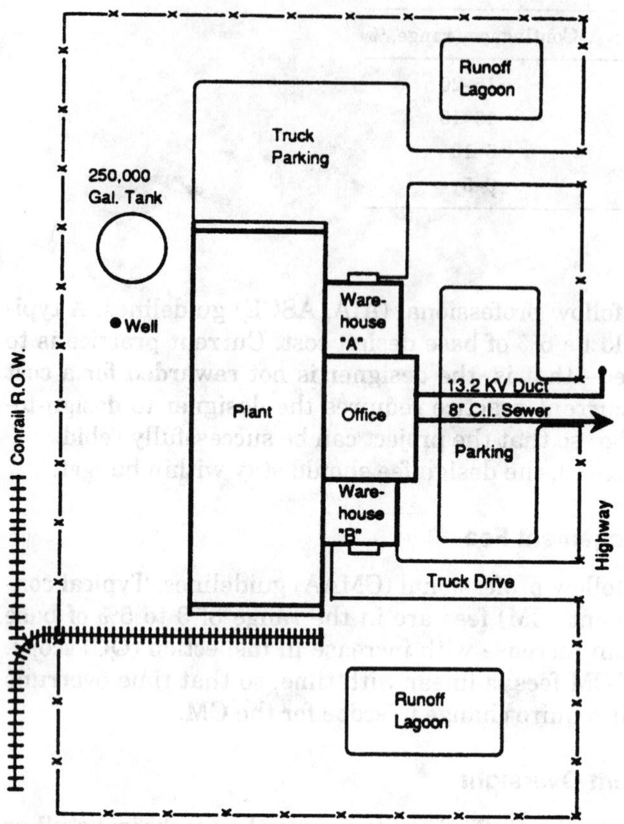

Figure 11.1 WHATCO site plan.

TABLE 11.1 Summary of Quantities and Costs, Site Work

Item	CSI No.	R.S. Means ID No.	Quantity	Unit	1995 Unit Price	Cost
Site clearing, trees to 12 in, cut and chip	02110	021 104 0200	5	acre	4,530.00	22,650
Grub stumps and remove	02110	021 040 0250	5	acre	2,780.00	13,900
Earthwork (lagoons)	02210	022 242 2020	6000	cy	2.15	12,900
Trenching for piping	02210	022 258 0750	1500	lf	1.35	2,025
Backfill for trenches	02210	022 258 1750	1500	lf	2.60	3,900
Base course for paved areas	02230	022 308 0304	4700	sy	11.90	55,930
RR turnout, No. 8	02452	024 524 2300	1	each	24,750.00	24,750
RR car bumper	02452	024 524 0100	1	each	3,950.00	3,950
Fence, 6 ft high, 3 strands of wire, aluminized steel	02830	028 308 0300	1900	lf	17.80	33,820

TABLE 11.1 Summary of Quantities and Costs, Site Work (*Continued*)

Item	CSI No.	R. S. Means ID No.	Quantity	Unit	1995 Unit Price	Cost
Gates (3), double swing, 6 ft high, 12 ft open	02830	028 308 5060	3	each	860.00	2,580
Motor operators for 3 double swing 12 ft	02830	023 308 27103	each	2,500.00	7,500	
Electrical service for swing gates	NA	Quote	1	ls	Quote	8,000
RR gate, 20-ft opening	02830	028 308 5070	1	each	1,050.00	1,050
Motor operator for 20-ft double swing	02830	028 308 2800	1	each	4,350.00	4,350
RR siding includes ties and ballast	02452	024 524 0810	220	lf	98.00	21,560
Curb, concrete, 6 in × 18 in, straight	02525	025 254 0300	1200	lf	8.55	10,260
Curb, concrete, 6 in × 18 in, radius	02525	025 254 0400	200	lf	15.15	3,030
Paving, bituminous, 2 in binder course	02510	025 104 0120	4700	sy	4.88	22,936
Paving, 2 in wearing course	02510	025 104 0380	4700	sy	5.26	24,722
Industrial well	02670	Quote	1	ls		25,000
Piping, rain spouts to lagoon, RCP, 24 in	02715	027 162 1080	700	lf	21.30	14,910
Lawn hydroseed turf mix	02930	029 308 5000	120	msf	50.00	6,000
Sod	02930	029 316 1000	32	msf	500.00	16,000
Water tower, elevated, 250,000 gal	13210	132 101 3400	1	each	390,000.00	390,000
8-in steel pipe	15170	151 701 0680	150	lf	81.25	12,188
4-in steel pipe	15170	151 701 0650	50	lf	32.50	1,625
4-in steel (cast) angle valve, 150 lb, flanged	15195	151 980 0860	1	each	2,500.00	2,500
8-in steel (cast) angle valve, 150 lb, flanged	15195	151 980 0880	1	each	6,560.00	6,560
8-in cast-iron pipe	15130	151 301 2220	200	lf	39.50	7,900
High-voltage entry duct, 13.2 kV	16375	NA	1	ls	Quote	9,600
Total						$772,096

mary of the site quantities and costs (manual estimate). The unit costs include 21.5% overhead, bond, and profit. The total of $772,096 is within 2% of the budget figure.

Following is an I.C.E. (Management Computer Controls, Inc.) site work estimate showing a site work total of $873,052 broken down as follows:

1. Cost breakdown by 11 categories
2. Detailed cost by 11 categories

File ID. 256S
Project No.
5 ACRE

MANAGEMENT COMPUTER CONTROLS, INC.
2881 DIRECTORS COVE
MEMPHIS, TENNESSEE 38131
WHATCO PROJECT - SITE WORK

Page 1

Description	Qty	Unit	Unit Price	Total Price	5 $/ACRE
GENERAL REQUIREMENTS				$154,491	30,898
CLEARING OF SITE				$32,506	6,501
EXCAV, GRADING & BACKFILL				$17,002	3,400
SITE DRAINAGE & UTILITIES				$51,738	10,347
ROADS & WALKS				$103,289	20,657
SITE IMPROVEMENTS				$42,168	8,433
LAWNS & PLANTING				$16,440	3,288
RAILROAD WORK				$45,992	9,198
SPECIAL CONSTRUCTION				$363,825	72,765
PLUMBING SYSTEMS & EQUIP				$26,307	5,261
ELEC SYSTEMS & EQUIPMENT				$19,294	3,858
***** ESTIMATE TOTAL				$873,052	174,610

File ID. 256S
Project No.
5 ACRE

MANAGEMENT COMPUTER CONTROLS, INC.
2881 DIRECTORS COVE
MEMPHIS, TENNESSEE 38131
WHATCO PROJECT - SITE WORK

Page 1

Description	Qty	Unit	Unit Price	Total Price	5 $/ACRE
GENERAL REQUIREMENTS					
OTHER GENERAL REQUIREMENTS					
OH/PROFIT/GC/BOND @ 21.5%	1	LS	154,491.00	154,491	30,898
** TOTAL OTHER GENERAL REQUIREMENTS				154,491	30,898
*** TOTAL GENERAL REQUIREMENTS				154,491	30,898
CLEARING OF SITE					
SITE CLEARING,CUT & CHIP	5	ACRE	3,880.80	19,404	3,880
GRUB STUMPS & REMOVE	5	ACRE	2,620.40	13,102	2,620
*** TOTAL CLEARING OF SITE				32,506	6,501
EXCAV, GRADING & BACKFILL					
SITE GRADING					
EARTHWORK, LAGOONS	6,000	CUYD	1.95	11,712	2,342
TRENCHING FOR PIPING	1,500	LNFT	1.29	1,946	389.20
BACKFILL FOR TRENCHES	1,500	LNFT	2.22	3,344	668.80
** TOTAL SITE GRADING				17,002	3,400
*** TOTAL EXCAV, GRADING & BACKFILL				17,002	3,400
SITE DRAINAGE & UTILITIES					
SITE DRAINAGE					
24"RCP,RAINSPT TO LAGOON	700	LNFT	18.78	13,150	2,630
** TOTAL SITE DRAINAGE				13,150	2,630

Item	Quantity	Unit Cost	Total	
MISC SITE UTILITIES				
INDUSTRIAL WELL	1 LS	38,588.00	38,588	7,717
** TOTAL MISC SITE UTILITIES			38,588	7,717
*** TOTAL SITE DRAINAGE & UTILITIES			51,738	10,347
ROADS & WALKS				
ASPHALT PAVING				
12" BASE COURSE	4,700 SQYD	11.06	52,020	10,404
PAVING, 2" BINDER COURSE	4,700 SQYD	4.01	18,885	3,777
PAVING, 2" WEARING COURSE	4,700 SQYD	4.46	20,971	4,194
** TOTAL ASPHALT PAVING			91,876	18,375
CURB & GUTTERS				
6X18 CONC CURB, STRAIGHT	1,200 LNFT	7.33	8,807	1,761
6X18 CONC CURB, RADIUS	200 LNFT	13.03	2,606	521.20
** TOTAL CURB & GUTTERS			11,413	2,282
*** TOTAL ROADS & WALKS			103,289	20,657
SITE IMPROVEMENTS				
FENCES				
6' FENCE	1,900 LNFT	14.61	27,760	5,552
GATE-12' OPNG, 6' HT	3 EACH	770.33	2,311	462.20
RR GATE-20' OPNG	1 EACH	939.00	939	187.80
MOTOR OPERATOR @ GATE	3 EACH	2,453.00	7,359	1,471
MOTOR OPR @ RR GATE	1 EACH	3,799.00	3,799	759.80
** TOTAL FENCES			42,168	8,433
*** TOTAL SITE IMPROVEMENTS			42,168	8,433
LAWNS & PLANTING				
LAWN HYDROSEED	120 MSF	32.27	3,873	774.60
SOD	32 MSF	392.71	12,567	2,513
*** TOTAL LAWNS & PLANTING			16,440	3,288

File ID. 256S
Project No.
5 ACRE

MANAGEMENT COMPUTER CONTROLS, INC.
2881 DIRECTORS COVE
MEMPHIS, TENNESSEE 38131
WHATCO PROJECT - SITE WORK

Description	Qty	Unit	Unit Price	Total Price	5 $/ACRE
RAILROAD WORK					
#8 RR TURNOUT	1	EACH	22,492.00	22,492	4,498
RR CAR BUMPER	1	EACH	3,272.00	3,272	654.40
RR SIDING	220	LNFT	91.94	20,228	4,045
*** TOTAL RAILROAD WORK				45,992	9,198
SPECIAL CONSTRUCTION					
MISC SPECIAL CONSTRUCTION					
250,000 GAL WATER TOWER	1	EACH	363,825.00	363,825	72,765
** TOTAL MISC SPECIAL CONSTRUCTION				363,825	72,765
*** TOTAL SPECIAL CONSTRUCTION				363,825	72,765
PLUMBING SYSTEMS & EQUIP					
PIPE & FITTINGS					
8" STEEL PIPE	150	LNFT	71.84	10,777	2,155
4" STEEL PIPE	50	LNFT	31.26	1,563	312.60
8" STL ANGLE VALVE,150#	1	EACH	4,530.00	4,530	906.00
4" STL ANGLE VALVE,150#	1	EACH	1,384.00	1,384	276.80
8" CAST IRON PIPE	200	LNFT	40.26	8,053	1,610
** TOTAL PIPE & FITTINGS				26,307	5,261
*** TOTAL PLUMBING SYSTEMS & EQUIP				26,307	5,261
ELEC SYSTEMS & EQUIPMENT					
ELECTRICAL SYSTEMS					
ELECTRICAL SERVICE @ GATE	1	LS	2,756.00	2,756	551.20
13.2KV HIGH VOLT ENTRY DU	1	LS	16,538.00	$16,538	3,307
** TOTAL ELECTRICAL SYSTEMS				$19,294	3,858
*** TOTAL ELEC SYSTEMS & EQUIPMENT				$19,294	3,858
***** ESTIMATE TOTAL				$873,052	174,610

Following is the site work estimate ($882,200) by Composer Gold (Building Systems Design):

1. Summary with markups and escalation
2. Basis of estimate
3. Summary of indirects on subcontracts
4. Detailed estimate broken down by units, labor, equipment, subcontracts, total cost

Building Systems Design

PROJECT WHATCS: Whatco Mfg-Whse-Office Sitework

Pre-Bid Design Review Cost Estimate

** PROJECT DIRECT SUMMARY - LEVEL 1 (Rounded to 100's) **

SUMMARY PAGE 1

	QUANTITY UOM	Labor	Equipmnt	Material	SubContr	TOTAL COST	UNIT COST
40 Site Support	5.00 ACR	82,000	36,000	189,900	375,900	683,900	136783.57
Whatco Mfg-Whse-Office Sitework	5.00 ACR	82,000	36,000	189,900	375,900	683,900	136783.57
Overhead @ 8.0 %						41,000	8207.01
SUBTOTAL						725,000	144990.58
Profit @ 7.0 %						36,200	7249.53
SUBTOTAL						761,200	152240.11
Bond @ 1.5 %						6,000	1192.54
TOTAL INCL INDIRECTS						767,200	153432.66
Escalation @ 5.0 % / Annum						115,100	23014.90
TOTAL INCL OWNER COSTS						882,200	176447.55

BASIS OF ESTIMATE

This estimate was produced from Design Development information received from James J O'Brien, P.E. of O'BREIN - KREITZBERG & ASSOCIATES, INC. on January 27th, 1993.

Sketch Plans & Designs	: Dated	"NOT DATED"
Original Specifications	: Dated	"NOT DATED"

The estimate has been compliled using the 1993 R.S.Means Facilities Data and Davis Bacon wage rates for Philadelphia, Pennsylvania.

BASIS FOR PRICING

Pricing shown reflects probable construction costs obtainable in the Southeastern Pennsylvania area on the date of this statement of costs. This estimate is a determination of fair market value for the construction of this project. It is not a prediction of low bid. Pricing assumes competitive bidding for every portion of the construction work for all subcontractors, as well as the general contractor; that is to mean 6 to 7 bids. If less bids are received, bid results can be expected to be higher.

Length of construction is assumed to be 12 months. Any costs for excessive overtime to meet stringent milestone dates are not included in this estimate.

Bid date is assumed to be April 1995. A value of 5% per annual escalation is added to the cost for the construction which is assumed to be completed March 1996.

The General Contractor's Overhead is set at 8% and his Profit margins are set at 7% on all of his work, and 5% on all of his Subcontractors. Bond is set at 1.5%. The Subcontractors have been properly teired for this project and this is shown in the Indirect Cost Summary Report.

PROJECT DESCRIPTION

This project consists of : The sitework package associated with the construction of a new Plant-Office-Warehouse on a site in Philadelphia, Pennsylvania.

STATEMENT OF PROBABLE COST

Building Systems Design has no control over the cost of labor and materials, the general contractor's or any subcontractor's method of determining prices, or competitive bidding and market conditions. This opinion of probable cost of construction is made on the basis of experience, qualifications, and best judgement of Building Systems Design estimators familar with the construction industry. We cannot and do not guarantee that proposals, bids or actual construction costs will not vary from this or subsequent cost estimates.

Building Systems Design
PROJECT WHATCS: Whatco Mfg-Whse-Office Sitework
Pre-Bid Design Review Cost Estimate

** CONTRACTOR INDIRECT SUMMARY (Rounded to 100's) **

	TOTAL DIRECT	Overhead	Profit	Bond	TOTAL COST	UNIT COST
GM General Markup / Prime Contract						
SW Sitework Contractor						
PV Paving Subcontractor	85,700	6,000	5,500	0	97,100	97146.84
FE Fencing Subcontractor	35,000	2,800	3,800	0	41,600	41619.25
LS Landscaping Subcontractor	16,300	1,000	1,400	0	18,600	18648.34
Subtotal Subcontract Work	137,000	9,800	10,700	0	157,400	157414.43
Indirect on Subcontracts	157,400	9,400	8,300	1,600	176,800	176800.52
Indirect on Own Work	86,700	6,900	6,600	900	101,100	101085.21
SW Sitework Contractor	244,100	16,400	14,900	2,500	277,900	277885.72
TK Water Tank Contractor	260,600	26,100	28,700	4,700	320,000	319994.48
EL Electrical Contractor	17,600	1,300	1,500	0	20,400	20433.60
PL Plumbing Contractor						
WE Well Drilling Subcontractor	25,000	2,500	4,100	0	31,600	31625.00

Subtotal Subcontract Work	25,000	2,500	4,100	0	31,600	31625.00
Indirect on Subcontracts	31,600	1,900	1,700	300	35,500	35518.15
Indirect on Own Work	26,300	1,800	1,700	300	30,100	30085.88
PL Plumbing Contractor	57,900	3,700	3,400	600	65,600	65604.04
Subtotal Subcontract Work	580,200	47,500	48,400	7,800	683,900	683917.84
Indirect on Subcontracts	683,900	41,000	36,200	6,000	767,200	767163.28
GM General Markup / Prime Contract	683,900	41,000	36,200	6,000	767,200	767163.28

Building Systems Design
PROJECT WHATCS: Whatco Mfg-Whse-Office Sitework
Pre-Bid Design Review Cost Estimate
40. Site Support

DETAILED ESTIMATE

DETAIL PAGE 1

40.10. Site Preparation	QUANTY UOM	Labor	Equipmnt	Material	SubContr	TOTAL COST
40. Site Support						
40.10. Site Preparation						
02110 40000 Clear And Grub						
RSM SW Grub Stumps & Remove		346.11	750.10	0.00	0.00	1096.21
	5.00 ACR	2,017	4,372	0	0	6,390
RSM SW Clear Medium Trees To 12" Dia.		1943.02	1489.43	0.00	0.00	3432.45
Cut & Chip	5.00 ACR	11,325	8,682	0	0	20,007
02224 20000 Excavating, Bulk, Dozer						
RSM SW Earthwork (Lagoons)		0.93	0.73	0.00	0.00	1.66
Excav Bulk 75Hp Dozer Push 50'	6000.00 CY	6,498	5,103	0	0	11,600
Common Earth						

Item	Quantity					
02225 80000 Excavating, Utility Trench						
RSM SW Trenching for Piping--Common Earth, Chain Trnchr 12Hp Operator walking 8"W 36"Dp	1500.00 LF	0.81 1,409	0.26 460	0.00 0	0.00 0	1.07 1,870
RSM SW Backfill Utility Trench By Hand W/Compctn 8"X36"D	1500.00 LF	1.60 2,799	0.43 759	0.00 0	0.00 0	2.03 3,558
Site Preparation		24,049	19,376	0	0	43,425
40.15. Site Improvements						
40.15.10. Paving						
02230 80000 Base Course						
RSM PV Base Course, Roadways & Large Areas, 1-1/2" Stone, 12" Deep	4700.00 SY	0.66 3,963	0.62 3,718	7.80 46,700	0.00 0	9.08 54,381
02510 40000 Asphaltic Concrete Pavement						
RSM PV Asphaltic Concrete Pavement, Paving, Binder Course, 2" Thick	4700.00 SY	0.40 2,393	0.25 1,510	2.67 15,986	0.00 0	3.32 19,888
RSM PV Asphaltic Concrete Pavement, Paving, Wearing Course, 2" Thick	4700.00 SY	0.44 2,642	0.29 1,721	2.95 17,662	0.00 0	3.68 22,025
02525 40000 Curbs						

Building Systems Design
PROJECT WHATCS: Whatco Mfg-Whse-Office Sitework

DETAILED ESTIMATE — Pre-Bid Design Review Cost Estimate — DETAIL PAGE 2

40. Site Support

40.15. Site Improvements	QUANTY UOM	Labor	Equipmnt	Material	SubContr	TOTAL COST
RSM PV Curbs,Concrete,Cast In Place		3.26	0.07	3.10	0.00	6.42
Straight,6"X18"	1200.00 LF	4,980	101	4,739	0	9,820
RSM PV Curbs,Concrete,Cast In Place		8.14	0.16	3.45	0.00	11.76
Radius,6"X18"	200.00 LF	2,075	42	879	0	2,996
Paving		16,052	7,092	85,967	0	109,111
40.15.20. Fencing						
02830 80000 Fence, Chain Link Industrial						
RSM FE Fence,Chainlink,6'H,Barbwire		3.53	1.71	6.90	0.00	12.14
2"Post,1-5/8"Rail,9Ga Wire,	1900.00 LF	8,957	4,328	17,493	0	30,778
Aluminized						
RSM FE Motor Operator For up to 15'Wide		441.65	213.40	1375.00	0.00	2030.05
Swing Gate, No Wiring	3.00 OPN	1,768	854	5,504	0	8,126

RSM FE Motor Operator For Gate,No Gate		441.65	213.40	2475.00	0.00	3130.05
Or Electric Wiring	1.00 OPN	589	285	3,302	0	4,176
RSM FE Double Swing Gates,Incl. Posts &		276.03	133.38	240.00	0.00	649.40
Hardware 6' High 12' Opening	3.00 OPN	1,105	534	961	0	2,600
RSM FE Rail Road Double Swing Gates,20'		339.73	164.15	294.00	0.00	797.88
Incl. Posts & Hardware 6' High	1.00 OPN	453	219	392	0	1,065
Fencing		12,873	6,220	27,652	0	46,745

40.15.30. Landscaping

02930 80000 Seeding

RSM LS Seeding;Fescue 5.5#,Hydro Or Air		8.68	6.74	23.00	0.00	38.41
Including Mulch&Fertilizer	120.00 MSF	1,339	1,040	3,549	0	5,927

02931 60000 Sodding,

RSM LS Sodding;Bent Grass Sod,On Level		55.31	4.69	305.00	0.00	365.00
Ground,Over 6M.S.F.	32.00 MSF	2,276	193	12,549	0	15,018
Landscaping		3,614	1,233	16,098	0	20,945

Building Systems Design
PROJECT WHATCS: Whatco Mfg-Whse-Office Sitework
DETAILED ESTIMATE — Pre-Bid Design Review Cost Estimate — DETAIL PAGE 3
40. Site Support

40.15. Site Improvements	QUANTY UOM	Labor	Equipmnt	Material	SubContr	TOTAL COST
40.15.40. Industrial Well						
OBK WE Industrial Well		0.00	0.00	0.00	25000.00	25000.00
	1.00 LS	0	0	0	35,518	35,518
Industrial Well		0	0	0	35,518	35,518
40.15.50. Plumbing						
02716 20000 Piping, Drainage & Sewage, Concrete						
RSM SW Piping, Rain Drains to Lagoon		5.44	0.74	9.50	0.00	15.68
Concrete Non Reinforced Extra	700.00 LF	4,440	601	7,752	0	12,793
Strong,B&S Or T&G Joint, 24"Dia.						
15130 10000 Pipe, Cast Iron						
RSM PL 8" Pipe CI Soil Service Weight		19.18	0.00	15.85	0.00	35.03
	200.00 LF	4,389	0	3,627	0	8,017

15170 10000 Pipe, Steel

Description	Quantity					
M RSM PL 4" Black Steel Pipe, Schedule 40		14.82	0.00	16.50	0.00	31.32
Threaded, W/Couplings 10'OC	50.00 LF	848	0	944	0	1,792
This is placed underground and						
is coated and wrapped.						
M RSM PL 8" Pipe Steel Threaded Schd 40		30.74	0.00	36.50	0.00	67.24
Black W/Couplings 10'OC Coated	150.00 LF	5,276	0	6,265	0	11,541
and Wrapped.						

15198 00000 Valves, Steel

Description	Quantity					
RSM PL Valve Steel Cast Angle 150# 4"		177.83	0.00	1925.00	0.00	2102.83
Flanged	1.00 EA	203	0	2,203	0	2,406
RSM PL Valve Steel Cast Angle 150# 8"		331.95	0.00	5200.00	0.00	5531.95
Flanged	1.00 EA	380	0	5,950	0	6,330
		-----------	---------	-----------	----------	-----------
Plumbing		15,537	601	26,742	0	42,879

40.15.60. Water Storage Tank

13210 10000 Tanks

Description	Quantity					
RSM TK Tanks;Elevated Water Tanks,100'		0.00	0.00	0.00	260550.00	260550.00
Bottom Capacity Line,250,000 Gal	1.00 EA	0	0	0	319,994	319,994

Building Systems Design
PROJECT WHATCS: Whatco Mfg-Whse-Office Sitework
Pre-Bid Design Review Cost Estimate
40. Site Support

DETAILED ESTIMATE DETAIL PAGE 4

40.15. Site Improvements	QUANTY UOM	Labor	Equipmnt	Material	SubContr	TOTAL COST
Water Storage Tank		0	0	0	319,994	319,994
40.15.70. Site Electrical						
OBK EL Electrical Service Swing Gates		0.00	0.00	0.00	8000.00	8000.00
	1.00 SUM	0	0	0	9,288	9,288
OBK EL High Voltage Entry Duct, 13.2 KV		0.00	0.00	0.00	9600.00	9600.00
	1.00 SUM	0	0	0	11,146	11,146
Site Electrical		0	0	0	20,434	20,434
40.15.80. Rail Road						

02452 40000 Railroad

Description	Quantity					
RSM SW Railroad;Car Bumpers, Heavy Duty		669.17	100.30	2225.00	0.00	2994.47
	1.00 EA	780	117	2,594	0	3,491
RSM SW Railroad;Siding,Yard Spur,Level		23.48	3.52	50.00	0.00	77.00
Grade,100Lb Rail New Mat.On	220.00 LF	6,022	903	12,823	0	19,748
Wood Ties						
RSM SW Railroad;Turnout#8 Incl 100#		2655.43	398.02	15500.00	0.00	18553.45
Rail,Plate,Bar,Timber&Ballast	1.00 EA	3,096	464	18,069	0	21,629
6"Below Tie						
Rail Road		9,897	1,483	33,486	0	44,867
Site Improvements		57,973	16,629	189,945	375,946	640,493
Site Support		82,022	36,004	189,945	375,946	683,918
Whatco Mfg-Whse-Office Sitework		82,022	36,004	189,945	375,946	683,918

TABLE 11.2 Comparison of Base Construction Cost Estimates, Site Work

Composer Gold	$882,200
I.C.E.	$873,052
Manual	$772,096

Base Construction Cost

Table 11.2 compares the bottom line site work estimates. To review the estimates item by item, Table 11.3 lists the 12 major categories and the cost of each. However, the category costs must be normalized for comparison. Each estimate has (or lacks) different factors in the category cost as shown in Table 11.4.

To normalize the costs, the I.C.E. estimate is the closest when general requirements are deleted. (The $154,491 is really the 21.5% overhead and profit markup.)

To normalize the Composer Gold, add 10% to each category cost to escalate the 1993 database number to 1995.

To normalize the 1995 manual estimate, divide each category cost by 1.215. The results are listed in Table 11.5. Figure 11.2 is a plot of the results. Table 11.6 compares the three site work estimates: Examination of Table 11.5 shows two categories with substantial shortfalls vs. the average of the two computerized estimates: the well is $18,200 low and the water tank $37,100 low. Adding these amounts to the manual estimate brings its total to $690,807, and the average to $720,553. The results are shown in Table 11.7.

TABLE 11.3 Site Work Costs by Category

	I.C.E. (MC^2)	Gold (BSD)	Manual
General requirements	154,491	0	0
Clearing	32,506	26,397	36,550
Excavation and backfill	17,002	17,028	18,825
Site drainage	13,150	12,793	14,910
Well	38,588	35,518	25,000
Roads and walks	103,289	109,111	116,878
Fencing	42,168	46,745	49,300
Lawns	16,440	20,945	22,000
Railroad	45,992	44,867	50,260
Water tank	363,825	319,994	390,000
Plumbing	26,307	30,068	30,773
Electrical	19,294	20,434	17,600
	873,052	683,900	772,096

TABLE 11.4 Factors in Category Cost

	Manual estimate	I.C.E. estimate	Gold estimate
Base calendar	1995	1995	1995
Escalation	No	No	15% (to 1996)
Overhead and profit in base	21.5%	No (general requirement is 21.5%)	Subcontracts only
Below bottom line	0	0	16.5%

TABLE 11.5 Normalized Category Costs

	I.C.E. (MC²)	Gold (BSD)	Manual
General requirements	0	0	0
Clearing	32,506	29,037	30,082
Excavation and backfill	17,002	18,731	15,494
Site drainage	13,150	14,072	12,272
Well	38,588	39,070	20,576
Roads and walks	103,289	120,022	96,196
Fencing	42,168	51,420	40,576
Lawns	16,440	23,040	18,107
Railroad	45,992	49,354	41,366
Water tank	363,825	351,993	320,900
Plumbing	26,307	33,095	25,328
Electrical	19,294	22,477	14,486
	718,561	752,311	635,383

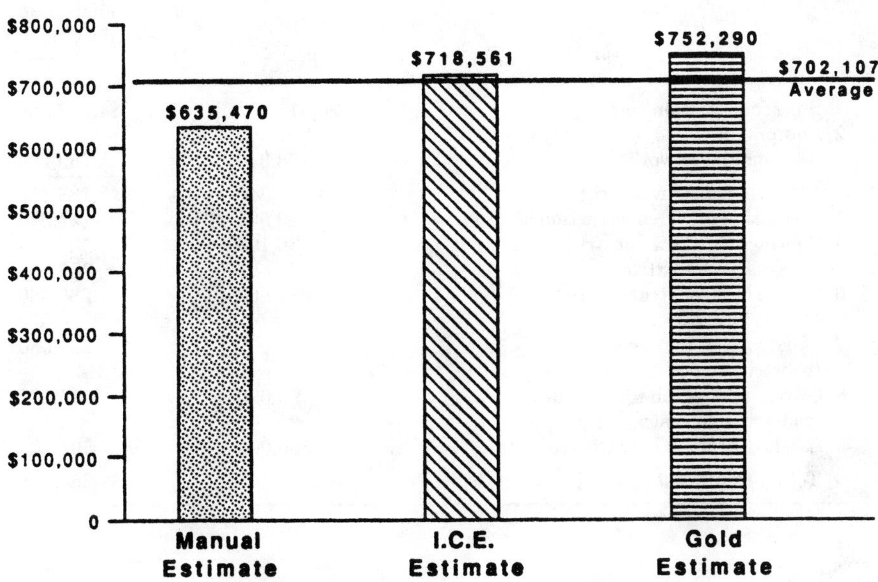

Figure 11.2 Normalized category cost totals.

TABLE 11.6 Comparison of Normalized Category Cost Totals

Gold estimate	7.1% above average
I.C.E. estimate	2.4% above average
Manual estimate	9.5% below average

TABLE 11.7 Comparison of Adjusted Normalized Category Cost Totals

Gold estimate	4.4% above average
I.C.E. estimate	Average
Manual estimate	4.1% below average

The adjustment to the manual site work estimate is:

$$\text{Original estimate} = \$772{,}096$$

$$\text{Adjustment } (\$55{,}300)\ 1.215 = \underline{\quad 67{,}190}$$

$$\text{Adjusted manual estimate} = \$789{,}286$$

Reconciliation with Budget

	Budget	Construction document estimate
1. Base construction cost	$3,611,452	$4,993,826
2. Equipment	381,200	
3. Additional site work	788,340	789,286
Subtotal construction cost	$4,788,340	$5,783,112
4. Escalation 5%/2 (construction cost)	119,525	144,578
5. Contingency (20% construction cost)	956,198	
5% reserve for extras		289,156
6. Design 6% (construction cost) (cap at budget)	286,860	286,860
7. Construction management 5% (construction cost) (cap at budget)	239,050	239,050
8. Owner's project management 2% (construction cost)	95,620	115,662
9. Land cost (per WHATCO real estate)	500,000	500,000
Total project cost	$6,978,245	$7,358,418

Chapter

12

Second VE

Value-engineering review in the latter portion of the design phase should be of a limited nature. This it true for several reasons. First, a major change will usually cause major redesign. At this stage of design, major redesign requires coordination across many design disciplines. Second, a major redesign requires a major time impact. Also, redesign costs should be offset against any potential savings. Ideal VE suggestions at this design stage are those which can be implemented in the specification with no drawing changes.

The second VE still follows the precepts of the Society of American Value Engineers (SAVE) 40-hour workshop. However, because of the limited scope of the second study, it can be compressed in terms of time and/or scope.

WHATCO Project

The last estimate vs. budget reconciliation shows a contingency (including the 5% reserve for changes) of $380,000, or 6.6% of construction cost. WHATCO management decides on a second VE to try to build up the reserve and contingency account.

CVS (Certified Value Specialist) James J. O'Brien assembles the following team:

Wesley F. Mikes, VETC (VE team coordinator)

John D. Orr, P.E. (mechanical)

Vincent Pagliaro (electrical)

Alfred Keil (construction accountant)

Lillian Watson (senior estimator)

O'Brien briefs the VE team and instructs that there are several "givens" or "untouchables" in this second VE. First, the team will not restudy the areas reviewed in the first VE study for two reasons: all VE suggestions in the first study were acceptable; changes in structure, etc., would require major redesign. Second, the second VE study will focus on mechanical, electrical, and site work areas not previously reviewed.

Figure 12.1 is the speculation phase for the second VE.

Figure 12.2 is the electrical idea evaluation. Two ideas, E1 and E2, are scored high and will be costed out. E3 (plastic conduit) is dropped.

Figure 12.3 is the mechanical idea evaluation. Two ideas, M1 and M2, are scored high and will be costed. Of course, only one of these two can be selected (i.e., an "either or").

Figure 12.4 is miscellaneous idea evaluation (i.e., building and site). All three ideas (B1, S1, S2) are scored high enough to cost out.

Figure 12.5 is the electrical cost worksheet. All three ideas show savings:

E1. Using EMT in plant instead of rigid conduit will save $44,808.

E2. Using EMT in warehouse instead of rigid conduit will save $9,100.

E3. Using romex in office building instead of conduit and wire will save $39,120.

Figure 12.6 is the mechanical cost worksheet. The savings are:

M1. Fiberglass duct instead of aluminum, or $64,318.

M2. Galvanized steel duct instead of aluminum, $10,180.

Clearly, the M1 fiberglass option would be the choice.

Figure 12.7 is the site and building cost worksheet. Investigation of other building panels shows that a higher-quality aluminum panel will cost substantially more, $167,760. This is dropped as a VE option.

Both site ideas produce substantial savings:

S1. City water instead of well and elevated tank, $239,563.

S2. Reduce depth of roadbed, $27,965.

Figure 12.8 is a summary of the recommendations. These savings increase the reserve from $380,000 to $805,000, or 14% of construction cost.

If the decision were made to accept the more expensive panels at an increased cost of $167,760, the reserve would still be $637,240, or 11% of construction cost.

LIST ALL IDEAS-
EVALUATE LATER

SPECULATIVE PHASE	
ITEM:	
BASIC FUNCTION:	

Electrical
E1 Use romex instead of wire in conduit in the office building.
E2 Use EMT instead of rigid conduit in plant and warehouse.
E3 Use plastic conduit in plant and warehouse instead of rigid conduit.
Mechanical
M1 Use fiberglass ductwork instead of aluminum.
M2 Use galvanized ductwork instead of aluminum.
Building
B1 Review panel specification for wall panels (no change in support system).
Site Work
S1 Replace water well and tank system with city water.
S2 Reduce roadway base to 6" depth of stone.

OKA FORM 124

Figure 12.1 Speculative phase form.

IDEA EVALUATION

ITEM:	
BASIC FUNCTION:	

IDEA	ADVANTAGES	DISADVANTAGES	IDEA RATING*
E1 Romex instead of conduit in office building.	o Lower cost. o Protection not necessary.	o Less protection.	10
E2 EMT instead of rigid conduit plant and warehouse.	o Lower cost. o Higher level of protection unnecessary	o Less protection.	10
E3 Plastic conduit instead of rigid conduit plant and warehouse.	o Lower cost.	o Provides almost no impact protection.	4

* 10 = MOST DESIRABLE, 1 = LEAST DESIRABLE

Figure 12.2 Electrical idea evaluation form.

IDEA EVALUATION

ITEM:	
BASIC FUNCTION:	

IDEA	ADVANTAGES	DISADVANTAGES	IDEA RATING*
M1 Use fiberglass ductwork instead of aluminum.	o Lower cost. o Self insulated.	o Poorer transmission qualities.	8
M2 Use galvanized ductwork instead of aluminum.	o Lower cost o Can use ligher gage.		10

* 10 = MOST DESIRABLE, 1 = LEAST DESIRABLE

Figure 12.3 Mechanical idea evaluation form.

IDEA EVALUATION

ITEM:	
BASIC FUNCTION:	

IDEA	ADVANTAGES	DISADVANTAGES	IDEA RATING*
Building			
B1 Review panel specification.	o Possible lower cost.	o None	10
Site			
S1 Replace water well and tank with city water.	o Possible lower first cost. o Delete cost of treating well water.	o Higher operating cost.	8
S2 Reduce road subbase from 12" to 6" stone.	o Lower cost.	o Possible shorter life span.	8

* 10 = MOST DESIRABLE, 1 = LEAST DESIRABLE

Figure 12.4 Miscellaneous idea evaluation form.

COST WORKSHEET

Item	Units	No. Units	Original Estimate Cost/Unit	Original Estimate Total	New Estimate Cost/Unit	New Estimate Total
El Plant 2" Rigid	LF	1500	15.60	23,400		
3" Rigid	LF	1800	30.06	54,108		
2" EMT	LF	1500			6.50	9,750
3" EMT	LF	1800			12.75	22,950
				77,508		32 700
SAVINGS						$44,808
El Warehouse 2" Rigid	LF	1000	15.60	15,600		
2" EMT	LF	1000			6.50	6,500
SAVINGS						$9,100
El Romex - office						
1¼" Rigid CT	LF	2000	10.56	21,120		
600V Wire 2 #12	CLF	40	650	26,000		
Romex 2 #12	CLF	40			200	8,000
				$47,120		$8,000
SAVINGS						$39,120

Figure 12.5 Electrical cost worksheet.

COST WORKSHEET

Item	Units	No. Units	ORIGINAL ESTIMATE Cost/Unit	ORIGINAL ESTIMATE Total	NEW ESTIMATE Cost/Unit	NEW ESTIMATE Total
M1 Fiberglass in place of aluminum	#	8,920	10.00	89,200		
Fiberglass	#	6,380			3.90	24,882
SAVINGS						$64,318
M2 Galvanized Steel in place of aluminum	#	8,920	10.00	89,200		
Steel	#	18,000			4.39	79,020
SAVINGS						$10,180

Figure 12.6 Mechanical cost worksheet.

COST WORKSHEET

Item	Units	No. Units	ORIGINAL ESTIMATE Cost/Unit	ORIGINAL ESTIMATE Total	NEW ESTIMATE Cost/Unit	NEW ESTIMATE Total
B1 Panels	SF	24,000	14.88	357,120		
	SF	24,000			21.87	524,880
SAVINGS						NONE
S1 City Water						
Water Tank	EA	1		435,077		
Well	EA	1		47,113		
Piping	EA	1		22,873		
City Service	EA	1				250,000
Site Pipe 4"	LF	400			32.50	13,000
Valve 4"	EA	1				2,500
				$505,063		$265,500
SAVINGS						$239,563
S2 Road Bed 12"	SY	4,700	11.90	55,930		
6"	SY	4,700			5.95	27,956
SAVINGS						$27,965

Figure 12.7 Site and building cost worksheet.

V.E. RECOMMENDATION

ITEM:

PROPOSED CHANGE:

Summary of Recommendations

E1	Romex in Office Building	39,120
E2	EMT Plant/Warehouse	53,908
M1	Fiberglass Ductwork	64,318
S1	City Water	239,563
S2	Reduce Subbase	27,965
		$424,874

COST SUMMARY	Electrical	Ductwork	City Water	Subbase
INITIAL - ORIGINAL	140,228	89,200	505,063	55,930
INITIAL - PROPOSED	47,200	24,882	265,500	27,965
INITIAL SAVINGS	93,028	64,318	239,563	27,965
TOTAL ANNUAL COSTS - ORIGINAL				
TOTAL ANNUAL COSTS - PROPOSED				
ANNUAL SAVINGS				
PRESENT WORTH - ANNUAL SAVINGS				

Figure 12.8 Summary of VE recommendations.

Chapter

13

Construction Bid

There are a variety of ways in which an owner (or owner's agent) can request construction bids. The most common way is to request a sealed fixed-price bid based upon the completed construction documents (plans and specifications). Depending upon the complexity and size of the project, a bidding period of 4 to 6 weeks is normal.

The estimating by contractors, subcontractors, and suppliers during the bidding period is very similar to the procedures discussed previously. However, the estimates are at different tiers of responsibility—and range from macro to micro in detail.

General Contractors

The general contractor (GC) provides the general conditions, supervision, and home office support. Some general contractors perform their own concrete work. The GC performs a macro estimate to provide a framework for the final bid. This macro (or overview) estimate can be generated in a number of ways, such as the computer-generated estimates described in earlier chapters. The GC usually inputs their own experience factors in the form of unit prices and/or labor costs.

Another approach to generating the macro estimate is to use a modeling approach to apply factors to adjust prior project cost information for project size, type, and escalation. The GC also prepares a macro schedule either "from scratch" or by modeling it on similar prior projects.

The macro estimate and schedule have various purposes, including:

- General conditions costs are lineal with time.
- General conditions scope varies at certain milestone points.
- Supervision scope (number and/or type) varies at certain milestone points.

- Supervision costs are lineal with time.
- Can be used as a check on the realism of subcontractor bids.
- Can be used to be certain that all required scope is included in the bid.

Figure 13.1 is a spreadsheet input for general conditions estimating by a GC. Work to be done by the GC, such as concrete, is taken off

General Conditions	Monthly Cost	Months	# Units	Total
1. Trailers				
1.1 Office Type				
1.2 Storage				
1.3 Change				
1.4 Move-in Cost				
1.5 Move-out Cost				
2. Portable Toilets				
3. Services				
3.1 Electric				
3.2 Water				
3.3 Telephone				
4. Office Equipment				
4.1 Furniture				
4.2 Xeroxes				
4.3 P.C.s				

General Conditions	Initial Cost	Cost/Month	Months	Total
5. Security				
5.1 Fence				
5.2 Lighting				
5.3 Watchmen				
6. Access				
6.1 Roads				
6.2 Dust Control				
6.3 Parking				
6.4 Signage				
7. Survey Team(s)				

Figure 13.1 General conditions input sheet.

and micro estimated in the same fashion as subcontractor pricing. The GC sets up a bid sheet to be sure that all items are priced. Prices are received up to literally the last minute, so the bid sheet is usually on a PC spreadsheet.

The GC identifies areas where the bidding subcontractors may miss significant work. For instance, where the HVAC subcontractor supplies air handling units (AHU), are the motors included? Also, if the HVAC subcontractor supplies the motors, who hooks them up?

Specifications may call on the contractors to identify and report any omissions or errors during the bidding process. Any discovered should be reported so that the design professional can issue an addendum so that all bidders are on an even basis. However, bidding contractors are not obligated to discover all errors and/or omissions (*Blount Brothers Construction Co. v. United States,* 346F. 2d 963). The United States Court of Claims made the following comments:

> Where specifications can and should but do not expressly provide for a desired condition in an end product such omission increases the duty of the drawings to be clear and free from ambiguity in portraying the condition.
>
> Contractors are business men, in the business of bidding on Government contracts. They are usually pressed for time and are consciously seeking to underbid a number of competitors. Consequently, they estimate only on those costs which they feel the contract terms will permit the Government to insist upon in the way of performance. They are obligated to bring to the Government's attention major discrepancies or errors which they detect in the specifications or drawings, or else fail to do so at their peril. But they are not expected to exercise clairvoyance in spotting hidden ambiguities in the bid documents, and they are protected if they innocently construe in their own favor an ambiguity equally susceptible to another construction, for as in Peter Kewit Sons Co. v. United States, 109 Ct. Cls 390,419 (1947), the basic precept is that ambiguities in contracts drawn by the Government are construed against the drafter.

Major Subcontractors

The mechanical and electrical subcontractors have duties similar to those of the GC. Traditionally, the GC relies on these two to handle their own scope of work. Under some public law, these two are separate prime contractors. Because of the need for the mechanical and electrical subcontractors to direct and coordinate their own work, it really makes little difference whether they are prime or subcontractor.

Again, similar to the GC, the mechanical contractors, in particular, often subsubcontract parts of their work. The subcontracts would typically be ductwork, air conditioning, heating, plumbing, and controls.

The structural steel subcontractor is also in the major category. Once the foundations are in place, this subcontractor takes control of the job progress. The structural steel subcontractor must also coordinate the shop drawing, fabrication, shipping, and shakeout and staging preceding the erection phase.

Specialty Subcontractors

Specialty subcontractors include trade contractors (masonry, glazing, roofing, painting, curtain wall, etc.) and smaller specialty contractors (tile, carpeting, hardware, hung ceilings, etc.). The GC (and major subcontractors) invite quotations (by phone, mail, or fax). The specialty subcontractors typically do not take out sets of plans and specifications because they are bidding so many projects. To accommodate them, the GCs (and major subcontractors) provide plan rooms where the specialty subs can visit to take off their scope of work.

Performance Contracts

Most of the construction project is described in a prescribed manner. Certain elements, however, are described by their performance requirements. The subcontractor provides design and/or manufacturing as well as installation. Examples of this type of subcontract include elevators, escalators, sprinklers, alarm and detection systems, and HVAC control systems.

Suppliers

Microlevel information is solicited from suppliers by the subcontractor estimators. Each subcontractor will have a network of suppliers. Based on criteria such as volume of order, prior business, and payment terms, each subcontractor will have their own discount from list price. The estimators get quotations by mail or fax, and enter the information into the appropriate part of their draft estimate.

Long-Term Bidding Strategies

Long-term bidding strategies should be established well before the decision to prepare a bid. In his article, "Why Some Prosper and Others Fail,"[1] Joseph P. McDonough asks:

> Have you every wondered why some contractors prosper during a recession and others fail? How, in a time when productive estimating is critical, one company with a small staff seems to outperform their competi-

---

[1]"An Effective Bidding Strategy: Why Some Prosper and Others Fail," Joseph P. McDonough, ASPE (American Society of Professional Estimators) July/August, 1992.

tors with large departments? What makes these contractors different? Why do they win so many bids? Do they have an *unfair advantage?*

He offers some examples of "unfair advantage."

Example A

A small GC building single and multifamily dwellings felt the effect of the shrinking housing market. Instead of laying off his team or bidding jobs at a loss, he looked for an "unfair advantage." He located a medium-sized mall project where his nearby location and trained team made him more competitive than the big downtown contractors.

Example B

A large northeast electrical contractor saw problems looming as his profit margin on successful bids was shrinking, with no real prospects of improving in the current market. Looking for an "unfair advantage," he identified Key West as a unique area for him to bid projects because of a building (and rebuilding) boom in southern Florida. The closest major electrical contractor was in Miami, several hours away. That meant that competing bidders would have to pay housing and expenses for their men—an equalizing factor. Coming from a very competitive bust market, he could *raise* his margins and still be lower than the not-so-hungry competition.

Short-Term Bidding Strategies

When the base estimate is assembled, the bidder's management (at all tiers) make adjustments to enhance their chance to win (i.e., to gain an "unfair advantage"). Some of the decisions and/or strategies include the following:

Profit. Reducing profit percentage enhances chances of a winning bid.

Home office overhead. Instead of a flat percentage allocation of home office overhead (3 to 5% range), identify costs of specific home office costs to be charged to the job.

Equipment. If your company owns equipment needed for the project, and it is part or fully depreciated, bid less than market value for using it on this job.

Contingency. If there are exposures inherent in the project scope (such as liquidated damages), reduce or delete contingency for same.

Unbalanced bid. On a unit price bid, identify your own estimate of quantities. For instance, on an excavation category where the owner's base bid quantity is 10,000 cy common and 2000 cy rock, your estimate suggests that rock is closer to 5000 cy and common excavation 7000 cy, unbalance your unit bid to high on rock and low on common excavation.

Penny bid. This is a variation of the unbalanced bid. For instance, where unit prices are called for excavation, under a pipeline, and separately for support piling as directed by the engineer, a penny bid is given for the piling. The bidder studies the borings and estimates reasonable piling costs. The excavation and piling costs are added and divided by the cubic yards of excavation. If the soil conditions are more favorable than the engineer expected, the contractor keeps whatever piling cost was built into the bid.

Of course, the bidder can put zero piling cost in the bid to enhance chances of winning.

Bidding procedures

In 1991, the American Society of Professional Estimators recognized "a need to establish more uniform guidelines for bidding competitively priced construction projects." Working with a number of national construction industry associations, they addressed 37 separate issues (and subissues) and published a 53-page booklet identifying each issue, discussing it, and offering recommendations.[2] Following are excerpts (with permission) of 10 issues and subissues which would directly affect the preparation of a bid:

Advertisement for bids

The Issue: Should a bid be advertised? If so, what information should be included in the advertisement?

Discussion: A typical bid advertisement is issued prior to the issuance of bid documents for the purpose of attracting prospective bidders. It may be mailed to individual firms, published in an appropriate newspaper or magazine of general circulation in the construction industry, and/or posted in plan rooms.

Recommendations: A bid advertisement should:

1. Be circulated sufficiently in advance of the distribution of bidding documents to allow prospective bidders to include the project in their respective bid calendars. The minimum advance notice should be 30 days prior to bid document issue.

---

[2]"Recommended Bidding Procedures for Competitively Bid Construction Projects," prepared by the American Society of Professional Estimators, April 8, 1991, 11141 Georgia Ave., Suite 412, Wheaton, MD, 20902.

2. Contain a short description of the project including bid date, time, approximate contract amount, approximate size (or capacity), project location, licensing requirements, and bid and performance and payment bond requirements.
3. Be circulated to both individual prime and major subcontractor and material supplier prospective bidders, published in construction-oriented magazines and newspapers, and posted in plan rooms.
4. State date of document availability, location to obtain documents, and deposit and refund provisions.

Estimating and bidding time

The Issue: How much time for the issue of bidding documents should be allowed for bid preparation?

Discussion: Bid preparation is a function of the number of sets of bid documents issued, the efficiency with which they are dispersed, the completeness of the documents, their complexity, and the complexity of the bid form and of the project.

Simple projects (usually of lesser value, for example, under $1,000,000) with simple bid forms, complete documents, and which involve only bidders in a small geographic area do not require estimating periods as long as more complex projects with more complex bid documents, a wider geographic appeal, and/or incomplete documents.

Of whatever size and complexity, it is essential that bidding documents be complete before they are issued. This results in less wasted time in bid preparation and lower bids for the owner.

Recommendations:

1. Complete all bidding documents before issuing them for bid.
2. Allow 3 weeks minimum preparation for small (under $1,000,000) or simple projects.
3. Allow 4 to 6 weeks for large and/or complex projects.
4. For extremely large projects ($50,000,000 or more), extremely complex projects, or extremely complex bid requirements, the estimating time allowed should be determined after consultation with several potential prime bidders.

Bidding document availability. This issue involves bidding document availability, the plan deposit, and the prime bidder.

The Issue: Under what conditions should prime bidders be required to pay a plan deposit when bidding a project?

Discussion: Many owners and architects charge a deposit fee for plans used during bidding their projects. In most cases, these deposits are refundable to the prime bidder after bid time if certain requirements are met.

The owners and architects use this policy to ensure the return of their plans. In this way, they may be released to the prime bidder awarded the project, thus reducing reproduction costs.

Recommendations: When plan deposits are used:

1. The apparent low bidder and other bidders whose bid security is retained by the owner should be allowed to retain bid documents until the contract has been awarded. If other bidding documents are returned to the owner or architect within ten (10) days of bid, in good condition, their deposit should be fully refunded.
2. A bidder which decides against bidding and which notified the owner or architect two days prior to bid and which returns its bid documents in good condition should have its deposit fully refunded.
3. The deposit amount should closely approximate reproduction costs of plans and specifications, and deposits for all sets returned in good condition should be fully refunded.
4. If a plan deposit system is not used, bid documents should be available for purchase at the cost of reproduction and handling.
5. The successful prime bidder's deposit should also be returned.

This issue involves bidding document availability, plan deposit, subcontractors, and material suppliers.

The Issue: Under what conditions should subcontractors and material suppliers be required to pay a plan deposit when bidding a project? Should subcontractors and suppliers be afforded the same opportunity to obtain bid documents upon the same basis as prime bidders?

Discussion: Many major subcontractors (e.g., mechanical and electrical subcontractors) have a great a need for complete sets of bidding documents, as does the prime bidder.

In many cases owners do not allow major subcontractors and material suppliers to use the plan deposit system to receive plans for bidding. Subcontractors and material suppliers should be afforded the opportunity to obtain bidding documents upon the same basis as prime bidders.

Recommendations: Where plan deposits are used:

1. The apparent low subcontractor and material supplier should be allowed to retain documents until the contract has been awarded. In other cases, if the bidding documents are returned in good condition to the owner within ten (10) days of bid date the deposit should be fully refunded.

2. A subbidder which decides against bidding and which notifies the owner or architect two days prior to bid and which returns its plans in good condition should have its deposit fully refunded.
3. The deposit amount should closely approximate the reproduction costs of plans and specifications, and deposits for all sets returned in good condition should be fully refunded.
4. If plan deposit system is not used, or if a bidder requires less than a full set of documents, documents should be available for reproduction in blueprint shops. In this way, subcontractors and material suppliers may purchase individual sheets or full sets directly from the printer. The cost for such documents should not exceed the cost of their reproduction and handling.
5. As an alternative the architect may have plans and specifications available for purchase by subcontractors and/or material suppliers. The cost for such documents should not exceed the cost of reproduction.
6. Bid documents which allow the purchase of partial document sets should emphasize the potential hazards of their use.

This issue involves bidding documents availability and quantity issued for bidding.

The Issue: How many sets of plans and specifications should be issued to each prime bidder on a project? How many to subcontractors and material suppliers?

Discussion: Frequently, limits upon the number of sets of documents issued to prime bidders and subcontractors and material suppliers are enforced. This limits the total number of sets of documents available to the construction community, resulting in fewer and less competitive bids.

Recommendations:

1. Each prime bidder should be allowed at least four (4) sets of bidding documents for its use and the use of some of its subcontractor and material suppliers.
2. Major subcontractors (e.g., mechanical and electrical) should be issued one set of bidding documents.
3. Both prime bidders and subcontractor and material supplier bidders should have bidding documents available through the plan deposit system.
4. Any prospective bidder, whether or not a major bidder, should either be afforded the opportunity to obtain complete sets of documents, through the deposit system, and/or to purchase partial or complete sets directly from a printer or the architect.

5. Any prospective bidder who has been provided a copy of partial or complete bid documents should be notified of any addenda issued and be provided an opportunity to obtain copies of them.

This issue involves bidding documents availability, and plan rooms and plan services.

The Issue: Should the architect or owner supply plans to the local plan rooms and plan services?

Discussion: With the increasing use of plan rooms and plan services, the architect and owner have a means of reaching a larger number of interested bidders for their projects.

These facilities enable a large number of bidders to have access to the documents, many of whom might not have had the opportunity to review them otherwise. The availability of bid documents to a larger number of bidders benefits the owner by increasing competition.

Recommendations:

1. Each local plan room should receive, free of charge, two (2) to four (4) complete sets of bidding documents at the time of initial issue.
2. If the projects are large or complex, additional sets of documents should be furnished to the plan rooms and services, in order to offer more time for the bidders in their bid preparation.
3. Local plan rooms or services should notify the architect if additional sets of plans and specifications are needed.
4. The architect should furnish the plan rooms and services any addenda or changes to bid documents, as well as names of prime bidders.
5. Plan rooms and plan services should not be required to provide a bid deposit or purchase bid documents.

Bid forms for prime bidders

The Issue: Are bid forms becoming overly complex?

Discussion: It is increasingly difficult to complete bid forms for projects. Bid forms are asking for more information in the form of alternates, unit prices, listing requirements, and general information.

Complex bid forms add to the degree of possible error at bid time. The closing hours prior to bid time are filled with prime bidders receiving bids from subcontractors and material suppliers and trying to confirm that each bid is complete.

Overly complex bid forms add to the prime bidder's work load and unduly encumber the bidder during the critical hours prior to bid time.

Recommendations:

1. Bid forms should be as simple as possible, with the minimum pertinent information included at bid time. General information

required from the prime bidders should be requested prior to or after bid time, prior to award.

2. Alternates and unit prices should be kept to a minimum. Where possible, these should be negotiated with the successful bidder. Open all bids upon receipt of alternates.

Alternates

The Issue: Should bids have required alternates?

Discussion: It is becoming increasingly popular for architects and owners to include alternates with the base bid for the project. This offers the owner a multiple number of building features, making determination of the low bid at bid time more tenuous and increasing the amount of time and work needed to compile a bid for the project. These increased costs must be passed on to the owner in increased bids.

The need for alternates from the original bidding documents can be reduced or eliminated through competent estimating during design development. Each alternate required increases the probability of an error in the bid. How each alternate affects the base project must be determined and clearly defined in the bid documents.

Recommendations:

1. Where alternates cannot be eliminated from the bid, they should be the absolute minimum in number, and as simple as possible. Multiple discipline alternates (those that impact the work of many trades) in fact require the preparation of two (or more) complete mini-bids and should be avoided.

2. Pricing for alternates should be submitted 2 hours after the base bid is submitted. At that time, both base and alternates bids should be opened and read aloud.

3. Alternates should be clearly defined in both plans and specifications including adequate specification description of work areas and drawing exposition of interface requirements.

4. Determination of the apparent low bidder should be made upon the base bid, precluding the appearance of arbitrary low bidder selection by manipulation of alternate selection. If alternates are used to determine the apparent low bidder, the project budget should be announced prior to opening bids, and the alternates should be evaluated in the order of priority. This order, as well as the basis of award, should be clearly set out in bid documents.

5. Where alternate prices result in listed subcontractors or material suppliers changes, such changes in the listing of subcontractors or

material suppliers should be recorded when alternate prices are submitted.

Prebid conference

The Issue: Should prebid conferences be held?

Discussion: Prebid conferences are a vehicle for the owner and design team to communicate information about project procurement, financing, and administrative procedures. They may also be used to present technical questions related to bidding documents, though the response to such questions, after investigation, is usually issued in an addendum.

Many members of the industry question the value of prebid conferences, believing them to be an unnecessary restatement of information contained in the bidding documents, and a dilution of bid preparation efforts, rather than an enhancement of them.

Recommendations:

1. Prebid conferences, if necessary, should be held where possible at the project site at the approximate midpoint of the bidding period, but no later than ten (10) days prior to bid date.
2. Information communicated should be a substantive addition to the bidding process; restatement of information already available in bidding documents should be avoided.
3. Prebid conference minutes and answers to technical questions should be issued to all plan holders as an addendum.
4. The source and amount of funds available for construction should be discussed.
5. Mandatory attendance at prebid conferences by prospective bidders should be avoided.

Day and time for receiving bids

The Issue: Upon what days of the week, times of the day, and under what conditions should bids be received?

Discussion: Owners who wish to receive the most competitive bids must recognize that the construction industry has certain days and times to tender bids which are much preferred. The preferred days are Tuesday, Wednesday, and Thursday. The preferred time is a time specific between 2:00 and 4:00 P.M. local time. Bids tendered at times and on days other than these will receive substantially less coverage and very likely will be higher.

Recommendations:

1. Bid should be received at a specific time between 2:00 and 4:00 P.M. on Tuesday, Wednesday, or Thursday.

2. Bids should not be received on holidays, on the day before or after holidays, or during the week between Christmas Day and New Years Day.
3. Prime bidders should be afforded telephones in close proximity to the place for receiving bids for receiving and recording last-minute changes to their bids.
4. The time of receipt should be clearly stated and strictly enforced. An official clock should be displayed in the place receiving bids, and the person receiving bids should stamp bids received with the time of receipt and publicly state when the time for receipt of bids has elapsed. Bids received after that time should not be accepted.
5. The owner's construction budget for the project and basis for selecting the low bidder should be announced prior to opening bids.
6. Bids should be publicly opened and read aloud. Obvious problems with a bidder's responsiveness (e.g., no bid bond, failure to acknowledge addenda, failure to comply with listing requirements, etc.) should be noted. If protests are allowed, the mechanism for them should be set out in the bidding documents.
7. Other pertinent bid information (alternates, subcontractor listing, etc.) should also be read aloud.
8. Interested parties should be allowed to review the bidding documents of all bidders.
9. The schedule for award should be announced. Bids should not be held longer than sixty (60) days without an award announcement being made.
10. Complete bid results should be formally published in a timely way.

Chapter

14

Estimating During Construction

During the construction phase, the major role of estimators is in change order negotiations. Two other roles are: (1) approval of the payment breakdown at the outset; and (2) review of VECPs (value-engineering change proposals). Chronologically, payment breakdown review comes first.

Payment Breakdown

The successful bidder submits a proposed payment breakdown by CSI division (and subdivision). The submittal will be turned over to estimators for review. The thrust of the review is to identify the "front-end loading." This (front-end loading) overprices early work for the purpose of offsetting retention. The typical construction contract retains 10% of progress payments (up to 50% of total). At 50% the owner (typically) can continue 10% retention if there is any problem with the contractor. If progress is satisfactory, the owner (or construction manager) can waive additional retention. If the payment breakdown overprices by 10% (or more), the impact of retention on cash flow is offset.

The estimator(s) identify the cost by CSI division in their estimate and compare it with the matching account in the contractor's proposed payment breakdown. If the variance (in terms of percentage of the total figure) in chronologically early work (sitework, excavation, concrete) is higher by more than 5 to 10%, the contractor should be challenged to support the costs. Similarly if later work (Divisions 10 to 16) is underpriced by more than 5 to 10%, there should be a similar support to demonstrate that the costs will be sufficient to put the work in place.

Change Orders

Change orders in value of 5% or less of the original contract value are usual and should be anticipated by the contractor. No similar number of appropriate changes is available as an accepted statistic, principally because the number of changes can be disguised by packaging smaller changes into single change orders. Packaging of change orders in this fashion is a poor practice, since it makes the negotiation of change orders more difficult.

In virtually all contracts, the owner retains the prerogative of making changes to the work by the use of clauses, such as:

> The director, at his discretion, may, at any time during the progress of the work, authorize additions, deductions, or deviations from the work described in the specifications as herein set forth; and the contract shall not be vitiated or the surety released thereby.... Additions, deductions, and deviations may be authorized as follows at the director's option:
>
> 1. On the basis of unit prices specified
> 2. On a lump-sum basis
> 3. On a time and material basis
>
> Additional work is work that can be imposed within the contract documents. It is a change or alteration to the plans or specifications for a number of reasons implicit in the original agreement. These reasons could include, but not be limited to, omissions in the design documents, recognition of better methods or materials to achieve the required effect, resolutions of problems recognized, resolution of unforeseen conditions not anticipated, and similar adjustments within the intent of the original contract.

Identification of change conditions may emanate from the architect-engineer, field engineer at the jobsite (unforeseen site conditions), and/or the contractor.

The question of whether a change order will result in a credit or payment to the contractor is a matter of interpretation of the plans and specifications. The field staff should exercise great care to avoid casual comment or agreement with purported change items, and should research a change or requirement in the plans and specifications to determine that it is not included in the contractual scope, either partially or completely.

Change orders are initiated through various methods. An RFI (request for information) usually precedes a change order. Or a bulletin can be issued requesting a price from the contractor and subcontractor for proposed revisions. Sketches often accompany this bulletin. Direction can also be given verbally to make changes in the field. Revised drawings can also be issued.

The most important consideration of a change order is to completely understand what the change involves. A clear definition of the change and the scope of work will enable the estimator to develop a list of materials and labor, and an estimate for the work to be credited and/or added to the original contract. It is usually not difficult to reach agreement between the owner (CM) estimators and contractor estimators as to the scope of the change order.

Further, the typical contract limits the total for overhead and profit to 20%. The contractor has two avenues to cover the cost of the change order: unit pricing and special conditions.

Unit pricing

The cost of units (such as cubic yard of concrete in place, or cubic yard of common backfill) must be negotiated. These agreed-upon units are then used on later change orders. To support the negotiation, the contractor's estimators must provide backup such as invoices. The owner's (CM) estimators use databases (such as R. S. Means) to evaluate the reasonability of the negotiated unit pricing.

Special conditions

For individual change orders, the contractor's estimators must identify special items not included in the standard overhead. This would include items such as engineering of temporary structures, mobilization and remobilization costs, and special (low) productivity costs. For instance, if additional materials had to be purchased, increased unit prices can be incurred for purchasing lesser quantities. Other considerations can involve rescaffolding an area, loss of productivity, out-of-sequence work. stacking of trades, congested installations of other trades, and mobilization costs.

Value Engineering Change Proposals (VECP)

Many contracts encourage VECPs and offer to share 50% of any *net savings* initiated by the contractor. In this negotiation, the owner's (CM) estimators have somewhat of a role reversal, in that their goal must be to be sure the savings figure is large enough.

For instance, if a contractor has a VECP worth $100,000 and manages to negotiate it as an $80,000 savings, the official share to the contractor will be $40,000. However, when added to the undisclosed $20,000, the result would be a 60% share for the contractor.

WHATCO Project

The GC submits a VECP for an alternate to the cross bracing in the plant (refer to Figs. 6.18 and 6.19). The VECP would use 1-in-diame-

COST WORKSHEET

Item	Units	No. Units	ORIGINAL ESTIMATE Cost/Unit	ORIGINAL ESTIMATE Total	NEW ESTIMATE Cost/Unit	NEW ESTIMATE Total
Cross Bracing	Tons	60.1	$2,240	134,400		
Wire Rope	LF	627			12.00	7,524
Clamps	EA	48			30.00	1,440
Connectors	EA	32			75.00	2,400
Turnbuckles	EA	16			50.00	800
						$10,724
SAVINGS						$123,676

Figure 14.1 VECP, wire rope instead of cross bracing.

ter galvanized wire rope, tensioned by turnbuckles. Figure 14.1 shows a total cost savings of $123,676. At 50%, the contractor's saving will be $61,838. Note that the VECP does not reduce the OH&P (plus bond) related to the contract work (nor does it add OH&P plus bond to the cost of the VECP alternate).

Index